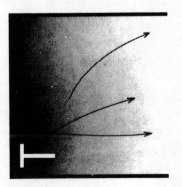

THE UNIVERSITY OF WESTERN ONTARIO SERIES IN PHILOSOPHY OF SCIENCE

A SERIES OF BOOKS

ON PHILOSOPHY OF SCIENCE, METHODOLOGY,

AND EPISTEMOLOGY

PUBLISHED IN CONNECTION WITH

THE UNIVERSITY OF WESTERN ONTARIO

PHILOSOPHY OF SCIENCE PROGRAMME

VOLUME 10

FOUNDATIONAL PROBLEMS IN THE SPECIAL SCIENCES

PART TWO OF THE PROCEEDINGS
OF THE FIFTH INTERNATIONAL CONGRESS OF
LOGIC, METHODOLOGY AND PHILOSOPHY OF SCIENCE,
LONDON, ONTARIO, CANADA–1975

Edited by

ROBERT E. BUTTS

The University of Western Ontario

and

JAAKKO HINTIKKA

The Academy of Finland and Stanford University

D. REIDEL PUBLISHING COMPANY

DORDRECHT-HOLLAND/BOSTON-U.S.A.

Library of Congress Cataloging in Publication Data

International Congress of Logic, Methodology, and Philosophy of
    Science, 5th, University of Western Ontario, 1975.
    Foundational problems in the special sciences.

    (Proceedings of the Fifth International Congress of Logic,
Methodology, and Philosophy of Science, London, Ontario, Canada,
1975 ; pt. 2) (University of Western Ontario series in philosophy of
science ; v. 10)
    Bibliography: p.
    Includes index.
    1. Science–Congresses. 2. Science–Methodology–Con-
gresses. 3. Science–Philosophy–Congresses. 4. Social sciences
–Congresses. I. Butts, Robert E. II. Hintikka, Kaarlo Jaakko
Juhani, 1929– III. Title. IV. Series: University of Western Ontario. The University of Western Ontario series in philosophy of science ;
v. 10.
Q174. I58 1975a pt. 2   [Q101] 501s [501]   77–22431
ISBN 90–277–0710–3

The set of four volumes (cloth) ISBN 90 277 0706 5

Published by D. Reidel Publishing Company,
P.O. Box 17, Dordrecht, Holland

Sold and distributed in the U.S.A., Canada, and Mexico
by D. Reidel Publishing Company, Inc.,
Lincoln Building, 160 Old Derby Street, Hingham,
Mass. 02043, U.S.A.

Printed in The Netherlands

TABLE OF CONTENTS

PREFACE

The Fifth International Congress of Logic, Methodology and Philosophy of Science was held at the University of Western Ontario, London, Canada, 27 August to 2 September 1975. The Congress was held under the auspices of the International Union of History and Philosophy of Science, Division of Logic, Methodology and Philosophy of Science, and was sponsored by the National Research Council of Canada and the University of Western Ontario. As those associated closely with the work of the Division over the years know well, the work undertaken by its members varies greatly and spans a number of fields not always obviously related. In addition, the volume of work done by first rate scholars and scientists in the various fields of the Division has risen enormously. For these and related reasons it seemed to the editors chosen by the Divisional officers that the usual format of publishing the proceedings of the Congress be abandoned in favour of a somewhat more flexible, and hopefully acceptable, method of presentation.

Accordingly, the work of the invited participants to the Congress has been divided into four volumes appearing in the University of Western Ontario Series in Philosophy of Science. The volumes are entitled, *Logic, Foundations of Mathematics and Computability Theory, Foundational Problems in the Special Sciences, Basic Problems in Methodology and Linguistics,* and *Historical and Philosophical Dimensions of Logic, Methodology and Philosophy of Science.* By means of minor rearrangement of papers in and out of the sections in which they were originally presented the editors hope to have achieved four relatively self-contained volumes.

The papers in this volume consist of all those submitted for publication by invited participants working on foundational problems in the physical sciences, biology, psychology and the social sciences. Contributed papers in these fields appeared in the volume of photo-offset preprints distributed at the Congress. The full programme of the

Congress appears in *Historical and Philosophical Dimensions of Logic, Methodology and Philosophy of Science.*

The work of the members of the Division was richly supported by the National Research Council of Canada and the University of Western Ontario. We here thank these two important Canadian institutions. We also thank the Secretary of State Department of the Government of Canada, Canadian Pacific Air, the Bank of Montreal, the *London Free Press,* and I.B.M. Canada for their generous support. Appended to this preface is a list of officers and those responsible for planning the programme and organizing the Congress.

THE EDITORS

February 1977

OFFICERS OF THE DIVISION

A. J. Mostowski	(Poland)	President
Jaakko Hintikka	(Finland)	Vice President
Sir A. J. Ayer	(U.K.)	Vice President
N. Rescher	(U.S.A.)	Secretary
J. F. Staal	(U.S.A.)	Treasurer
S. Körner	(U.K.)	Past President

PROGRAMME COMMITTEE

Jaakko Hintikka (Finland), Chairman
R. E. Butts (Canada)
Brian Ellis (Australia)
Solomon Feferman (U.S.A.)
Adolf Grünbaum (U.S.A.)
M. V. Popovich (U.S.S.R.)
Michael Rabin (Israel)

Evandro Agazzi (Italy)
Bruno de Finetti (Italy)
Wilhelm Essler (B.R.D.)
Dagfinn Føllesdal (Norway)
Rom Harré (U.K.)
Marian Przełęcki (Poland)
Dana Scott (U.K.)

CHAIRMEN OF SECTIONAL COMMITTEES

Y. L. Ershov (U.S.S.R.)	Section I:	Mathematical Logic
Donald A. Martin (U.S.A.)	Section II:	Foundations of Mathematical Theories
Helena Rasiowa (Poland)	Section III:	Computability Theory
Dagfinn Føllesdal (Norway)	Section IV:	Philosophy of Logic and Mathematics
Marian Przełęcki (Poland)	Section V:	General Methodology of Science
J.-E. Fenstad (Norway)	Section VI:	Foundations of Probability and Induction
C. A. Hooker (Canada)	Section VII:	Foundations of Physical Sciences

Lars Walløe   Section VIII: Foundations of Biology
   (Norway)

Brian Farrell   Section IX: Foundations of Psychology
   (U.K.)

J. J. Leach   Section X: Foundations of Social Sciences
   (Canada)

Barbara Hall Partee Section XI: Foundations of Linguistics
   (U.S.A.)

R. E. Butts   Section XII: History of Logic, Methodology
   (Canada)   and Philosophy of Science

LOCAL ORGANIZING COMMITTEE

R. E. Butts (Philosophy, the University of Western Ontario),
Chairman

For the University of Western Ontario:
Maxine Abrams (Administrative Assistant)

R. N. Shervill (Executive Assistant to the President)
G. S. Rose (Assistant Dean of Arts)
R. W. Binkley (Philosophy)
J. J. Leach (Philosophy)
C. A. Hooker (Philosophy)
J. M. Nicholas (Philosophy)
G. A. Pearce (Philosophy)
W. R. Wightman (Geography)
J. D. Talman (Applied Mathematics)
J. M. McArthur (Conference Co-ordinator)

For the City of London:

Betty Hales (Conference Co-ordinator)

For the National Research Council of Canada:

R. Dolan (Executive Secretary)

I

FOUNDATIONS
OF THE PHYSICAL SCIENCES

JOHN ARCHIBALD WHEELER

GENESIS AND OBSERVERSHIP*

ABSTRACT. Four lines of evidence out of physics would seem to have direct relevance to the question, how did the universe come into being? (1) Gravitational collapse speaks for the mutability of the laws and structure of physics. (2) No way is evident how physics can bottom out in a smallest object or most basic field or continue on to forever greater depths; and the third possibility presents itself that the observer himself closes up full circle the links of interdependence between the successive levels of structure. (3) What is the explanation for the 'anthropic principle' of Dicke and Carter? Carter's 'ensemble of universes', in only a very small fraction of which life and consciousness are possible? Or is the very mechanism for the universe to come into being meaningless or unworkable or both unless the universe is guaranteed to produce life, consciousness and observership somewhere and for some little time in its history-to-be? (4) The quantum principle shows that there is a sense in which what the observer will do in the future defines what happens in the past – even in a past so remote that life did not then exist, and shows even more, that 'observership' is a prerequisite for any useful version of 'reality'. One is led by these considerations to explore the working hypothesis that 'observership is the mechanism of genesis'. Testable, 'falsifiable', consequences may never flow from this concept unless and until one discovers how to *derive* from it the existence and structure of quantum mechanics.

HOW DID THE UNIVERSE COME INTO BEING:
FOREVER BEYOND EXPLANATION?
IS GENESIS UNFATHOMABLE?

Less than four years after the Nov. 24, 1859 publication of *The Origin of Species*, Charles Darwin (1863) wrote to Joseph Dalton Hooker, "It is mere rubbish, thinking at present of the origin of life; one might as well think of the origin of matter". Today, thanks not least to Darwin himself, we possess an attractive and actively investigated scenario [Oparin (1938)] for the origin of life. Will we ever know anything about the still deeper issue, what is the origin of matter?

Leibniz put it in his famous words, "Why is there something rather than nothing?" William James (1911), translated the "why" to the more meaningful "how": "How comes the world to be here ..."

We ask today, "How did the universe come into being?", realizing full well that how properly to ask the question is also a part of the

Butts and Hintikka (eds.), *Foundational Problems in the Special Sciences*, 3–33.

question. One can even believe that one can only then state the issue in the right words when one knows the answer. Or is there an answer? Is the mystery of genesis forever beyond explanation?

The investigator of today is not content to let a major question remain forever in the air, the football of endless indecisive games. Either it can be ruled out or it must be answered: that is his credo. Something may rule out the question as meaningless, as quantum mechanics rules out any possibility to find out simultaneous values for the position and momentum of an electron. Or something may establish the issue to be undecidable, as Gödel has proved certain propositions to be undecidable. But in the absence, as here, of some clear indication that the question is meaningless or undecidable, the question must be faced and the relevant evidence sought out.

FOUR LINES OF EVIDENCE

If no two detectives turn up the same clues in a murder case, perhaps no two investigators regard the same pieces of evidence as relevant to a major issue that is still far from resolution. However, one inquirer's search of the literature and questioning of many colleagues ends up, at least today, always with the same four points at the center of consideration:

(1) Einstein's general relativity theory of cosmology leads to big bang or collapse or both–not periodicity, not big bang, collapse and reexpansion, not big bang, collapse and 'reprocessing' [Wheeler (1971); see also chapter 44 in Misner, Thorne, and Wheeler (1973)], but big bang and collapse; and big bang or collapse or both argue for the mutability of the laws of physics.

(2) There is no law of physics that does not lend itself to most economical derivation from a symmetry principle. However, a symmetry principle hides from view any sight of the deeper structure that underpins that law and therefore also prevents any immediate sight of how in each case that mutability comes about. Moreover, no search has ever disclosed any ultimate underpinning, either of physics or of mathematics, that shows the slightest prospect of providing the rationale for the many-storied tower of physical law. One therefore

suspects it is wrong to think that as one penetrates deeper and deeper into the structure of physics he will find it terminating at some nth level. One fears it is also wrong to think of the structure going on and on, layer after layer, *ad infinitum*. One finds himself in desperation asking if the structure, rather than terminating in some smallest object or in some most basic field, or going on and on, does not lead back in the end to the observer himself, in some kind of closed circuit of interdependences. The final two points fuel this thought:

(3) A book of Henderson (1913) and papers of Dicke (1961), of Carter (1974), and of Collins and Hawking (1973a,b) give evidence that substantial changes in certain of the constants or initial data of physics would rule out, not only life and consciousness as we know them, but even the planets and elements-heavier-than-hydrogen that would be needed for almost any other imaginable form of life. This line of reasoning raises a central question. Could the universe only then come into being, when it could guarantee to produce 'observership' in some locality and for some period of time in its history-to-be? Is 'observership' the link that closes the circle of interdependences?

(4) David Hume (1779) asked, "What peculiar privilege has this little agitation of the brain that we call *thought* that we must make it a model for the entire universe?" In conformity with this assessment it was long natural to regard the observer as in effect looking at and protected from contact with existence by a 10 cm slab of plate glass (Figure 1a). In contrast, quantum mechanics teaches the direct opposite. It is impossible to observe even so minuscule an object as an electron without in effect smashing that slab and reaching in with the appropriate measuring equipment (Figure 1b). Moreover, the installation of apparatus to measure the position coordinate, x, of the electron automatically prevents the insertion in the same region at the same time of the equipment that would be required to measure its velocity or its momentum, p; and conversely. The act of measurement typically produces an unpredictable change in the state of the electron. This change is different according as one measures the position or the momentum. This is the strange feature of quantum mechanics that led Niels Bohr to say [Weizsäcker (1971)], "If a man does not feel dizzy when he first learns about the quantum of action, he has not understood a word". The choice one makes about what he observes makes

machinery that makes it mutable. No source is evident, either in physics or mathematics, for any 'ultimate underpinning' at the bottom of it all. Is this a clue that the structure of physics, rather than having any 'bottom', returns in the end full circle to the observer with which it began?

(3) No reason has ever offered itself why certain of the constants and initial conditions have the values they do except that otherwise anything like observership as we know it would be impossible.

THE FOUR THESES IN FOUR PHRASES;
AND ONE CENTRAL THEME AND THESIS

Look again in a moment at (1) 'mutability', (2) 'no ultimate underpinning', (3) 'observership as prerequisite for genesis', also to be known in what follows as the 'anthropic principle' of Dicke and Carter, and (4) 'observer-participator as definer of reality'. However, pause here to ask if these four central points together suggest any still more central theme and question. If so, question though it were and must for long remain, it could bind the points together and bring a certain helpful unity to the discussion.

No other way has disclosed itself to bring the four assortments of evidence into tight connection except to ask, is the universe a 'self-excited circuit'? Does the universe bring into being the observership, and the observership give useful meaning (substance, reality) to the universe? Can one only hope some day to understand 'genesis' via a proper appreciation of the role of the 'observer'? Is it his indispensable role in genesis that is some day to explain the otherwise so mysterious centrality of the 'observer' in quantum mechanics? Is the architecture of existence such that only through 'observership' does the universe have a way to come into being?

The collective knowledge of mankind makes an overwhelming pyramid but it is not evident that any of it bears more directly than physics on the question 'how the universe came into being'. Even physics itself is a gigantic structure of observation, theory and experiment; but out of it all it is not clear that any evidence lays more claim to attention in connection with genesis than what we have out of the four named areas and out of the active work going on today in every one of them. To bind these results and this work together into any coherence demands a central theme and thesis. The search for such an architectural plan can and will continue. Up to now, however, no pattern suggests itself from the available clues except this, to interpret quantum mechanics as evidence for the tie between genesis and observership.

EXPLORATION AS FIRST STEP TOWARDS
REVISION AND ADVANCE

To advocate the thesis 'genesis through observership' is not the purpose here; nor is it the purpose to criticise the thesis. It is too frail a reed to stand either advocacy or criticism. The purpose here is rather to explore the thesis.

At least four objections offer themselves against investigating the question 'How did the universe come into being?', along lines such as these: (1) the question is meaningless, therefore is beyond answer, and therefore should be rejected; (2) thinkers have debated for centuries before and after Leibniz and Berkeley between 'realist' and 'idealist' views of existence and the debate is as far as ever from being ended today, so what profit is there in re-raking these stale issues? (3) the laws of physics persist forever – so it makes no sense to ask how they 'came into being'; (4) any exploration of the concept of 'observer' and the closely associated notion of 'consciousness' is destined to come to a bad end in an infinite mystical morass. Every one of these objections has the same counsel about the present exploration: 'Drop it!' Moreover, we have to be open to the possibility that any one of them may be right – or all of them. However, it is not the way of science to sit inactive in the face of mystery. More acceptable to the many active in this field today is a continuing search for more evidence and a continuing attempt to bring it into order. Nothing better suggests the outlook of science in the search for a plan than the motto of the engine inventor John Cris, "Start 'er up and see why she don't run". That 'start-up' is beyond our power here because we are not even sure we have before us all the parts of the 'engine', still less a plan of how they fit together. Handicappéd though we are by these circumstances, it is not obvious that we are in any worse position than any engine inventor. He does not know that his engine will ever go. We know that ours 'runs'.

No better way is evident to get on with the exploration of this 'engine' than to review again the four central points in greater detail.

THESIS 1. *Gravitational Collapse Argues for the Mutability of all Physical Law*

About the conditions of extraordinary density and temperature that prevailed in the first few minutes of the universe we have today a wealth of evidence that one would hardly have dared hope for twenty years ago. Radiotelescopes less than a meter long operating in the range of wavelengths of a few centimeters have brought evidence of the so-called primordial cosmic fireball radiation, or 'relic radiation' from the time when the universe was enormously smaller than it is today and enormously hotter. Measurements of the abundances of the elements and their isotopes and analysis of the mechanism of formation of these nuclear species reveal more about what went on around $\sim 10 \times 10^9$ yr ago than we know about the biochemistry in our own interiors in the here and the now. The prediction of a big bang at the hands of Friedmann out of Einstein's 1915 and now standard geometrical theory of gravitation is now strongly supported by the evidence. Estimates of galactic distances as they stood 20 years ago were wrong, we realize today, by a factor of the order of 6. With the correction of this error interest has dwindled in such alternative views of cosmology as 'the continuous creation of matter' and a 'steady state universe'.

The simplest cosmology compatible with Einstein's theory of gravitation is characterized by a 'big stop' as well as by a 'big bang'. In this model the fluctuations in density from galaxy to galaxy are idealized as smoothed out and the geometry of space is treated as homogeneous and isotropic, curved equally in all directions and everywhere by the same amount. Einstein gave arguments [collected in Wheeler (1975b)] that this bending should be great enough to curve space up into closure, making the geometry that of a '3-sphere', the 3-dimensional analog of the 2-sphere of the familiar geographic globe. If he had stated this requirement for closure as an equation, 'closure'

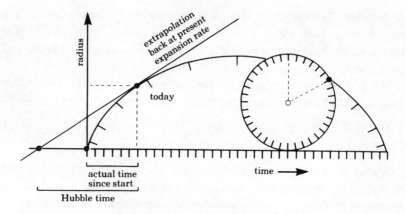

Illustrative values all derived from

Time from start to now	10	x 10⁹ yr
Hubble time now	20	x 10⁹ yr

Hubble expansion rate now	49.0 $\dfrac{\text{km/sec}}{\text{megaparsec}}$
Rate of increase of radius now	0.66 lyr/yr
Radius now	13.19 x 10⁹ lyr
Radius at maximum	18.94 x 10⁹ lyr
Time, start to end	59.52 x 10⁹ yr
Density now	14.8 x 10⁻³⁰ g/cm³
Amount of matter	5.68 x 10⁵⁶ g
Equivalent number of baryons	3.39 x 10⁸⁰

FIG. 2. Sample figures for Friedmann model of homogeneous isotropic 'stardust filled' universe [from Misner, Thorne and Wheeler (1973)]. The last 8 numbers are derived, via Einstein's standard gravitation theory, from the first two numbers treated as 100% accurate, although the present uncertainty in each is of the order of 30%: (1) the actual age of the universe as estimated from the rate of evolution of stars and star clusters; and (2) the extrapolated or 'would-be' or 'Hubble' age of the universe; that is, the time it *would* have taken for galaxies to get to their presently observed separations at their presently observed rates of recession (slower, by reason of the continued pull of gravity, than their speeds of recession in the intervening time).

could have been regarded then and thereafter as a built-in boundary condition and necessary part of general relativity. Instead he left it in the form of words and the alternative view that the universe is 'open' is often explored.

Space bending is proportional to matter density. However, the amount of matter *seen* in galaxies today falls short by a factor of the order of 30 compared to that required by Einstein's theory for closure

[Oort (1958); Gott, Gunn, Schramm, and Tinsley (1974)]. Therefore it is a remarkable development that Ostriker and Peebles (1973, 1974) find reason to believe that galaxies contain in their outer fringes 3 to 20 times as much matter as one had previously attributed to the visible parts of the galaxies. An intense search is now under way for direct observational evidence of faint stars or other objects in galactic halos. It is conceivable that insufficient matter is present; that the universe is open; and that the universe goes on expanding forever. However, if the searches now under way reveal the 'missing matter' and the conditions for closure are verified this will be another impressive confirmation of Einstein's conception of general relativity. In this event Figure 2 gives a quantitative impression of the predicted variation with time of the radius of the universe. Collapse or 'big stop' is symmetric to big bang.

Extreme models of the singularity and big bang and big stop have been proposed, including a so-called 'whimper' singularity [Ellis and King (1974); Criss et al. (1975)] and "closed-in-space to closed-in-time transition" [Misner and Taub (1968)] very different in character from the singularity of simple Friedmann cosmology. However, all available evidence [Hawking and Penrose (1969)] suggests that the Friedmann description of singularity is closer to describing the extreme conditions encountered in the generic case [Belinsky, Lifshitz, and Khalatnikov (1971)]. Moreover, when a cloud of matter collapses to a black hole and gives rise to a singularity in the here and now, this singularity is not topologically distinct from the final cosmological singularity but [Penrose (1974)] part and parcel of it (see Figure 3). Thus if the universe is closed – as Einstein argued, and as available evidence allows one to believe or disbelieve – one has a choice whether to rocket in to the nearest black hole and encounter the singularity in the near future or wait some tens of 10^9 years down the road to encounter the singularity. But escape? It is not evident how. Big bang, black hole and big stop mark *The Gates of Time*.

How can one see the lesson of gravitational collapse – whether big bang or black hole or big stop – in a larger framework? No concept puts itself forward with greater force than 'mutability' [Wheeler (1973)]. To summarize the message of gravitational collapse and the rest of physics, Figure 4 symbolizes the progress of physics by a

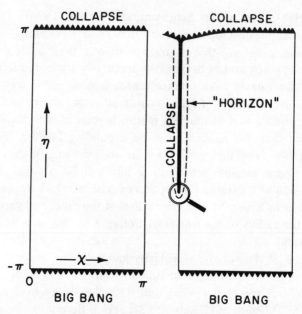

F<small>IG</small>. 3. Black hole singularity as part of, and inseparable from, the final cosmological singularity. Left, the spacetime of the ideal Friedmann model universe. Right (schematic and illustrative only), spacetime as modified by the formation of a black hole (by the collapse of a star or cloud of gas) about $\frac{1}{3}$ of the way through the history of the universe. (In the diagram at the left, space points are distinguished for convenience by an angle coordinate rather than a distance coordinate; $\chi = 0$ at the 'N-pole' of the 3-sphere; $\chi = \pi$ at the 'S-pole'; and the time parameter, η, is so defined that light rays – not shown – run at 45° angles in the diagram. At the limits marked in heavy black, according to classical general relativity, the spacetime geometry becomes singular and the tide-producing components of the gravitational field become infinite.).

staircase. Each tread marks a new law. Each riser symbolizes the achievement of conditions sufficiently extreme to overpower that law. Density (step one) seemed a conserved quantity until one discovered that sufficiently great pressures alter density (next riser in Figure 4). Chemical valence (step two) won even wider fields of application until one found that temperatures can be raised so high (next riser) that the concept of valence loses all usefulness. Physics advanced to the point of assigning a charge number and a mass number to each atomic nucleus, crystallizing the age-old idea of the fixity of atomic properties. But then came the day of nuclear transmutations under conditions never before matched, altering the charge number and mass number.

With the advent of still higher energies it was possible to break into the world of elementary particle physics. There no principles proved themselves more reliable than the laws of conservation of particle number: in every reaction the total number (particles minus antiparticles) of heavy particles or baryons is conserved; and conserved also is the net number of light particles or leptons (indicated tread in Fig. 4). However, black hole physics 'transcends' these laws of conservation of particle number [Wheeler (1971); Hartle (1971, 1972); Bekenstein (1972a, b); Teitelboim (1972a, b, c)]. According to all available evidence a black hole of macroscopic dimensions has properties uniquely characterized by its mass, charge, and angular momentum, and no other parameter (Fig. 5). To distinguish between three black holes of the same mass, charge and angular momentum, one built from matter, one from antimatter, and one primarily from radiation, no method, even of principle, has ever disclosed itself. The law of conservation of baryons is not violated. Rather it has lost all possibility of application. It is transcended. We have passed (Fig. 4) the next tread and the next riser of the staircase of law and law transcended. The laws of conservation of mass-energy, electric charge and angular momentum (next

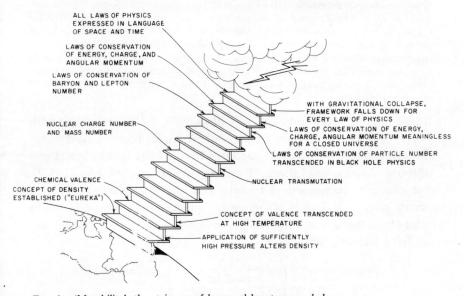

FIG. 4. 'Mutability': the staircase of law, and law transcended.

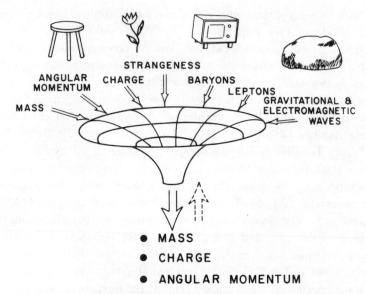

Fig. 5. A black hole of macroscopic dimensions extinguishes all the qualities of everything dropped in. The resulting object is characterized at the classical level by mass, charge, and angular momentum and by no other parameter. (A black hole of small dimensions evaporates particles [Hawking (1975)] and thereby has the possibility to give evidence of additional, non-classical parameters, or 'chemical potentials' [Wheeler (1974)]. However, no possibility offers itself in either case to compare total number of particles in with total number of particles out [Wheeler (1975b)]. A large black hole requires an essentially infinite time to evaporate. A small black hole sends *out* particles at a countable rate. However, no way is evident to form such an object except primordially, in the big bang, an event depriving the term, 'number of particles that went *in*', of all observable meaning.)

tread) rule the transformations of a black hole and its interaction with other black holes. However, none of these would-be conserved quantities has an application when one turns from a finite system to a closed universe [Landau and Lifshitz (1951) and Weber and Wheeler (1957)]. For example, every method to determine the total mass-energy of a system goes back to or is in some way equivalent to placing a planet in orbit about that system, measuring the period and size of that orbit and applying Kepler's law,

$$\text{mass (in cm)} = \left(\frac{2\pi}{\substack{\text{period, in cm} \\ \text{of light-travel time}}} \right)^2 (\text{radius, in cm})^3$$

Surrounding the closed universe, however, there is nowhere to put a planet in orbit. Space offers no asymptotically flat region for defining a 'translation in space' or a 'translation in time'. Or in mathematical terms the law of conservation of energy applied to a closed universe degenerates to the vacuous identity, $0 = 0$. All three familiar conservation laws are transcended in the physics of a closed universe (next riser in Fig. 4). Or, to state the point in still more vivid terms, the last thing one has to worry about is where the energy 'comes from' that starts the big bang.

Finally we come to the last level of the staircase. There never has been a law of physics that did not demand 'space' and 'time' for its statement. However with gravitational collapse we come to the end of time. Never out of the equations of general relativity has one been able to find the slightest argument for a 'reexpansion' or a 'cyclic universe' or anything other than an end. Moreover in gravitational collapse – whether big bang or black hole or big stop–it is not the matter alone that collapses. The space in which that matter is located also collapses. With the collapse of space and time the framework falls down for everything one ever called a law of physics.

Not law but mutability has the last word. It is not evident how to put the lesson of gravitational collapse in any more pregnant form than this. The physics seems to cry out, 'Find a way to state each law that lets that law fade from being!'

THESIS 2. *Law Derived from Symmetry; but Symmetry Hides the Mechanism of Mutability*

Of all laws of physics, among the oldest are the laws of elasticity; and none illustrate more vividly the dual themes of law derived from symmetry, and symmetry as concealment. A homogeneous isotropic solid, subjected to a small deformation, stores an amount of energy equal to the sum of two terms. One term is proportional to the trace of the square of the tensor of deformation; the other, to the square of the trace. The constants of proportionality that multiply these terms – the two 'elastic constants' – completely characterize the behavior of the substance under small deformations; there is no place for any other constants or parameters. This is elasticity in brief, as it follows from the symmetry of the solid by standard arguments based on the mathematical theory of groups of transformations (here rotations and translations).

Not from elasticity does one learn the origin of the elastic constants, but from a level of structure that elasticity conceals from view: atoms and molecules. Each interaction is described by a complicated molecular potential energy curve. The second derivatives of all these interaction potentials have to be taken and multiplied by appropriate direction cosines and added to reproduce the elastic constants. Those two constants give far too little information to allow one to work backward even to the existence of atoms, let alone the interactions between them.

'Elasticity' is transcended, its laws fade, and a new way of description is forced on us when we tear the solid or pluck out atoms here and there or load it with dislocations or vaporize it. We are driven to a new mode of analysis of what goes on, based on the twin ideas of atoms and interatomic forces or 'chemical physics'. The many regularities and symmetries of chemical forces are too well known to bear recounting here. Important for our theme is only this, that the hundred regularities of chemistry in their turn completely shield from view the deeper structure underpinning them. The mechanism by which valence fades from relevance at high temperature is concealed by the laws of valence. And even more deeply hidden by those laws and symmetries is the electronic structure in which valence takes its origin. Lewis, Langmuir, Heitler, London, Pauling, and others taught us that it was in no way required nor right to try to explain each complication of chemical bonds with a corresponding complication of principle. All have their origin in something so fantastically simple as a system of positively and negatively charged masses moving in accordance with Schrödinger's equation. Yet only with the help of Schrödinger's equation could one understand how the laws of valence fade when temperatures or pressures rise without limit.

A hundred years of the study of elasticity would never have revealed chemical bonds; and a hundred years of study of chemical bonds would never have revealed Schrödinger's equation. The direction of understanding ran, not from the upper levels of structure to the deeper ones but from the deeper ones to the upper ones. The foundations for the mutability of a law are hidden by the structure of that law.

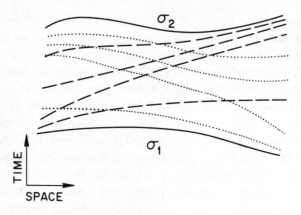

FIG. 6. The 'history of deformation' indicated by the dashed hypersurfaces leads from initial-value hypersurface σ_1 to final-value hypersurface σ_2. So does the history indicated by the dotted hypersurfaces. The physics on σ_2 resulting from a complete specification of the initial value data on σ_1 must be independent of the history one chooses to integrate along in passing from σ_1 to σ_2 via the Hamiltonian equations of motion. This heavy but simple requirement suffices to fix the form of the Hamiltonian both for the dynamics of a vector field (giving Maxwell theory) and for the dynamics of the 3-geometry itself (giving Einstein's geometrodynamics) [Hojman, Kuchař, and Teitelboim (1973)].

Space, like elasticity, Einstein taught us, is a participant in physics and not only an arena for physics. Its geometry is dynamic and changes with time according to a geometrodynamic law of beautiful symmetry. Many derivations of Einstein's field equation have been given over the years but none more compact and none more penetrating than that of Hojman, Kuchař, and Teitelboim (1973) [see Figure 6; see also Teitelboim (1973a, b) and Kuchař (1973, 1974)]; but the very perfection of these derivations again denies us any insight into how this law fades at the beginning of the big bang and in the final stage of collapse.

Given the mutability of other physical laws and given gravitational collapse, it is difficult to see how one can escape two conclusions for space itself: (1) the law for its dynamics must also be mutable, must also be superseded, must also be transcended under sufficiently extreme conditions; and (2) underneath must lie some deeper structure, call it 'pregeometry' or call it what one will.

For envisaging that deeper structure one is tempted to try as guide one by one all the structures one has ever seen in the worlds of physics and mathematics, from crystal lattices to standing waves and from

Borel sets to the calculus of propositions. However, nowhere in all this hunting around has any one ever found any structure that compels acceptance by its simplicity and scope [Patton and Wheeler 1975a); Finkelstein *et al.* (1974)]. Physics offers no sufficient model. Mathematics offers no unique structure as central. It is conceivable that all those who have searched through physics and mathematics for guide to the 'pregeometric' structure underpinning space – and all physics – have been wanting in imagination and that the necessary model stands ready at hand. However, it is also conceivable that one has so far looked in the wrong domain for the central point. Could it be that the quantum is trying to tell us the answer? Could it be that the *observership* of quantum mechanics is the ultimate underpinning of the laws of physics – and therefore of the laws of time and space themselves?

THESIS 3.   *Observership as Prerequisite for Genesis; or the 'Anthropic Principle' of Dicke and Carter*

No consideration argues more forcibly that the 'observer' has nothing to do with the scheme of physics that the disparity in size of 26 powers of 10 between man and the universe; and none argues more strongly that life and consciousness are a rather unimportant development in a faraway and not particularly relevant part of space. Quite another assessment of the situation develops when one turns one's attention from scales of distance to scales of time. In no way is one led more directly into this altered approach than through the considerations of Dicke (1961).

Paraphrased, Dicke asks what possible sense it could make to speak of "the universe" unless there was someone around to be aware of it. But awareness demands life. Life in turn, however anyone has imagined it, demands heavy elements. To produce heavy elements out of the primordial hydrogen requires thermonuclear combustion. Thermonuclear combustion in turn needs several times 10^9 years cooking time in the interior of a star. But for the universe to provide several times 10^9 years of time, according to general relativity, it must have a reach in space of the order of several times 10^9 lyr. Why then is the universe as big as it is? Because we are here!

Rephrase the argument to make it seem less paradoxical. Why should it need such an enormous universe to allow consciousness and observership in a limited region of space for a limited length of time? Why $\sim 10^{11}$ galaxies, and $\sim 10^{11}$ stars per galaxy, and all this to bring about life on one planet of one star of one galaxy? Why not economize! It may be too much to strike for the whole potential economy factor of $\sim 10^{22}$; but at least introduce an economy factor of 10^{11}! That 'budget cut' would still leave an enormous amount of matter, enough to make one galaxy. The difficulty is not in the resulting size but in the resulting reduction in time scale. Instead of having $\sim 10^{11}$ yr from big bang to big stop the universe would have only ~ 1 yr. This is not long enough to produce stars or planets, let alone life. From this point of view it is not at all obvious that there is any extravagance in the scale of the universe.

The 'anthropic' requirement on the universe, that it should last long enough to be able to give rise to life, Dicke (1961) and Carter (1974) translate into numbers. The central quantity connecting the observer and the universe in their considerations is the lifetime of a typical 'main-sequence' star. As Carter puts it, "at times much later than this [characteristic star life,] the ... [universe] will contain relatively few (and mainly very weak) energy producing stars, whereas at times much shorter than this the heavy elements (whose presence seems necessary for life) could not have been formed." One typical main–sequence star is the sun, with mass 1.989×10^{33} g. The mass-energy available from it through thermonuclear reactions is about one percent of the total, or $\sim 2 \times 10^{52}$ erg. The luminosity of the sun is 3.9×10^{33} erg/sec. Thus the lifetime of the sun is of the rough order of $\sim (2 \times 10^{52}$ erg$)/(4 \times 10^{33}$ erg/sec$) \sim 5 \times 10^{18}$ sec or $\sim 10^{11}$ yr. On the 'anthropic principle' of Dicke and Carter it is not unreasonable that the time from big bang to the development of life should be of the same order, $\sim 100 \times 10^9$ yr. Whether the universe lives that long or much longer or lives forever they leave open; and leave open also whether it is open or closed. However if we take seriously Einstein's arguments for closure, and the concept of 'economy', we arrive at a figure of the same order for the time from big bang to collapse. This consideration thus supplies an order-of-magnitude way to estimate the size of the wheel that generates the cycloid of Figure 2.

Carter applies the anthropic principle to a second feature of the universe, this time not an 'initial condition' (the size of the generating wheel in Figure 2) but one of the most famous of the 'physical constants', the so-called inverse fine-structure constant,

$$\alpha^{-1} = \frac{\text{(quantum of angular momentum) (speed of light)}}{\text{(elementary unit of electric charge)}^2} = \frac{\hbar c}{e^2} = 137.036.$$

His line of reasoning runs from α^{-1} to convection in the atmosphere of a star, and from convection to the formation of a planet as platform for life. Carter notes that the physics of the formation of a planet [see especially Alfvén and Arrhenius (1975)] is not yet fully elucidated, but that angular momentum is an important factor. Thus in the solar system the planets carry only a fraction of one percent of the mass, but about 99 percent of the angular momentum [Allen (1964)]. How is a star to spin off most of its angular momentum and form planets? Have for some time in its evolutionary history – probably just before it joins the main sequence – a strongly convective atmosphere: that is Carter's proposal in brief. In support of this proposal he notes that it is an "empirical fact that the larger stars [in which convective transfer of energy from interior to surface takes second place to radiation flow] ... retain much more of their angular momentum than those which ... [like the sun, transfer energy to the surface primarily by convection]". He goes on to show that the relative importance of convection and radiation flow is governed by a high power of the factor $\alpha^{-1} = \hbar c/e^2$. Thus a value for this physical constant only a few percent higher than '137' would make the stars in the main sequence all radiative blue stars, with relatively little convection and therefore no obvious possibility to produce planets.

What about other constants and initial conditions of physics and their influence on the possibilities for life? Regge (1971) discusses the possibility that the ratio of mass between proton and electron has a critical bearing on some subtle step of biological replication – if not upon its possibility (one knows that deuterated DNA will 'work'), at least upon its reliability. Other consequences of physics for the possibility of life, from the special properties of water to the special features of CO_2 and from regulatory process of the ocean to the chemistry of carbon, hydrogen and oxygen, are summarized in the long known book of Henderson (1913) (see especially the reprint updated by the very interesting introduction of Wald).

It is one thing to say that changes in the constants and initial conditions would have changed the character of life. It is quite another to establish clearly that a substantial change in one or another direction in one or another parameter would have ruled out any plausible possibility for life whatsoever as in the case of a model universe that is too small and lives too short a time; or as may be the case for a value of $\hbar c/e^2$ substantially larger than 137. The anthropic principle of today is no conjecture too vague to stand analysis, but a hypothesis that admits and demands investigation on a dozen fronts. Already the state of the subject today raises with renewed force a great question.

If an anthropic principle, *why* an anthropic principle? Envisage as Carter does 'an ensemble of universes' in only a very small fraction of which life and consciousness are possible? Or ask as we do now if no universe at all could come into being unless it were guaranteed to produce life, consciousness and observership somewhere and for some little length of time in its history-to-be?

THESIS 4. *Participatory Observership as the Source of All Useful Meaning*

'The universe, meaningless or not, would still come into being and run its course even if the constants and initial conditions forever ruled out the development of life and consciousness. Life is accidental and incidental to the machinery of the universe'. Or, going beyond the anthropic principle, is the directly opposite view closer to the truth, that the universe, through some mysterious coupling of future with past, required the future observer to empower past genesis?

Nothing is more astonishing about quantum mechanics than its allowing one to consider seriously on quite other grounds the same view that the universe would be nothing without observership as surely as a motor would be dead without electricity.

Is observership the 'electricity' that powers genesis? What does quantum mechanics have to say about the role of the observer in defining 'reality'? This newly developing science of 'electricity' may still be in the Leyden jar and galvanic cell stage but it also has the equivalent of Franklin's kite and key experiment in the Einstein-Podolsky-Rosen experiment (1935).

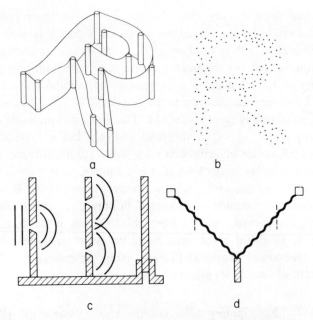

FIG. 7. Four ways to symbolize the role of observation in the definition of 'reality' ('R'). (a) The results of observations and experiments [regarding which we can 'communicate to others' ... "in plain language" ... "what we have done and what we have learnt" (Bohr 1963, p. 3)] serve as the iron posts of reality; all else is *papier maché* plastered in between them. (b) Image of 'R' on the retina envisaged as a pattern of dots made by individual photons. (c) Whether the photographic plate at the right [see Bohr (1949)] records interference between the waves coming from two slits or tells through which slit the photon has come depends upon the choice – which is principle can be delayed until the last moment – whether one will hold fixed the receptor screen at the right or measure the vertical component of momentum imparted to it by the photon. No means is open to determine both. But doing the one observation is incompatible with doing the other ('complementarity'). (d) The EPR experiment depicted in a spacetime diagram (space coordinate measured to the right; time advancing upward). Lower hatched world line, segment of history of (e^+, e^-) atom up to the point of disintegration into two photons. Their world lines run at 45° in the diagram. Each is recorded by a detector (square block) provided it makes its way through the polarization analyzer (dashed line) (see text for discussion).

In the conceptually simplest version of the EPR experiment (Fig. 7d) the very light atom composed of one positive and one negative electron in a ground state of zero angular momentum annihilates. Two photons fly off with equal energy and opposite momentum and polarization. 'Opposite polarization' is the heart of the lesson. Whatever the polarization (up and down, or perpendicular thereto, or right-handedly circular, or left-handedly circular) recorded by the right hand detector it is guaranteed that the left hand analyzer, when adjusted to the opposite polarization, will pass the other photon. It does not matter whether the right hand detector is 5 lyr away on the one side and the left hand one 5 lyr away on the other. It does not matter whether one waits up to the very last picosecond to decide whether to turn that right hand polarizer to accept up-down polarized photons or right-handedly polarized photons. Whichever the choice, if the arriving photon gets through one is immediately guaranteed that the other photon, by now 10 lyr away, has the opposite polarization. To Einstein, Podolsky, and Rosen (1935) this result was contradictory to any reasonable concept of reality. In their view, "physics is an attempt conceptually to grasp reality as it is thought, independently of its being observed". In contrast Bohr insisted that what counts in any proper definition of reality is not the system (here the two photons) alone but a larger unity composed of the system plus the measuring equipment. He went on to stress (1935) that " ... these [experimental] conditions constitute an inherent element of the description of any phenomenon to which the term 'physical reality' can be properly attached". Inescapable, he concluded, is "a final renunciation of the classical ideal of causality and a radical revision of our attitude towards the problem of physical reality" [for more on the great debate over the years between Bohr and Einstein, perhaps the greatest debate in the history of science, see Bohr (1949) for a summary, the relevant chapters of Jammer (1974) for details, and Bohr (1963) for Bohr's most developed statement of the situation; and for a summary of several experiments done to date to check the predictions of quantum mechanics, see Freedman and Holt (1975)].

What was 'reality' during 5 years while the photons were flying from source to receptor? What polarization did they have all that time? Meaningless, utterly meaningless, that question is, we have to say. Try another question: When did the polarization of the photons 'come into being'? At the moment of emission? At the moment when the one or the other of them 'first' (not a well defined idea, according to special relativity!) passes through an analyzer? At the instant when one of the detectors 'first' records a photon? Not one of these answers can be defended, nor even the question itself. The biggest question about those photons is, what question does it make sense to ask about them? Happily the mathematical formalism of quantum mechanics is set up in just such a way as to allow one to ask all physically meaningful questions. Moreover in the interpretation of that formalism one has learned enough by now not to be misled by appearances. The appearance of the wave function is strange in a Lorentz or inertial frame of reference, moving with such a velocity that the photon is registered in the right hand counter a year before it is registered in the left hand counter, because formalistically the probability amplitude for this or that state of the left hand photon appears instantaneously to have adjusted itself to the newly gained information. One at first sight gains the impression that the basic principle has been violated, that no information can be propagated at a speed in excess of the speed of light. Deficit though this is in the intermediate stages of that particular version of the quantum mechanical formalism, no such difficulties appear in the final predictions about the outcome of the experiment. Moreover Wigner (1973) has given the beginnings of a formalism which in its predictions is identical to that of the usual wave mechanics but which nowhere brings into evidence any nonphysical effects. This development reemphasizes Bohr's (1963, pp. 5, 6) words about the Schrödinger state function, "that we are here dealing with a purely symbolic procedure, the unambiguous physical interpretation of which, in the last resort, requires a reference to a complete experimental arrangement". He adds "that all departures from common language and ordinary logic are entirely avoided by reserving the word 'phenomenon' solely for reference to unambiguously communicable information".

There is no more remarkable feature of this quantum world than the strange coupling it brings about between future and past, between (1) the way one will choose to orient his polarization analyzers two years down the road and (2) what one can then say about the photons that already – as judged by us now – had their genesis 3 years ago. There is a sense in which the polarization of those photons, already on their way, 'was' brought into being by the disposition of the polarization analyzers that the observer has yet to make and will make. Moreover there is no difference in principle whether the time between the emission of a photon and its detection is 5 years; or a nanosecond, as it is when one is looking at an object a foot away; or 10^{10} years, as it is when one is receiving direct or indirect radiative evidence of what went on in the first few seconds after the big bang. In each case what one chooses to ask (as for example by the setting of the polarization analyzer) forms an inseparable part of a phenomenon that in earlier thinking one would have said had 'already happened'. In this sense it is incontrovertible that the observer is participator in genesis.

It was a long step from the kite and key experiment to the electromagnetism of a later era with its quantitative measurements, its units, and its all-embracing theoretical structure. Even in elucidating the EPR experiment the quantum mechanics of today has only gone a small fraction of the way, one can believe, down the corresponding long road. If it took years to go from such qualitative notions as 'intensity currents' and 'quantity currents' to such quantitative concepts as inductance and capacity, it may well take even longer to arrive at a theoretical framework [see for example Wigner (1973)] deep enough to define in and by itself what is to be meant by 'observer' and 'observership'. Wigner remarks that a measurement is only then complete when the result has entered the consciousness. Bohr says more: that an observation is only then an observation when one can communicate the result to another in plain language. Or to turn to the role of observation in defining meaning, that central topic in the philosophical thought of our time, we have these words of Føllesdal (1975) in the great Oxford Lectures on Mind and Language, "Meaning is the joint product of all the evidence that is available to those who communicate." How is one to translate these qualitative notions into quantitative ones?

It is appropriate to make a distinction between quantum mechanics and the quantum principle; between the formalism that predicts the properties of matter and radiation, and the larger framework of ideas that defines both what can be measured and how the observer participates in bringing about that which he measures. The one field is well developed. It is the other that would seem to be in the primordial state of electricity 200 years ago. It provides one small illustration of the many confusions that can arise to recount a conversation overheard between two students of quantum theory. A: "The wave function does not describe the electron. It describes the state of the observer's knowledge about the electron". B: "With that I agree and everyone agrees". A: "And therefore there are as many wave functions (or density matrices) as there are observers of that electron". B: "Oh, no. Nobody can accept that". A: "Oh, yes. The logic of the situation leaves no alternative". B: "What state function the electron is in is determined by the last measurement on the electron. It is determined by an experiment. But an experiment is only an experiment when the outcome is expressed in the form of communicable knowledge, knowledge that can be shared. That is why there is only one wave function for the electron, not many". A: "Now I see; and now I agree". And so do we!

So then what is 'observership'? It is too early to answer. Then why the word? The main point here is to have a word that is not defined and never will be defined until that day when one sees much more clearly than one does now (except in the foregoing obvious instance) how the observations of all the participators, past, present and future, join together to define what we call 'reality'. It is an inspiring circumstance for future developments along this line that the formalism of quantum mechanics as it already stands has furnished us and continues to furnish us with such a trustworthy guide.

One does not have to know Maxwell electrodynamics to perceive a thunderbolt; and one does not have to trace out the unbelievably many ties of 'observership' with the deepest questions of knowledge, meaning and quantum theory to accept willingly or unwillingly – but accept – the strange sense in which observation now participates in defining the reality of the past: direct involvement of observership in genesis.

The question is often asked, how can the existence or non-existence of an observer have the slightest consequence for the existence of the universe? Do we not know that the universe went through many 10^9 years of astrophysical evolution before there was any life around? We will not give the answer of the old limerick (Knox, undated):

> There was a young man who said, "God
> Must think it exceedingly odd
> If He finds that this tree
> Continues to be
> When there's no one about in the Quad."

REPLY

Dear Sir:
    Your astonishment's odd;
I am always about in the Quad.
    And that's why the tree
    Will continue to be,
Since observed by
               Yours faithfully,
               God.

Rather – changing the words of the limerick a little – we will say that 'observership' allows and enforces a transcendence of the usual order in time. However difficult it may be today to spell out the conception of genesis as taking place through observership, it is difficult to see any other line that lends itself to exploration. What other way of genesis is there?

Quite the opposite view is often taken: that the development of life is accidental and unimportant in the scheme of things. It is therefore interesting to read old views of the spontaneous generation of life [see for example Roger (1963)]: Take a glass jar, fill it half full with grain, stuff its mouth with a piece of a dirty shirt. Place it in the corner of the room for 21 days. Life will develop in the shape of a mouse! Compare with today's view of 'life as an accident': Take a universe, stuff it with $\sim 10^{80}$ particles, let it run $\sim 10^{10}$ yr, and life will develop! Is it conceivable that this new prescription for life is on no better footing than the old one? Quantum mechanics has led us to take seriously and explore the directly opposite view that the observer is as essential to the creation of the universe as the universe is to the creation of the observer.

WHAT IS MISSING FROM THE QUANTUM STORY

Today the quantum is recognized as the central principle of every branch of physics. However, in its many features of indeterminism, complementarity, and interference of probability amplitudes, it has sometimes seemed something strange, incomprehensively imposed from outside on an unwilling world of physics. If it were fully understood would its inevitability in the construction of the universe not

stand out clearly for all the world to see? And could it not then be
derived, along with all its mathematical superstructure, from some
utterly simple first principle? Until we shall have arrived at this basic
idea we can even say that we have not understood the first thing about
the quantum principle. That rationale is what is missing from the
quantum story.

If the situation poses a challenge and a question, is not the central
role of the observer in quantum mechanics the most important clue we
have to answering that question? Except it be that observership brings
the universe into being what other way is there to understand that
clue?

"Everything that happens in the world is like a great game in which
nothing is fixed in advance except the rules", say Eigen and Winkler
(1975) in their beautiful new book on the role of chance in biological
evolution. But unless the blind dice of mutation and natural selection
lead to life and consciousness and observership at some point down the
road, the universe could not have come into being in the first place,
according to the views under exploration here; there would be nothing
rather than something.

CONCLUSION

We have reviewed (1) the evidence out of the big bang that the
universe did come into being and the evidence that not only space but
all the structures and laws of physics are mutable. (2) In the domain in
which each law is applicable that law states itself most compactly as the
consequence of a symmetry; but that symmetry hides any view of the
next layer of structure. Not the slightest hint presents itself of any
'ultimate' layer of structure either of mathematics or physics. One is
led to ask whether the interconnections between one level of the
structure and the next, rather than continuing forever, do not close up
full circle on the observer himself. Nothing speaks more strongly for
this thesis than (3) the anthropic principle of Carter and Dicke and (4)
the indispensable place of the participating observer – as evidenced in
quantum mechanics – in defining any useful concept of reality. No way
is evident to bring these considerations together into a larger unity
except through the thesis of 'genesis through observership'.

Despite the many penetrating considerations that have been advanced over the centuries about how the universe came into being it is not evident that anyone has the requisite private judgment to declare out of his inner resources which story, if any, is the right one; nor is it evident that many would believe him if he did. What is required in the analysis of genesis is not private judgment, but public judgment – which is to say, science. If judgment is an awareness of all the factors in a situation and an appreciation of their relative importance, science evidently includes that and more: falsifiability. Not only should the individual steps along the way open themselves to detailed analysis. In addition the final overall picture should, in the words of Mendeléev, "expose itself to destruction". It should make predictions that are new and testable. Both analysis and testing are public. Any qualified analyst from anywhere has the democratic right to trace out and verify for himself one by one the logical links in a chain of analysis. Any qualified observer from anywhere has the democratic right to do an experiment, or make an observation, and see for himself whether it agrees with a prediction.

To such logical analysis and observational test there lend themselves many pieces of cosmology, and many parts of the quantum story, but up to now no overall picture of genesis whatsoever. Will 'genesis takes place through observership' ever start yielding testable, falsifiable, consequences? First then, we can believe, if and when one discovers how to *derive* the quantum principle from this concept.

Acknowledgment

The author expresses thanks to several colleagues for discussions of this report, and especially to John Bell, Miroslav Benda, Brandon Carter, David Deutsch, Robert H. Dicke, B. Kanitscheider, Gary Miller, John Moffitt, Karl Popper, and Eugene P. Wigner.

Joseph Henry Laboratories   Now: *Physics Department*
Princeton University   *University of Texas*
Princeton, New Jersey 08540   *Austin, Texas* 78712

NOTES

* Preparation for publication assisted in part by National Science Foundation Grant GP30799X to Princeton University.
* Invited contribution given August 30, 1975 at the Fifth International Congress of the Division of Logic, Methodology and History of Science of the International Union of History and Philosophy of Science at the University of Western Ontario, London,

Ontario, originally under the title, "How Did the Universe Come Into Being: Forever Beyond Explanation?", as updated for publication March 15, 1976 in the light of many much appreciated discussions both at the London meeting and after similar presentations (also under the original title) at the University of Florida, Gainesville, January 26, 1976; the University of Waterloo, Ontario, January 29; and the Columbia University Seminar on the Philosophy of Science, New York City, January 30. A portion of the present report was presented (in reoriented form) at the annual meeting of the American Association for the Advancement of Science, February 23, 1976, under the title, "What is Missing from the Quantum Story?".

BIBLIOGRAPHY

Alfvén, H., and G. Arrhenius, 1975, *The Structure and Evolutionary History of the Solar System*, Reidel, Dordrecht.

Allen, C, W., 1964, *Astrophysical Quantities*, The Athlone Press, University of London, London (second edition 1964).

Bastin, T., ed., 1971, *Quantum Theory and Beyond*, Cambridge University Press [cited under Weizsäcker (1971)].

Bekenstein, J., 1972a, "Nonexistence of baryon number for static black holes", *Phys. Rev.* **D5**, 1239–1246.

Bekenstein, J., 1972b, "Nonexistence of baryon number for static black holes, II", *Phys. Rev.* **D5**, 2403–2412.

Belinsky, V. A., E. M. Lifshitz and I. M. Khalatnikov, 1971, "Oscillatory mode of approach to a singularity in homogeneous cosmological models with rotating axes", *Zh. Eksp. & Teor. Fiz.* **60**, 1969–1979. English trans. in *Sov. Phys. – JETP* **33**, 1061–1066 (1971).

Bohr, N., 1949, "Discussion with Einstein on epistemological problems in atomic physics", pp. 201–241 in Schilpp (1949); reprinted in Bohr (1958), pp. 32–66.

Bohr, N., 1958, *Atomic Physics and Human Knowledge*, John Wiley and Sons, New York.

Bohr, N., 1963, *Essays 1958–1962 on Atomic Physics and Human Knowledge*, Wiley-Interscience, New York.

Carter, B., 1974, "Large number coincidences and the anthropic principle in cosmology", in M. S. Longair (1974), pp. 291–298.

Collins, C. B., and S. W. Hawking, 1973a, "The rotation and distortion of the universe", *Mon. Not. Roy. Astron. Soc.* **162**, 307–320.

Collins, C. B., and S. W. Hawking, 1973b, "Why is the universe isotropic", *Astrophys. J.* **180**, 317–334.

Criss, T. B., R. A. Matzner, M. P. Ryan, Jr. and L. C. Shepley, 1975, "Modern theoretical and observational cosmology", in Shaviv, G. and J. Rosen (1975), 33–107.

Darwin, C., 1863, March 29 letter from Down to Joseph Dalton Hooker; in Darwin, F. (1919), p. 203. Warm appreciation is expressed to Frederick Burckhardt for locating this source of the cited passage.

Darwin, F., ed., 1919, *The Life and Letters of Charles Darwin, Including an Autobiographical Chapter*, Appleton, New York and London.

Debever, R., ed., 1974, *Astrophysics and Gravitation; Proceedings of the Sixteenth Solvay Conference on Physics*, Editions de l'Université de Bruxelles, Brussels.

Dicke, R. H., 1961, "Dirac's cosmology and Mach's principle", *Nature* **192**, 440–441.

Eigen, M., and R. Winkler, 1975, *Das Spiel; Naturgesetze steuern den Zufall*, München, Piper.

Einstein, A., B. Podolsky, and N. Rosen, 1935. "Can quantum-mechanical description of physical reality be considered complete", *Phys. Rev.* **47**, 777–780.
Ellis, G. F. R. and A. R. King, 1974, "Was the big bang a whimper?", *Comm. Math. Phys.* **38**, 119–156.
Finkelstein, D., G. Frye and L. Susskind, 1974, "Space-time code V", listed as "Phys. Rev., to be published, 1974"; June 1974 preprint of D. Finkelstein, "Space-time code, VI"; and prior papers of D. Finkelstein cited in V.
Føllesdal, D., 1975, "Meaning and experience", pp. 25–44 in S. Guttenplan, ed., 1975.
Freedman, S. J., and R. A. Holt, 1975, "Test of local hidden-variable theories in atomic physics", *Comments Atom. Mol. Phys.* **5**, 55–62.
Gott, J. R. III, J. E. Gunn, D. N. Schramm and B. M. Tinsley, 1974, "An unbound universe?," *Astrophys. J.* **194**, 543–553.
Guttenplan, S., ed., 1975, *Mind and Language*, Clarendon Press, Oxford [cited under Føllesdal (1975)].
Hartle, J. B., 1971, "Long-range neutrino forces exerted by Kerr black holes", *Phys. Rev.* **D3**, 2938–2940.
Hartle, J. B., 1972, "Can a Schwarzschild black hole exert long-range neutrino forces?" in Klauder 1972.
Hawking, S. W., 1975, "Particle creation by black holes", *Commun. Math. Phys.* **43**, 199.
Hawking, S. W., and R. Penrose, 1969, "The singularities of gravitational collapse and cosmology", *Proc. Roy. Soc. London* **A314**, 529–548.
Henderson, L. J., 1913, *The Fitness of the Environment*, Macmillan, New York; 1958 reprint, with an introduction by George Wald, Beacon Press, Boston.
Hojman, S. A., K. Kuchař and C. Teitelboim, 1973, "New approach to general relativity", *Nature Physical Science* **245**, 97–98.
Hooker, C. A., ed., 1973, *Contemporary Research in the Foundations and Philosophy of Quantum Theory*, Reidel, Dordrecht [cited under Wigner (1975)].
Hume, D., 1779, *Dialogues Concerning Natural Religion*, Robinson, London; 1947 reprint by Thomas Nelson, London; U.S. edition thereof, Bobbs-Merrill, Indianapolis and New York; p. 148 in latter edition. I thank Wesley Salmon for this reference.
Isham, C. J., R. Penrose and D. W. Sciama, eds. 1975, *Quantum Gravity; an Oxford Symposium*, Clarendon Press, Oxford [cited under Patton and Wheeler (1975)].
Israel, W., ed. 1973, *Relativity, Astrophysics and Cosmology*, Reidel, Dordrecht and Boston [cited under Kuchař (1973)].
James, W., 1911, *Some Problems of Philosophy*, New York, pp. 38–40. I owe this quotation to Gary Miller.
Jammer, M., 1974, *The Philosophy of Quantum Mechanics*, Wiley, New York.
Klauder, J., ed., 1972, *Magic without Magic: John Archibald Wheeler*, W. H. Freeman, San Francisco [cited under Hartle (1972)].
Knox, R., undated, limerick quoted in Bertrand Russell, *A History of Western Philosophy*, p. 648.
Kuchař, K., 1973, "Canonical quantization of gravity", in Israel (1973).
Kuchař, K., 1974, "Geometrodynamics regained: a Lagrangian approach", *J. Math. Phys.* **15**, 708–715.
Landau, L. and E. Lifshitz, 1951, *The Classical Theory of Fields*, translated by M. Hamermesh, Addison-Wesley, Reading, Mass.

Longair, M. S., ed., 1974, *Confrontation of Cosmological Theories with Observational Data*, I.A.U., Reidel, Dordrecht. [cited under Carter (1974) and Penrose (1974)].

Misner, C. W., K. S. Thorne and J. A. Wheeler, 1973, *Gravitation*, Freeman, San Francisco.

Misner, C. W. and A. H. Taub, 1968, "A singularity-free empty universe", *Zh. Eksp. & Teor. Fiz.* **55**, 233. English original in *Sov. Phys. – JETP* **28**, 122, 1969.

Mehra, J., ed., 1973, *The Physicist's Conception of Nature*, Reidel, Dordrecht [cited under Wheeler (1973)].

Oort, J. H., 1958, "Distribution of galaxies and the density of the universe", in *Onzième Conseil de Physique Solvay: La Structure et l'évolution de l'univers*, Editions Stoops, Brussels.

Oparin, A. I., 1938, *The Origin of Life*, ... with annotations by Sergius Morgulis, New York, Macmillan.

Ostriker, J. P., P. J. E. Peebles, and A. Yahil, 1974, "The size and mass of galaxies and the mass of the universe," *Astrophys. J. Lett.* **193**, L1–4.

Ostriker, J. P., and P. J. E. Peebles, 1973, "A numerical study of the stability of flattened galaxies; or, can cold galaxies survive?", *Astrophys. J.* **186**, 467–480.

Patton, C. M., and J. A. Wheeler, 1975, "Is physics legislated by cosmogony?," pp. 538–605 in Isham, C. J. et al., eds., 1975. This paper lays out for study the idea that the universe is 'a self-excited circuit' in which the universe gives rise to the observer and the observer – in development of the ideas of quantum mechanics – gives a meaningful existence to the universe. It recalls the concept of Parmenides of Elia that "what is ... is identical with the thought that recognizes it" and the concept of George Berkeley "To be is to be perceived". However, through the kindness of Prof. B. Kanitscheider and Prof. Baumgartner of the University of Giessen, I have since learned that already in the writings of the philosopher F. W. J. von Schelling (1775–1854) there is an earlier statement, in older language, of the present thesis – as Prof. Kanitscheider puts it – "dass das Universum von vornherein ein ihm immanentes Ziel, eine teleologische Struktur, besitzt und in allen seinen Produkten auf evolutionäre Stadien ausgerichtet ist, die schliesslich die Hervorbringung von Selbstbewusstsein einschliessen, welches dann aber wiederum den Entstehungsprozess reflektiert und diese Reflexion ist die notwendige Bedingung für die Konstitution der Gegenstände des Bewusstseins." For details he cites M. Schröter (ed.), 1958–59, *Schellings Werke, nach der Originalausgabe in neuer Anordnung herausgegeben*, München, Beck, in 6 vols. See especially, in Vol. 5, his 'Darstellung des Naturprocesses' (fragment of a series of lectures on the principles of philosophy given at Berlin the winter of 1843–1844, from the handwritten literary remains), particularly these words from pp. 428–430: "... liessen wir die Idee auseinander treten in ihre Momente, damit sie durch Wiederkehr in die Einheit sich verwirkliche. Das Auseinandergehen und successiv Wiedereinswerden dieser Momente ist die Natur. Die Wiederherstellung der Einheit ist ihr Ende und der Zweck der Natur. Die Wiederherstellung der Einheit ist die Verwirklichung der Idee. Die verwirklichte Idee ist der Mensch, und er ist die Intention nach nur diese ... der Mensch hat keinen Zweck, denn er ist selbst Zweck, er ist nur, um Bewusstsein zu sein, und das Bewusstsein ist der Zweck; der Mensch ist also nichts als Bewusstsein, und nicht noch etwas anderes ... Zu dem Menschen hat das gesammte Weltall mitgewirkt ... Weil er das Existirende ist, so waren alle Potenzen des Universums, alle diese getrennten Momente bestimmt, in ihm als in der letzten Einheit zusammenzugehen ... der

Mensch ist...nicht speciell ein Produkt der Erde–er ist ein Produkt des ganzen Processes–nicht die Erde allein, das ganze Weltall ist bei ihm betheiligt, und wenn aus der Erde, so ist er,...doch nicht ausschliesslich für sie, er ist für alle Sterne, denn er ist für das Weltall, als Endzweck des Ganzen, erschaffen. Wenn er als locales Wesen erscheint, so ist er diess nicht ursprünglich, er ist localisirt worden: wie? diess muss durch die Folge sich zeigen. Er ist, wie gesagt, das universale Wesen, und sollte daher nicht an einem bestimmten Punkt, er sollte im Ganzen wohnen..."

Penrose, R., 1974, "Singularities in cosmology", in Longair, M. S. (1974), 263–272.

Regge, T., 1971, "S-matrix theory of elementary particles", pp. 395–402 in Verde, M., ed. 1971; see p. 398 for the cited proposal.

Roger, J., 1963, *Les Sciences de la Vie dans la Pensée Française*, Armand Colin, Paris. See p. 101 for the reference to Jean-Baptiste von Helmont, 1648, *Ortus medicine* and the story of the mouse. I owe this reference to the kindness of Thomas Hankins.

Schilpp, P. A., ed., 1949, *Albert Einstein: Philosopher-Scientist*, Library of Living Philosophers, Evanston, Illinois and subsequent paperback reprints elsewhere; cited under Bohr (1949).

Shaviv, G. and J. Rosen, eds., 1975, *General Relativity and Gravitation; Proceedings of the Seventh International Conference (GR7), Tel-Aviv University, June 23–28, 1974*, Wiley, New York [cited under Criss et al. (1975) and under Wheeler (1975b)].

Teitelboim, C., 1972a, "Nonmeasurability of the baryon number of a black hole", *Let. al Nuovo Cimento* **3,** 326–328.

Teitelboim, C., 1972b, "Nonmeasurability of the lepton number of a black hole", *Let. al Nuovo Cimento* **3,** 397–400.

Teitelboim, C., 1972c, "Nonmeasurability of the quantum numbers of a black hole", *Phys. Rev.* **D5,** 2941–2954.

Teitelboim, C., 1973a, "How commutators of constraints reflect the spacetime structure", *Ann. Phys.* **79,** 542–557.

Teitelboim, C., 1973b, "The Hamiltonian structure of spacetime", Ph.D. thesis, Princeton University.

Verde, M., 1971, *Atti del Convegno Mendeleeviano; "Periodicità e Simmetrie nella Struttura elementare della Materia"*, Accademia delle Scienze di Torino, Torino. [cited under Regge (1971)].

Weber, J. and J. A. Wheeler, 1957, "Reality of the cylindrical gravitational waves of Einstein and Rosen", *Rev. Mod. Phys.* **29,** 509–515.

Weizsäcker, C. F. von, 1971, "The Copenhagen interpretation", pp. 25–31 (cf. p. 26 for the cited quote) in T. Bastin, ed., 1971.

Wheeler, J. A., 1971, "Transcending the law of conservation of leptons", *Atti del Convegno Internazionale sul Tema: The Astrophysical Aspects of the Weak Interactions*, Roma, Accademia Nazionale dei Lincei, 133–164.

Wheeler, J. A., 1973, "From relativity to mutability", pp. 202–247 in Mehra, ed. (1973).

Wheeler, J. A., 1974, "The black hole", pp. 279–316 in Debever, ed. (1974).

Wheeler, J. A., 1975, "Conference summary: more results than ever in gravitation physics and relativity", pp. 299–344 in Shaviv and Rosen, eds. (1975).

Wigner, E. P., "Epistemological perspective on quantum theory", pp. 369–385 in Hooker, ed. (1973).

I. A. AKCHURIN

THE METHODOLOGY OF PHYSICS AND TOPOLOGY

I should like to call attention to certain recent advances in the methodology of physics; specifically, I wish to call attention to certain recent advances in the essentially new mathematical-conceptual formulation of physical laws.

Along with many others, I believe that several of these new mathematical-conceptual structures will soon play a role in our theory of empirical scientific knowledge similar to that played by the propositional and predicate calculi in the epistemological analysis of deductive knowledge. In particular, it seems that the explication of a materialistic dialectics in the language of modern category theory and non-trivial topologies in Grothendieck (1960) schemas may for the first time allow us to effect a theoretical synthesis of biological and physical knowledge similar to the synthesis of physics and chemistry achieved in quantum theory.

1. NEW ASPECTS OF THE UNITY OF PHYSICS FROM THE VIEWPOINT OF FIBER SPACES AND SPECTRAL SEQUENCES

I would suggest the adoption of a unified topological point of view in our attempt to attain a 'cohomological' methodology of concrete research in modern theoretical physics. Various important results of this approach have already been summarized in, for example, the following publications: with regard to mechanics, the review of S. MacLane (1970) and the books of C. Godbillon (1969) and R. Abraham (1967); with regard to relativistic electrodynamics, the papers of C. Misner and J. A. Wheeler (1957), E. C. Zeeman (1967) and A. Trautman (1973); with regard to quantum theory, the works of B. Kostant (1970), F. I. Berezin (1974) and A. S. Mischenko (1974). To put it very briefly and therefore very crudely, the methodology here

Butts and Hintikka (eds.), Foundational Problems in the Special Sciences, 35–45.

consists in considering certain generalized geometrical aspects of all three of these fundamental physical sciences – mechanics, relativistic electrodynamics and quantum theory – from the viewpoint of the mathematical theory of fiber spaces. Perhaps the first strikingly impressive example of fiber structures was given by Plato through his reference to shadows on a cave wall.

The fiber spaces of modern mathematics present a generalization of the products of mathematical structures with each other, where these latter no longer appear as simple quantities but more abstract characteristics of the objects studied – e.g. the Lorentz group or the groups of symmetry for various kinds of interactions of elementary particles. From an intuitive point of view, the simplest kind of topological fibered space is a structure both 'outside' of a given system of independent parameters (i.e. outside of a 'base') and at the same time 'attached' to it in a very stable way, 'rigidly' following the variation of these parameters (in any continuous transformation – i.e. any homotopic mapping – of the base, the mutual 'coordination' of its fiber points with those of the base remains invariant).

The simplest fiber structures are those of classical mechanics. The role of the base in this case, the independent variable determining the variation of all the other characteristics of the mechanical motion of bodies, belongs to time. The fiber, on the other hand, is the set of dependent variables of mechanics: i.e. the spatial (three-dimensional) coordinates of the moving points. The laws of mechanics – Newton equations – allow us to discover, on the basis of the second derivatives of these coordinates, new and more fundamental elements of physical reality. These are the forces which represent in mechanics the only cause of variation in the state of moving objects.

In the classical theory of fields, it is exactly these forces (i.e. the forces that act on unitary charges) which appear as the main object of investigation. The magnitudes of these field strengths are placed in the fiber of the fiber space, with the base (i.e. the set of independent variables) now being the whole four-dimensional continuum of space and time. The fundamental laws of field theory (e.g. the Maxwell equations in classical electrodynamics) allow us to employ certain combinations of the derivatives of the field strengths with respect to spatial and temporal coordinates in order to elucidate new and still more fundamental elements of physical reality; i.e. the charges and currents generating the force fields under consideration.

Quantum theory adopts as its main object of investigation these new elements of physical reality (i.e. charges and currents), but their variations (their 'motions') are now regarded as occurring (with a certain degree of probability) along all possible trajectories connecting the initial and final points of motion rather than as taking place only along a single trajectory (as in the case of classical mechanics). In quantum theory (at least in the formulation of it given by R. P. Feynman), the base of the fiber space of the theory is thus the space of all possible paths in the four-dimensional space-time continuum, and the fiber (i.e. the set of dependent variables) is the set of probabilities associated with these paths and expressed in a certain way through the wave functions.

The equations of motion in quantum theory (the Schroedinger, Dirac equations, etc., or Feynman's methods of functional integration, equivalent to the former from the mathematical point of view and accounting for the contribution of each possible trajectory of quantum motion to its total sum) reveal yet more fundamental elements of physical reality: namely, the most probable forms of the stable and quasistationary motions of micro-particles in atoms, molecules, crystals, etc., as well as their excited states (oscillatory, rotational, etc.).

The peculiar feature of all these three fundamental physical disciplines is that in each of them there is a special, dynamically most 'symmetric' state of variation (i.e. of motion) of their objects, corresponding in the simplest case to the inertial motion of bodies in mechanics. This state is such that the fundamental equations of motion of a given discipline (e.g. Newton's, Maxwell's, or Schroedinger's equations) can be now interpreted (from the point of view of algebraic topology) simply as cohomological characterizations of the degree of deviation in the behaviour of a given concrete physical system from its dynamically most 'symmetrical' state of motion. The latter is singled out by the fact that a certain generalized operation of taking the 'dynamic boundary' (always defined in these disciplines and with its properties in application to physical states being similar to those of the well-known 'static' operation of taking the boundary of a given set) results in zero if repeatedly applied to this state. The situation is simplest in mechanics; here the dynamically most symmetric state of motion – the inertial one – is defined by the fact that for it the second derivatives of the relevant coordinates with respect to time are equal to zero (the operation of taking the 'dynamic boundary' in mechanics is simply the taking of the temporal derivative). In electrodynamics, the role of inertial motion belongs to the field states, the four-dimensional potentials of which satisfy the wave equation with the right-hand side equal to zero: this can also be interpreted as a result of the repeated application to the field potentials of a certain abstract four-dimensional 'dynamic boundary' operation.

From the point of view of algebraic topology, the field states characterized by these kinds of potentials play the role of the so-called 'acyclic complex', a role corresponding dynamically to that of the inertial motion in mechanics. The forces in Newton's equations as well as the charges and currents in Maxwell's equations may be viewed simply as cohomological measures of the deviations of a given concrete physical system from this 'physical acyclic complex' (i.e. from the dynamical state of maximally 'simple' topological structure). If one could now provide a similar topological interpretation of the quantum mechanical theory of perturbation and could identify in the latter the corresponding operation of taking 'quantum dynamic boundaries', the problem of the theoretical unity of physics could be formulated in a quite different manner.

It would then be possible to formulate it as a problem concerning the existence of a certain spectral sequence (Spanier, 1966) which is unique for all physics. The sequence in question would characterize the uniform topological structure of the fiber spaces of mechanics, field theory, and quantum theory as though they were 'inserted' into each other. Indeed, spectral sequence is a special system of invariants determining the topological 'structure' of certain rather general fiber spaces in modern mathematics, and the relations among the first terms of such sequence appear similar to those obtained above for the dynamic-topological invariants of mechanics, field theory, and quantum mechanics.

In each of these disciplines, the system of invariants at a given level (mechanics, electrodynamics etc.) is defined as a measure of the deviation of the characteristic structures of moving objects at this level from a dynamical state of an especially high degree of topological simplicity; once again, this special state of motion is such that the repeated application to it of a certain 'dynamic boundary' operation (specified only at this level) results in zero. In the next term of the spectral sequence, the system of cohomological invariants of the preceding level becomes an object of investigation in respect of its topological properties and behaviour relative to the action on them of some other, more broadly defined differential operation (i.e. the 'taking' on them of another 'dynamic boundary'). The purpose at each of these successively higher levels of analysis is to discover new and more fundamental dynamical topological invariants of material motion.

Thus, our problem number 1 is this: is it possible to obtain a unified spectral sequence for 'all' physics and in this way to establish a deeper correspondence and mutual coherence among the respective operations of taking the 'dynamic boundary' in mechanics, field theory, and quantum mechanics?

2. Topos and the Existence of Isomorphisms between Biological and Physical Structures

All of the above, as well as the most recent results of relativistic cosmology, strongly suggests the conclusion that the most important place in physics now belongs to the concept of topos (Grothendieck, 1972). This concept will be defined here only in an intuitive way as the modern analog of a Riemannian space, but with the major novelty that, in topoi, the topology (not only metrics) may vary from one point to another. Topoi have already demonstrated their quite unique unifying efficiency in mathematics, bringing together into a single conceptual whole algebraic geometry, mathematical logic, and set theory, as well as the intuitionistic constructions. They have also shown, for example, that intuitionistic mathematics is first and foremost a mathematics of topologically non-trivial objects.

Through his concept of superspace, J. A. Wheeler (1968) appears already to have introduced the concept of topoi into physics; he has in fact done so only in a very special case, namely, in the case of topoi with metrics (i.e. with a well-defined concept of distance). At the same

time, however, topoi are perhaps the most variable objects yet discovered by man: a topos may represent an intuitionistic construction, a
set, an algebraic space, a scheme, or as was shown recently and quite
unexpectedly even for us, a magnificent Siamese cat of my colleague,
V. A. Smirnov.

It would seem that among the most promising applications of topoi
are those relating to the conceptual synthesis of modern physics and
biology. According to the fundamental results of the late N. N.
Rashevsky (1954), it is precisely variations in the topology of living
objects which represent the most salient features of their characteristic
forms of material motion. Topoi have finally allowed us to start a
concrete search for isomorphisms between biological and physical
structures (with the increase of time and energy, respectively), as has
been suggested by G. Cocconi (1970). We now believe that we are
even in a position to demonstrate that the biological principle of
evolution is just a 'four-dimensional generalization' of the 'three-
dimensional' principle of complementarity in quantum mechanics; the
latter need only be formulated in terms of topoi.

To summarize this matter very briefly, we may begin by observing
that, by analogy with the metric tensor of Riemannian spaces (which
tensor may vary from point to point), topoi are sometimes defined by
the initial (base) algebraic ring. Up to now, however, the topoi thus
generated have 'as a matter of course', been regarded as having only
the extremely simple structure of the number field. Operational measurement procedures in classical physics are essentially based on this
topological analog of the 'flat' Riemannian space, i.e. on a topos with
an everywhere identical and trivial topology.

However, as follows from quantum mechanics, the mere concept of
number fields is absolutely inadequate in this context for understanding of microscopic processes; indeed, it was exactly the non-classical
properties of the algebra of states of quantum objects that finally led to
the development of a detailed analysis of spaces with the base algebraic ring varying from one point to another. This kind of base algebraic
ring results from any operational procedure for fixing the states of
arbitrary quantum objects.

The diversity of the characteristic parameters of any object under
consideration as well as the variation of these parameters with time are
specified from the point of view of modern algebraic geometry by a

certain ideal of the base ring. The mathematical characteristics of the experimental device (e.g. in the investigation of the wave or corpuscular properties of microscopic objects – the coordinates of slits and openings, their width, etc.) realize as if 'automatically' the decomposition of this ideal into its so-called primary components.

Before Grothendieck, geometrical meaning (in terms of subspaces, components of curves etc.) was ascribed only to the simplest primary structures of abstract ideals – to the structures without any so called primary embedded components. The latter are apparently completely responsible for the unusual nature of the behaviour of essentially quantum objects. For when these are entered into the schemes used for the mathematical modelling of microscopic processes, the decomposition of the ideal into its primary components ceases to be unique. Only the 'automatic' fixation of some of these embedded primary components by the given experimental device allows us to restore uniqueness of the primary decomposition of the ideal; since this ideal characterizes the spatial and temporal parameters of the object under consideration, this restoration in turn leads immediately to a unique way of specifying, for example, the type of object (wave or corpuscle) that is being studied with the given experimental device.

This is the situation in the three-dimensional, that is, just the 'spatial' case, of quantum mechanics. If we now turn to the four-dimensional case of topos ('space + time'), similar mathematical characteristics of the external conditions surrounding the quantum-biological system under study are determining similarly embedded but now four-dimensional (space-time) primary ideals. The embedded primary components of ideals here characterize certain predominant and sometimes quite necessary trajectories in space-time of the individual quantum subsystems of living systems.

Thus, evolution is not interpreted here simply as a tautological 'survival of survivors', but it is shown to be conditioned by a fundamental physical principle which (when interpreted topologically) allows us to predict theoretically the inevitable appearance of certain complex, temporally coordinated trajectories of motion for the individual components of living systems. These trajectories, which the late C. Waddington (1968) calls 'chreods', seem to play a role in modern theoretical science similar to that which was played by the quantized (stable) electron orbits in atoms in the initial formulation of quantum theory.

Topoi also permit us, by means of the general duality theorems of Grothendieck-Hartshorne (1966) to consider in a more systematic way the definability of the most essential properties of an individual body through reference to the properties of its interaction with objects in the whole of the rest of the space (I am referring here to the topological generalization of Mach's principle). Crudely stated, the generalized dynamic characteristics of an object (its mass, charge, etc.), may be viewed as describing the 'force' of its interaction with the external world, as well as, in some sense, the 'speed' of its response to changes in the latter. Certain abstract duality theorems of A. Grothendieck (and P. Deligne) allow us to approximate all physically interesting effects of the external world on an object by means of a special set of operations leading to the construction of particular 'coverings' of the external world – the so-called 'dual topological complex'.

The idea behind this approach is, intuitively, to obtain all of the most essential properties of a given object from the properties of the topological complex dual to it by passing in the latter to the limit with respect to a series of 'ever-diminishing coverings' of the external world. For a rather large class of the base algebraic rings – for the so-called Gorenstein rings – the limiting dual complex of the 'external world' proves to be isomorphic to our usual Eudoxus space of one or another number of dimensions, while all of the connections of the studied object with the world outside of it appears summarized in a special 'complex of residues'. This complex of residues, due to the theory of functions of (many) complex variables, allows us to introduce a measure of, among other things, the 'inertial' aspect of the responses of an object to the action of external forces. According to this analysis, inertial mass is just a certain residue on the complex plane, where this interpretation of mass is perfectly analogous to that given in the method of Regge poles.

From this preliminary analysis, it appears as though the complex of residues is, as it were, the most general dynamic characteristic of a physical object, a total 'summary' of all its possible interactions with the external world; as yet, however, this conclusion is justified only with respect to the best studied and simplest cases, where each characteristic under consideration may be reduced to a numerical magnitude (e.g. mass or charge). In other cases, the complex of residues may contain additional global characteristics of dual connections without necessarily being able to reduce any of them to a number. Here, the characteristics are of a more general algebraic nature (the groups of symmetry, etc.) and the situation is more complicated.

Thus, we arrive at our problem number 2: can we find in high energy physics structures responsible, for example, for the specific shape of the Canadian maple leaf?

3. THE SEARCH IN PHYSICS AND BIOLOGY FOR AN INFINITY INTERMEDIATE BETWEEN COUNTABLE INFINITY AND THE CONTINUUM

The conceptual scheme proposed here with a view to furthering a theoretical synthesis of physics and biology on the basis of the concept of topos provides some hope that we may find within its framework the solution of still another interesting problem in modern science; *problem number 3: what is the physical and biological meaning of Paul J. Cohen's (1966) results?*

Topos is so variable a structure that it can contain non-Cantorian (Lawvere, 1972) as well as Cantorian sets (i.e. it can contain sets with intermediate 'steps' of infinity between countable infinity and the continuum as well as sets without such 'steps'). What kinds of physical and biological factors could determine such variations of set-theoretic structures? We still do not know, but the approach to this problem taken by D. Scott and R. Solovay seems to indicate at least one promising direction of research.

It seems that in this case we should think primarily in terms of a search for new operational tools for localizing the objects under study-tools which will have a very different nature from our classical method of attempting to 'catch the lion in the desert'. The latter now actually represents the most universal and general operational 'archetype' of any object identification and individuation. But it is based on and implies only a trivial topology: it is not applicable even on the torus, let alone in general topoi. We may say that it is just because in physics and biology topology offers this broad kind of variability that it has proven by no means less fundamental than logic and set theory. It may well prove even more fundamental than they are.

In the final analysis, it seems to be precisely our topology which determines our choice of objects and our most general, 'ontological' categorizations of their properties. Ever since the time of our III'rd Congress of LMPS in Amsterdam, our colleague B. d'Espagnat (1973) has conducted a very thorough and profound analysis of the philosophical meaning of the topological properties of non-separability of quantum objects. According to him, even the concept of atomism is based essentially on an hypothesis to the effect that there is in the world only one, trivial topology (Eudoxus' topology); thus, atoms are seen to be, as it were, but the 'artefacts' of a trivial, flat topology. The postulation of such a topology, however, seems at present to be altogether untenable, not only in microphysics, but also in relativistic cosmology and theoretical biology.

Specific theories concerning one or another concrete type of topology's taking place at a particular level of physical (or biological) reality should, in my estimation, today be regarded only as hypotheses concerning the most fundamental scientific basis of every physical and biological theory. Among other things, it seems quite possible that at some further stage in the development of the concepts discussed here, it will be precisely they which will allow us to answer the kinds of questions which I have raised in this paper, which will allow us to decide, for example, just which physical and biological objects are described by Cantor's set theory. Regarding this last question, one can already speculate on the basis of intuitive considerations that the boundary between Cantorian and non-Cantorian sets is passed somewhere in the area where essentially random (i.e. quantum) types of object behavior come into play. In order to establish hypotheses such as this in a rigorous fashion, however, we shall first have to concentrate upon 'operationalization', an effective manner of interpreting physically the 'localising categories' of modern algebra in the sense of C. Faith (1973).

ACKNOWLEDGEMENTS

It is a pleasure to thank Professor Robert S. Cohen and Blane Little for many discussions and help in correcting my English.

Academy of Science, Moscow

BIBLIOGRAPHY

Abraham, R.: 1967, *Foundations of Mechanics*, Benjamin, New York.
Berezin, F.: 1974, 'Kwantowanije', *Izwestija AN SSSR, ser. matem.* **38**, p. 1116.
Cocconi, G.: 1970, 'The Role of Complexity in Nature', in M. Conversi (ed.), *The Evolution of Particle Physics*, Academic Press, New York, p. 81.
Cohen, P. J.: 1966, *Set Theory and Continuum Hypothesis*, Benjamin, New York.
D'Espagnat, B.: 1973, 'Quantum Logic and Non-Separability', in J. Mehra (ed.), *The Physicist's Conception of Nature*, Reidel, Dordrecht, p. 714.
Faith, C.: 1973, *Algebra: Rings, Modules and Categories*, Vol. 1, Springer Verlag, Berlin.
Godbillon, C.: 1969, *Géométrie Différentielle et Mécanique Analytique*, Hermann, Paris.
Grothendieck, A.: 1960, *Éléments de Géométrie Algébrique*, t.t. I–IV, Presses Universitaires, Paris.
Grothendieck, A.: 1972, *Théorie des Topos et Cohomologie Étale des Schémas*, v.v. 1–3, Springer Verlag, Berlin.
Hartshorne, R.: 1966, *Residues and Duality*, Springer Verlag, Berlin.
Kostant, B.: 1970, 'Quantisation and Unitary Representations', in: *Lecture Notes in Mathematics*, Vol. 170, Springer Verlag, Berlin, p. 87.
Lawvere, F.: 1972, *Toposes, Algebraic Geometry and Logic*, Springer Verlag, Berlin.
MacLane, S.: 1970, 'Hamilton Mechanics and Geometry', *American Mathematical Monthly* **77**, p. 570.
Misner, C., Wheeler, J.: 1957, 'Classical Physics as Geometry', *Annals of Physics* **2**, p. 525.
Mišchenko, A.: 1974, *Metod kanoničeskogo operatora Maslowa*, Moscow.
Rashevski, N.: 1954, 'Topology and Life', *Bulletin of Mathematical Biophysics*, **16**, p. 317.
Spanier, E.: 1966, *Algebraic Topology*, McGraw-Hill, New York, Ch. 9.
Trautman, A.: 1973, 'Theory of Gravitation', in J. Mehra (ed.), *The Physicist's Conception of Nature*, Reidel, Dordrecht, p. 179.
Waddington, C.: 1968, *Towards a Theoretical Biology*, v. 1, Aldine, Chicago.
Wheeler, J.: 1968, *Einstein's Vision*, Springer Verlag, Berlin.
Zeeman, E.: 1967, 'Topology of Minkovski Space', *Topology* **6**, p. 161.

AXIOMATICS AND THE SEARCH FOR THE FOUNDATIONS OF PHYSICS

ABSTRACT. The paper deals with the peculiarities of the axiomatic method in physics. The heuristic role of the axiomatic method is underlined; its ability to increase scientific knowledge. The ability of axiomatic systems for development is analysed. Attention is concentrated on the dialectical character of physical axiomatics.

That nature is uniform in its diversity and is matter in development is an idea of dialectical materialism that has been generally accepted in modern physics and is reflected not only in its content but also in its methodology and logic. Modern physics has been making fruitful use of the principle of development and the principle of the uniformity of nature in its search for new phenomena and new laws. To show this is one of the primary aims of this article.

Strictly speaking, a definite general view of nature (a Weltanschauung problem), accepted in the various epochs of the historical development of physics, has always been intrinsically connected with the logic of research (a methodological problem) characteristic of it in that same epoch. That was the situation before classical physics, when physical knowledge, based on everyday observation and – with rare exceptions[1] – lacking systematic methods of research, was in complete accord with the vague and general views of the philosophers of the period, with their occasionally brilliant nature-philosophy guesswork. That was also the situation in classical physics, when Newton's method of research, subsequently called the method of principles and constituting a peculiar modification of Euclid's axiomatics, was in accord with the atomistic view of nature (which Newton also shared).

The coherence of knowledge is a reflection of the uniformity of nature. This coherence of knowledge was first expressed in axiomatics, while geometrical knowledge – the first type of knowledge to exist in its day – became a science, being axiomatically constructed by Euclid.

Relatively speaking a complete closed system of a physical theory (classical mechanics being the first to be established on these lines)

Butts and Hintikka (eds.), Foundational Problems in the Special Sciences, 47–65.

consists of basic concepts and principles (called axioms in geometrical terminology) which establish definite relationships connecting these concepts, and also of corollaries derived from these by means of logical deduction. It is these corollaries that must be in accord with experimental data, that is, must be put to the empirical test. Otherwise, no physical theory can be a *physical* theory or, in other words, in physics it is experiment alone that is the criterion of the truth of its theories, that is, it is experiment that ultimately certifies that a theory reflects objective reality and, consequently, certifies the adequacy of the mathematical formalism of that theory.

In physics, every system of concepts and principles corresponds to the mathematical formalism appropriate to it; it is a description of a definite sphere of physical phenomena indicated by experiment. The limits within which the concepts of the system (in terms of their correspondence to nature) are applicable are also established experimentally.

The axiomatic method has changed since Euclid's time, being enriched with fresh potentialities for explaining and forecasting the phenomena under research. While this method in its initial or, one might say, Euclidean form may be treated, according to S. Kleene's expression, as 'material axiomatics' (Kleene, 1952, p. 28), today, after the works of the famous mathematician D. Hilbert, and the research into mathematical logic, axiomatics appears both as 'formal' and as 'formalised' axiomatics. The two latter differ from material axiomatics in that their concepts and relationships stand out in something like a pure form free of empirical content with the language of symbols (formalism) being used in formalised axiomatics instead of verbal language, whereas in material axiomatics deduction is not in effect separated from the empirical and the visual.

Mutatis mutandis, this also applies to axiomatic constructions in physics. The axioms, or principles (which are also basic laws) of Newtonian mechanics, deal with inert mass or force, acceleration, space and time, and the relationships between these concepts. These relationships and concepts are the points of departure within the limits of Newtonian mechanics and in themselves represent idealised expressions of experimental facts. They were first set forth in Newton's *Principia* and can serve as a specimen of material axiomatics in classical physics.

The development of the axiomatic method in physics has run along basically the same lines as the development of this method in geometry. In modern physics, with its highly involved and ramified mathematical formalism, we have good grounds for speaking of the existence of formal and especially formalised axiomatics (which in a sense represents the highest point in the development of the axiomatic method). Strictly speaking, this has been fully brought out since the establishment and construction of the theories of non-classical physics. Any strict or comprehensive analysis of pertinent problems would go well beyond the framework of this paper, which is why I shall try to present no more than a general notion of the essence of the problem.

Consider the equation

$$F = \frac{\mathrm{d}(mv)}{\mathrm{d}t}.$$

It expresses the second law of Newtonian mechanics, which assumes that mass is a constant. But this equation can also be seen as expressing the law of the special theory of relativity; in which case m here denotes

$$m = \frac{m_0}{\sqrt{1 - \dfrac{v^2}{c^2}}},$$

where m_0 is mass of motionless body ('rest mass'), v is velocity of body, and c is velocity of light. Thus, the equation expresses the law of relativistic mechanics, which assumes that the mass of a body tends to change with its velocity.

This equation can also express the law of motion in quantum mechanics. Magnitudes in quantum mechanics and in classical mechanics are known to be linked with the same equations, but in quantum mechanics these equations contain operators, that is, magnitudes of a different mathematical nature than that of the magnitudes in classical mechanics.

The reader would be quite right in asking this question: what is the ground for this kind of 'substitution' in the equations (that is, substitution of operators for numbers, and a more complex expression for m, etc.), and what in general is their logical meaning? To answer this question is to describe the very content of classical, relativistic and quantum mechanics, and the transition of a special theory with its concepts to a more general and profound theory with its more meaningful concepts than those of the special theory. In other words, in the context of what has been said above it would mean describing the concept of mass in relativity mechanics and its origins, it would mean saying that operators in quantum mechanics are mathematical depictions of physical facts which are never treated in classical theory, it would mean describing the very logic underlying the origination of the special theory of relativity, of quantum mechanics, etc.

I have said all this to emphasise that the formal and formalised axiomatic construction of physical science comprises the development of its content, promoting an ever more profound cognition of nature. Let us note at this point that in physics special significance attaches to the question of interpretation of its formalisms as compared with the same question in mathematics; of this more below.

In what way, then, is the axiomatic method essential for physics? In both the logical and the methodological aspects the importance of this method in physics – whether in the form of material axiomatics or in its higher forms, formal and formalised axiomatics – is not merely great, but, as I shall try to show, so essential that it can hardly be overrated. If we compare it with other methods of analysis, we must agree with Hilbert, who said this about the axiomatic method in mathematics: 'Trotz des hohen pädagogischen und heuristischen Wertes der genetischen Methode verdient doch zur endgültigen Darstellung und völligen logischen Sicherung des Inhaltes unserer Erkenntnis die axiomatische Methode den Vorzug'. (Hilbert, 1900, p. 181).

Let me repeat that what Hilbert said about the axiomatic method in mathematics, I think, also applies to physical axiomatics. Of course, here, as always, one should not go to an extreme in expressing Hilbert's profound idea.

Let us start with the *genetic* method mentioned by Hilbert in the above-quoted statement about the axiomatic method. Let us look at its content, but from a somewhat different angle than Hilbert does (see p. 315 of his work referred to above). I want to speak of the role of the genetic method in cognition while stressing, in contrast to Hilbert, that in a way this method 'fits' into the axiomatic method.

How is the concept of number introduced? Assuming the existence of zero and a situation in which the increase of a number by unity gives rise to the following number, we obtain a natural series of numbers, using these to develop the laws of counting. If we take a natural number, a and add to it a unity b times, we shall obtain the number $a + b$, thereby determining (introducing) the operation of *addition* of natural numbers (together with the result called the sum).

Now by adding the numbers a which are b in number, we shall determine (introduce) the operation of *multiplication* of natural numbers and shall call the result of this operation the product of a and b, designating it as ab. Similarly – omitting the description of the process – we determine the operation of *raising to a power* and the exponent itself.

Let us look at so-called inverse operations with respect to addition, multiplication and raising to a power. Assume that we have the numbers a and b, and that it is required to find a number x satisfying the equations $a + x = b$, $ax + b$, $x^a = b$. If $a + x = b$, then x is found by means of the operation of *subtraction:* $x = b - a$ (a result which is called the difference). Similarly we introduce the operations of *division*, *extraction of root* and the *taking of logarithm* (the two latter, are inverse operations with respect to the raising to a power).

On the strength of these definitions it is possible to construct the axiomatics of natural numbers. The pertinent axioms are arranged in groups: (a) axioms of connection, (b) axioms of computation, (c) axioms of order, and (d) axioms of continuity.

I have reached the central point of my reasoning. The practical search for the solution of equations in which the numbers considered above appear indicates that inverse operations – subtraction, division, extraction of root – cannot always be performed. Let us assume now that inverse operations can be performed in every instance. Indeed, that is an assumption arithmetic has realised in the course of its historical development and as a result, as a peculiar logical resumé of its development, *positive* and *negative* numbers, whole numbers and *fractions*, *rational* and *irrational* numbers have appeared in it.

This kind of dichotomy of natural numbers into the indicated opposites and the relationships between these opposites resulted in the appearance of the concepts of relative number, of number as ratio and of real number; which means that the latter developed from the simple concept of natural number through consecutive generalisations. In modern arithmetic, the concept of real number is further developed, but what has been said will suffice for our purpose here.

The use of the axiomatic method emphasises that axiomatics does not in any sense rule out the recognition of variability of basic concepts and logically closed theories but, on the contrary, implies the need for new basic concepts and principles to emerge. Everything that makes the axiomatic method so valuable for logical formalisation and full logical substantiation of scientific theories is given in this kind of application of axiomatics its true (and in no sense a formal-logical) consummation and an expression adequate to reality.

This was very well put in mathematics by N. Bourbaki, who wrote: "C'est seulement avec ce sens du mot 'forme' qu'on peut dire que la methode axiomatique est un 'formalisme'; l'unité qu'elle confère à la mathématique, ce n'est pas l'armature de la logique formalle, unité de squelette sans vie; c'est la sève nourriciere d'un organisme en plein développement, le souple et fécond instrument de recherches auquel ont consciemment travaillé, dépuis Gauss, tous les grands penseurs des mathématiques, tous ceux qui, suivant la formule de Lejeune-Dirichlet, ont toujours tendu à 'substituer les idées au calcul' ", (Bourbaki, 1962, p. 47).

The picture is virtually the same in physics. Thus, the principle of realitivity, constituting a corollary of the principles of Newtonian mechanics, that is, of the principle of relativity in its Galilean form, did not work when it came to the speed of light, a phenomenon based on the principles of the theory of electromagnetism. This produced the task of extending the sphere in which the principles of mechanics applied, by including within it electromagnetic phenomena. But this meant that the principles of Newtonian mechanics had to constitute a single, integrated system with the principles of the theory of electromagnetism. This kind of combination led to the emergence of new concepts, which were broader and more meaningful than the concepts of classical mechanics. The first concepts to be modified were those of space and time; the concepts of absolute space and absolute time disappeared; the concepts of relative space and relative time, which turned out to be aspects of a single four-dimensional space-time continuum, appeared. Accordingly, the Galilean transformation (connecting inertial systems of reference in Newtonian mechanics and assuming absolute space and time) gave way to the Lorentz transformation, which in connecting the inertial systems of reference implies relative space and time. Here, the principle of relativity already appears in generalised Einsteinian form, showing the emergence of relativity mechanics.

Quantum mechanics is another example. This theory – it is dealt with here in view of its logically closed form – has a fundamental postulate: every physical magnitude (dynamic variable) of classical mechanics corresponds to a definite linear operator in quantum mechanics which affects the wave function, the assumption being that the relationships between these linear operators are similar to those between the corresponding magnitudes in classical mechanics. In quantum mechanics, a fundamental role also belongs to the postulate establishing the connection between the operator and the value of the magnitude characterising the reading of the measuring device (by means of which knowledge of the microobject is gained).

These two examples are in a sense a logical summary of the state of things which had taken shape in theory of relativity and quantum mechanics when these theories were constructed. Like any summary, it fails to convey the whole diversity of the logical and actual situations which had taken shape in the origination of these theories, or the details of the combination of reflection and experiment which brought to life the principles of these leading theories of modern physics. In order to avoid any possible confusion in clarifying the method used to find new concepts by means of axiomatics as indicated above, special attention needs to be drawn to the fact that axioms, once introduced into the definitions of certain basic concepts, for their part become the basis for the derivation of new, broader and more meaningful basic concepts than the original ones. Equations expressing axioms now contain symbols without any real value. The whole point is to discover these real values, that is, to discover the new concepts (and, consequently, to construct a new theory). Theoretical methods in modern physics, like the method of mathematical hypothesis and the method of 'observability in principle' (printsipial'noi nabl'udaemosti), in effect help to solve this problem. In view of circumstances of this type it becomes clear that although (to cite a well known example) the structure of axioms of relative numbers or real numbers is similar to the structure of natural numbers, this isomorphy alone does not help to find out, say, how negative numbers are added or multiplied. In the same way, the similarity of structure of the principles of classical relativity and quantum mechanics does not in itself as yet guarantee a knowledge of the basic laws of relativity and quantum mechanics, once the laws of classical mechanics are known. One should recall here what Engels said about the law of the negation of the negation. The knowledge that this law of dialectics applies both to the development of grain and the calculation of infinitesimal numbers, he says, does not enable me either successfully to grow barley, or to differentiate and integrate. As we have seen, the same applies to axiomatics, but this does nothing to reduce the fruitful methodological role either of the laws of dialectics or of axiomatics.

It is believed, and with good reason, that the possibility of expressing a theory through a system of axioms in theory is a sign that it is logically complete (closed), but in the history of knowledge and science the logical completion of theories has been most frequently regarded as something like a synonym for universality and invariability. This was historically justified, say, by the 2000-year reign (up until the mid-19th century) of Euclidean geometry as the only geometrical system, or of the 200-year domination (up until the 20th century) of Newtonian mechanics as the ultimate and indefeasible theoretical system of physics. I have sought to show that this notion is illusory when axiomatic ideas are considered in the logical plane. The logical consistency of a theory, far from ruling out its development, in fact implies such development.

It was Maxwell's electromagnetic theory that dealt the first blow to the classical idea of the axiomatic construction in physics, but the substance of this view of axiomatics did not in effect change: when the electromagnetic picture of the world was at its fullest, many physicists substituted the electromagnetic field with Maxwell's equations for the bodies of mechanics with Newton's axioms (Newtonian mechanics itself appeared to be refuted and irrelevant to the principles on which the universe was based, etc.). The standpoint of dialectical materialism on this matter was expressed at the time by Lenin, who pointed out, as the electromagnetic picture of the world was just being shaped, that it was wrong to say that materialism "necessarily professed a 'mechanical' and not an electromagnetic, or some other, immeasurably more complex, picture of the world of moving matter" (Lenin, 1968a, p. 280).

The final blow to the classical view of axiomatics in physics was dealt by the theory of relativity, and especially by the development of quantum mechanics as it assumed its modern form.

It turned out – a fact mentioned above in other contexts – that Newtonian mechanics has its limits in the sphere of phenomena which it was its task to explain and predict, that is, limits of applicability: electromagnetic phenomena on bodies in motion and also phenomena on the atomic scale cannot be either described or explained by means of the concepts and principles of Newtonian mechanics. Experimental research into these phenomena, together with an analysis of the theoretical situations arising in classical physics resulted, on the one hand, in the theory of relativity, and on the other, in quantum mechanics. Today physicists are known to have accepted the idea that no closed physical theory is absolute, that each has limits of applicability and is in that sense approximate, etc.

But how is the limit of applicability to be found and what is this limit? Let us deal with the last question first. There are phenomena which cannot be described by means of the concepts of a given theory, and if they can be so described they cannot be explained in it. This theory leaves aside the sphere of these phenomena, and the sphere of its applicability is the sphere of phenomena which it explains and can explain. In other words, a fundamentally different theory must function (that is, describe, explain and, consequently, predict) beyond the limits of applicability of the given theory.

I have no intention of considering in depth the question of the limits within which a theory applies, but one aspect needs to be dealt with more closely. The 'limit of a theory's development' is an expression which is frequently used and apparently with good logical reason. What does this expression mean and what relation does it have to the expression 'limit of, a theory's applicability' dealt with above?

This may appear to be a contrived question. It is generally said that there is no point in speaking of the development of axiomatic systems. Indeed, all the theorems of an axiomatic system can be treated as being implicit in the axioms and rules of inference; only the activity of a mathematician (or a corresponding device) can make explicit any theorem contained in it, and the axiomatic system contains an infinite number of such theorems of differing degrees of ordering. But everyone also knows that in reality the deduction of theorems from axioms is by no means a standard procedure and the derivation of, say, a geometrical (or mechanical) fact and proposition from a corresponding system of axioms – as everywhere else in the field of knowledge – the solution of the problem of discovering an unknown on the strength of *known data*. In his day Engels used to say that *even formal logic is a method for discovering new results.*

The deductive method (which includes the axiomatic method proper), like any other method applying the propositions of formal and dialectical logic, cannot disperse imagination. It is worth while to recall once again that, as Lenin put it, even "in the most elementary general idea ... there is a certain bit of fantasy" (Lenin, 1968b, p. 372). Quite naturally the role of the imagination greatly increases with the scale and depth of the generalisations in science, without which it ceases to be a science. It is a rewarding task to look closer at the role of the imagination.

Consequently, insofar as the deductive method (or the axiomatic – its highest form) leads from the known to the unknown and multiplies scientific knowledge, the axiomatic system should be regarded as a theoretical system capable of development and as developing under the right conditions. The development of axiomatic theory is the derivation of new and earlier unknown assertions and propositions within the limits of its applicability. This development of theory, as its definition suggests, proceeds, you might say, within itself; in its development theory does not go beyond its own limits, from the standpoint of its fundamentals (system of axioms) it remains the same. How then is the limit of applicability or limits of development of axiomatised theory to be found?

The answer does not, of course, lie in showing that one constructed theory contains another constructed theory, so that the first finds the limits of the applicability of the second theory in its own way, demonstrating that the latter is a limiting case of itself. That is no solution to the problem but is rather an assumption of the existence of such a solution.

Can the limits of applicability of a theory or the boundaries of the sphere of phenomena which it explains be discovered empirically?

That depends on the circumstances. The phenomena which Michelson's experiment established or the so-called 'ultraviolet catastrophe' have indeed become the marginal points for the applicability of classical mechanics: from these two 'little clouds' in the clear skies of classical physics – as they were once called – emerged the 'special' theory of relativity and quantum mechanics. But the relatively well-known fact of the movement of Mercury's perihelion, which was not covered by Newton's theory of gravitation, did not become the marginal point for the applicability of that theory. Einstein's theory of gravitation, which determined the limits within which Newton's theory of gravitation applied, was not discovered along the methodological path where the theory of relativity and quantum mechanics emerged. The principle of equivalence, implying the identity of inertia and gravitation had a crucial role to play in the formulation of Einstein's theory of gravitation, that is, in essence an experimental fact showing that all falling bodies in a vacuum have one and the same acceleration, which was known to Newton, but which he did not include in the theoretical content of his theory of gravitation but merely accepted empirically.

Thus, when it happens that an established theory fails to explain some well known experimental facts, this becomes traditional but it ultimately transpires that the theoretical interpretation or explanation (substantiation) of these facts goes beyond the limits of established theory, and this may subsequently be noticed only by a genius. That is precisely how the general theory of relativity, or Einstein's theory of gravitation, was created, for when it was being formulated it rested on the same experimental material (the same experimental basis) as Newton's theory of gravitation, but added to it a complex of new ideas alien to classical conceptions.

A logically structured theory, or axiomatised theoretical system, correctly functioning within the framework of its applicability, must be complete and non-contradictory. The non-contradictoriness and completeness of a system, as K. Gödel showed, cannot be demonstrated theoretically with the instruments of the system itself. When it comes to physical theories the fact that a theory is non-contradictory and complete is usually accepted without proof, unless the contrary is especially required, just as a theory is accepted as universal without proof where the facts are not in the way.

It follows from the latter's assertions, that if a phenomenon which stands, one might say, to be explained by a given theory, far from being explained, on the contrary produces contradictions (paradoxes) in the process of its explanation which cannot be resolved by that theory, we are entitled to regard the existence of such paradoxes as a symptom that the theory is approaching its limits.

It is quite possible, of course, that following the necessary reflection produced by the contradiction, some propositions and concepts of the theory will be specified and the contradiction can be resolved on the basis of the theory. In that case, the contradiction and its resolution merely help in the logical improvement of the theory on the basis of its own principles. *Mutatis mutandis*, the same applies to the second question about the completeness of theory. Einstein, Rosen and Podolsky formulated propositions from which ostensibly followed the incompleteness of quantum mechanics in Bohr's probability view. It turned out, however – and Bohr proved this – that Einstein was wrong: his initial proposition in the paradox in application to the problems of quantum mechanics was not unambiguous (Bohr, 1935; Einstein *et al.*, 1935). We are not interested in cases of that kind because they relate to the problem of logical improvement of a given theory in accordance with its axiomatics and not to the question of the limits of its applicability.

The movement from classical to modern physics occurred in consequence of the emergence in (classical) theory of a number of paradoxes, which I dealt with above. This peculiarity is to some extent also characteristic of Maxwell's electromagnetic theory, the immediate precursor of non-classical theories. Having brought together all the experimental data on electricity and magnetism discovered by Faraday, and having expressed these in the language of mathematical concepts, Maxwell found a kind of contradiction between the equations he obtained. To correct the situation, he added without any experimental substantiation (which came later) one expression to the equation, and the theory of electromagnetism emerged. The method of mathematical hypothesis applied by Maxwell[2] turned out to be extremely fruitful in the future as well.

Another example is Einstein's (special) theory of relativity. It emerged at the conjunction of classical mechanics and classical electrodynamics as a result of the resolution of the paradox of the contradiction between Galileo's principle of relativity and the principle that the velocity of light in a vacuum is independent of the motion of the source being analysed when considered together. M. I. Podgoretsky and Y. A. Smorodinsky called these borderline paradoxes 'contradictions of encounter' (Podgoretsky and Smorodinsky, 1969). The method of 'observability in principle' has a most important role to play in the resolution of this paradox, that is, in the formulation of the theory of relativity.

Quantum mechanics also emerged largely as the result of the resolution of the 'contradiction of encounter', in this instance between classical *corpuscular* mechanics (the same as Newtonian mechanics) and the classical *wave* theory. But here the role of the wave theory was not played by the corresponding theory of *matter*, but by the electromagnetic theory, which is why the 'encounter' was not as 'simple' as in the emergence of the special theory of relativity. Quantum mechanics emerged as a result of the resolution not only of the 'contradiction of encounter' but also of a number of others. It is essential to note that the theory that emerges is, in respect to the original one (and this also applies to the theory of relativity) a kind of metatheory, to use terms of modern logic.

Of primary importance for an understanding of the origination of quantum mechanics was the problem which could be called the problem of stability of the structure of physical bodies, molecules, corpuscles or the atoms which, from the standpoint of Newtonian mechanics, lie in the foundation of matter and whose motion ultimately determines all the changes in the universe. Newton 'found a way out of the situation' by postulating an *infinite hardness* for the original atoms of divine origin, etc. (Newton, 1952, part I). The same problem arose, you might say, in its immediately visual form, when it turned out that the 'primeval' atom became a system consisting of electrically charged particles (a positive nucleus and negative electrons), and there arose the need to solve the problem of its stability from the standpoint of the classical theory of electromagnetism. 'Rutherford's atom' is known to have owed its 'stability' not so much to the laws of the physics of the day as to the optimism of Rutherford and his associates and their confidence that a positive solution would subsequently be found for the problem. It was indeed solved in 1913 by the young Danish physicist Niels Bohr, who constructed an atomic model, applying to 'Rutherford's atom' what then used to be Planck's quantum *hypothesis*. 'Bohr's atom' turned out to be a truly stable atom, and this stability was explained by the laws of nature, that is, the ancient atom at long last 'acquired' stability and did so not because someone had assured himself and others of this on his own behalf or in the name of the Almighty, but because this was so established by the quantum laws of the motion of matter.

However, when one ponders more deeply the way the problem of the stability of the structure of the atomic particles of matter was solved, one may even find the idea of any other possible solution to be quite strange. After all, it is possible to explain the properties and motion of macroobjects by the laws of motion and the properties of their constituent microobjects – to avoid the danger of *regressus ad infinitum* – only when the properties and motion of macroobjects are not ascribed to the latter. That is what quantum mechanics did, providing a brilliant demonstration that microobjects are subject to totally different laws from those to which macroobjects are subject. But then the hardness of macrobodies, the constancy of the standards of length and time, etc., that is, all the physical characteristics of macroobjects, without which measurements and, consequently, physical cognition, are impossible, need to be substantiated – are in effect so substantiated – in quantum mechanics as the mechanics of objects on the atomic level.

On the other hand, man – if I may say so – is a macroscopic being; he learns of the microworld only when microobjects impinge on the macroobjects that man adjoins to his sensory organs; these macroobjects (which become man's instruments) in effect enable him to learn about the microworld in an indirect way. Thus, in gaining a cognition of microobjects man must use classical concepts because through these alone he describes the readings of the instruments, that is, because in his measurements he cannot do without the use of classical theories.

Such in brief is the relationship between quantum and classical mechanics, and it takes us close to an understanding of the relationship between the fundamentals of the theory of physics which, I think, are characteristic of 20th century physics.

Let us note, first of all, that the mechanics of the atomic world (quantum mechanics), far from being reducible to the mechanics of macrobodies (classical mechanics), just as the theory of electromagnetism is not reducible to classical mechanics (and does not absorb the latter), in effect produces a relationship with a much greater content. Quantum mechanics, as I have said, is in a sense the basis of classical mechanics, substantiating some of its fundamental concepts which reflect the properties of macroscopic objects so that, in consequence, with respect to these concepts it has a similar stand to classical mechanics in which derivative concepts are substantiated by means of axioms.

The system of axioms of a theory contains the basic concepts of the connections which are not logically substantiated within the system but are postulated on the basis of various convincing considerations taken into account when structuring the system. In this aspect, a theory is called *incomplete* (open) but this incompleteness is in principle of a different character than, say, the incompleteness of quantum mechanics which Einstein had in mind in his debate with Bohr, as described above. The fundamental concepts in their connections, constituting the axiomatic system of a theory, can be substantiated by means of a theory, with a new axiomatics, that is, broader and deeper than the one under review, etc. In the logical plane, the status of 'substantiation' of fundamental concepts in their connections in axiomatic theory is similar to the status of 'non-contradiction', 'completeness', etc., of an axiomatic system, which, as Gödel has proved, 'cannot be substantiated by means of the instruments of that system'. Or, to put it in a more general form: the fundamental propositions of a theoretical system cannot be obtained *by its logical means*, but they can be found by the logical means of a broader and deeper theory.[3] To use the same logical terminology, one might say that quantum mechanics is a kind of metatheory of classical mechanics.

Thus, for instance, Newton's theory of gravitation, like classical mechanics, 'gave no thought' to the proportionality or (given due selection of units) the equality of gravitational and inertial mass of a body: classical mechanics merely stated and accepted as experimental fact that in the gravitational field different bodies had the same acceleration. The discovery of the substantiation of the equality of gravitational and inertial mass, or rather, the substantiation of the proposition that the gravitational and the inertial mass of a body are equal, would have meant going beyond the limits of Newton's theory of gravitation and construction of a theory that would have been a peculiar metatheory with respect to that theory of gravitation. That is exactly what Einstein did by formulating a new theory of gravitation or, as he called it, the general theory of relativity. Let us put this in Einstein's words, by quoting some extracts from his works, adding that for our purposes we can confine ourselves to hints alone.

Concerning the proposition that the 'gravitational and inertial mass of a body are equal', Einstein says that this was stated but not interpreted by classical mechanics (in this instance, I used the expression: classical mechanics did not *substantiate*, did not find any substantiation, etc.). Einstein ends with these words: "A satisfactory interpretation can be obtained only if we recognise the following fact: The same quality of body mainifests itself according to circumstances as 'inertia' or as 'weight' (lit, 'heaviness')". (Einstein, 1964, p. 65). Having formulated this idea, Einstein thereby provided a substantiation for the equality of gravitational and inertial mass empirically stated in classical theory, and spelled out the beginnings of his theory of gravitation. The following extracts from Einstein's book, *What Is the Theory of Relativity* will serve as an illustration of his fundamental idea: "Imagine a coordinate system which is rotating uniformly with respect to an inertial system in the Newtonian manner. The centrifugal forces which manifest themselves in relation to this system must according to Newton's teaching be regarded as effects of inertia. But these centrifugal forces are, exactly like the forces of gravity, proportional to the masses of the bodies. Ought it not to be possible in this case to regard the coordinate system as stationary and the centrifugal forces as gravitational forces? This seems the obvious view, but classical mechanics forbid it" (Einstein, 1954, p. 231).

To sum up what has been said about theory and its metatheory, the following conclusions are suggested. The paradoxes which arise in theory and which *cannot be resolved by its logical* means are a sign that the given theory is reaching the limits of its meaningfulness, and its axiomatics (axiomatic construction) the highest logical completion possible from the standpoint of the actual content of the theory and its axiomatic form. Such paradoxes are fundamentally different from the paradoxes arising in a theory and capable of *being resolved by its logical means*, that is, from paradoxes which indicate that a theory is logically incomplete (or that the reasoning is incorrect or the premises imprecise).

The existence of paradoxes in a theory which cannot be resolved by its logical means is evidence of the need to seek more general and more profound theories with instruments helping to resolve these paradoxes (the resolution of such paradoxes usually coincides with the construction of the desired general theory).

Thus, the existence of this type of paradox in essence means that the physical cognition of objects does not stop at the level of this or that theory, but develops, ranging over new aspects of material reality, without discarding the knowledge already achieved by this theory. The existence of this type of paradox also means that the theory which contains them, but which cannot resolve them by its means, potentially includes a more general and deeper theory than itself. From this standpoint, every axiomatised theory necessarily and inevitably contains knowledge which cannot be substantiated by means of this theory. Otherwise knowledge would have frozen at a definite point, while that which has been gained would have become a metaphysical absolute.

The development of the theory of modern physics is ensured by a genetic series of theoretical systems representing closed or logically building axiomatic structure linked in definite relationships, from which the more general theoretical systems grow out of the more special in the genetic series. Thus, the single axiomatic system of the whole of physics in the spirit of the mechanical ideals of the 18th and 19th centuries was overthrown by the development of physical science. As Gödel's theorems have essentially demonstrated, such a system also proved to be impossible from the standpoint of logic: the logical development of theory and of physical science as a whole is expressed in a genetic hierarchy of axiomatic systems, combining both the tendency of stability and the tendency of change which are inherent in the individual axiomatic systems and in their aggregate.

Although the single axiomatic system (structure) in the spirit of classical physics has been overthrown, in the sphere of ideas the dead hand seizes the living in many more senses than in any other sphere. A single axiomatic system is also being revived in modern physics but, it is true, in a form that would appear to be a far cry from its 'classical' model. In literature today we find the following notion of physical science: physics is structured as being in principle a strict and non-contradictory axiomatic system, which ranges over all its sections, where the earlier theory (with its axiomatics) is historically a limiting special case of the historically later theory because it turns out to be broader than the former. As time goes on, the same happens to the latter, etc.

When theory is generalised, that is, when transition takes place from a special theory to a general one, the special theory does not in any sense disappear altogether within the general one, while the general theory does not become the only true theoretical system in physics, as the notion of a single axiomatics in physical science would suggest. Actually, the special theory is preserved within the general one and in a modified form (this also applies to definite concepts of the special theory): it remains within the general theory as an approximate theory and its concepts are also preserved as approximate ones. From this standpoint it is also possible to speak of absolute simultaneity in Einstein's theory of relativity. Thus, a theory is not rejected with its transition to a general theory, but remains as a relative truth, that is, an absolute truth within circumscribed limits.

All this is connected with finding the answers to these questions (some of which were considered above): why is it necessary to use Euclidean geometry in the search for the 'non-Euclidean' nature of some spatial form; why do we learn about the properties of the space – time continuum from measurements of local space and time; why are the concepts of classical mechanics applied to descriptions of experiments which are the experimental basis of quantum mechanics, etc.?

We find, therefore, that the dialectical contradiction – the source of all development and vitality – also works in axiomatics.

Institute of Philosophy, Moscow, USSR

NOTES

[1] I mean the statics developed by Archimedes.
[2] Maxwell himself assumed that he was guided by the mechanical model of the ether, but in certain conditions illusions frequently appear to be quite real.
[3] In this instance, excellent material is contained in the work of Fock (1936).

BIBLIOGRAPHY

Bohr, N.: 1935, 'Can Quantum Mechanical Description of Physical Reality be Considered Complete', *Physical Review* **48,** 696.
Bourbaki, N.: 1962, 'L'Architecture des Mathématiques,' *Dans: Les grands courants de la pensée mathématiques* (Presentes par F. Le Lionnais), Paris.
Einstein, A.: 1954, 'What's the Theory of Relativity', in *Ideas and Opinions*, New York.
Einstein, A.: 1964, *Relativity, the Special and General Theory*, London.
Einstein, A., Podolsky, B., and Rosen, N.: 1935, 'Can Quantum Mechanical Description of Physical Reality be considered Complete', *Physical Review* **47,** 777.
Fock, V. A.: 1936, 'Fundamental Significance of Approximation Methods in Theoretical Physics', in *Advances in the Physical Sciences*, Vol. XVI, Issue 8 (in Russian).
Hilbert, D.: 1900, *Über den Zahlbegriff. Jahresbericht der Deutschen Mathematiker-Vereinigung*, Bd. 8, Leipzig.
Kleene, S. C.: 1952, *Introduction to Metamathematics*, Amsterdam.
Lenin, V. I.: 1968a, *Collected Works*, Vol. 14, Moscow.
Lenin, V. I.: 1968b, *Collected Works*, Vol. 38, Moscow.
Newton, I.: 1952, *Optics*. Third book of Optics, Part 1, New York.
Podgoretsky, M. I. and Smorodinsky, Y. A.: 1969, 'On the Axiomatic Structure of Physical Theories', in *Questions of the Theory of Knowledge* (in Russian), Moscow, p. 74.

II

THE INTERPRETATION
OF QUANTUM MECHANICS

WHAT IS PHILOSOPHICALLY INTERESTING
ABOUT QUANTUM MECHANICS?*

What puzzles philosophers about the transition from classical to quantum mechanics?

It is not the phenomenon of interference for particles, or the uncertainty principle, that in itself generates a conceptual puzzle. That is, the problem is not merely that the magnitudes of a quantum mechanical system form a non-commutative algebra.

For consider a quantum mechanical system whose magnitudes are represented by linear operators on a 2-dimensional Hilbert space, for example a photon, or a spin-1/2 particle such as an electron. I denote the magnitudes by $A, B, C, ...$, possible values by $a_1, a_2; b_1, b_2; c_1, c_2;$ ..., statistical states by ψ, α_1, α_2, etc.[1] The theory generates probabilities for the possible values of the magnitudes $a_1, a_2; b_1, b_2;$ etc.[2] In this case there is no problem in understanding *all* the magnitudes as having *determinate* values (at the same time), and the probabilities generated by the statistical states as representing distributions over these values. We can introduce a measure space $(X, \mathscr{F}, \mu)$, represent the magnitudes $A, B, C, ...$ by random variables $f_A, f_B, f_C, ...$, on X, and associate each statistical state ψ with a measure μ_ψ on X, so that the quantum mechanical probabilities specified by ψ for $a_1, a_2; b_1, b_2;$ etc. are generated by μ_ψ, i.e.

$$p_\psi(a_1) = \mu_\psi(\Phi_{a_1}), \text{ etc.,}$$

where Φ_{a_1} is the set of points in X on which the random variable f_A takes the value a_1.[3]

Such a construction does not, of course, explain the order-dependence of sequential probabilities involving non-commuting magnitudes. For example, the probability that a photon will pass a pair of crossed polaroids, representing a sequential measurement of the magnitudes A and $A^\perp$, say, is zero. The probability of passing a sequence $A, A^\perp, B$ or $B, A, A^\perp$ is also zero (where B does not commute with

Butts and Hintikka (eds.), *Foundational Problems in the Special Sciences*, 69–79.

A, e.g. where the axis of polarization corresponding to B lies between the axis of polarization corresponding to A and that of $A^{\perp}$), but the probability of passing a sequence A, B, $A^{\perp}$ is non-zero. The problem posed by this feature of the quantum statistics is resolved if we assume that the class of available measuring instruments (for A, B, C, ...) is such that the quantum system undergoes a disturbance during measurement, and that the sequential probabilities of the theory refer to selections made by these instruments. Thus, in passing through a sheet of polaroid (which performs a measurement by selecting for a particular property, say a_1) the photon is disturbed, in the sense that values of the magnitudes other than the magnitude measured are altered during the interaction.

We can see what this measurement disturbance must be by the following argument: Suppose there is no order-dependence of sequential probabilities. Then the probability of a sequence $a_i - b_j$ is computed as:

$$p(a_i - b_j) = \mu_\psi(\Phi_{a_i})\mu_{\psi,a_i}(\Phi_{b_j}),$$

where μ_{ψ,a_i} is the conditional probability measure defined by:

$$\mu_{\psi,a_i}(\Phi) = \frac{\mu_\psi(\Phi_{a_i} \cap \Phi)^4}{\mu_\psi(\Phi_{a_i})}$$

According to the quantum theory:

$$p_\psi(a_i - b_j) = p_\psi(a_i)p_{\alpha_i}(b_j).$$

On the classical measure space, then:

$$p_\psi(a_i - b_j) = \mu_\psi(\Phi_{a_i})\mu_{\alpha_i}(\Phi_{b_j}).$$

Thus, in order to recover the correct quantum mechanical sequential probabilities, we must assume that after a measurement of A yielding the result a_i, the measurement disturbance is characterized by the transition:

$$\mu_{\psi,a_i} \rightarrow \mu_{\alpha_i}.$$

Both these measures assign zero probability to sets outside the set Φ_{a_i}, and unit probability to Φ_{a_i}. The difference is that the relative probabilities of subsets in Φ_{a_i} defined by μ_ψ are preserved by μ_{ψ,a_i}, but not by μ_{α_i}. In this sense, the measurement disturbance *randomizes* the information contained in the initial measure μ_{ψ}.[5]

There is, therefore, no problem in accounting for the statistics of a quantum mechanical system associated with a 2-dimensional Hilbert space by a 'hidden variable' theory. That is, there is no fundamental problem which is philosophically interesting. A quantum mechanical system can be understood as having *all* its properties, even though the algebra of magnitudes is non-commutative. It is possible to introduce additional variables which parametrize a classical measure space X, and to represent the magnitudes of the system by random variables on X, i.e. to associate the properties of the system with subsets of X. The non-commutativity of certain magnitudes can be regarded as reflecting a feature of the class of measuring instruments: Sequential probabilities involving non-commuting magnitudes are order-dependent because they refer to sequences involving measuring instruments as selection devices which disturb the systems measured in a particular way.

A problem only arises for systems associated with Hilbert spaces of three or more dimensions. In a 3-dimensional Hilbert space, the algebra of idempotent magnitudes is not merely non-Boolean, it is not even imbeddable into a Boolean algebra.[6] Now, a realist interpretation of the 2-dimensional case, along the lines sketched above, involved imbedding the non-Boolean algebra of idempotents into a Boolean algebra, isomorphic to the Boolean algebra of (measurable) subsets of the measure space X.[7] On the basis of this imbedding, the non-Boolean algebra of idempotents could be interpreted as part of a Boolean algebra of idempotents, representing the structure of properties of the system. Classically, the Boolean property structure or possibility structure yields all possible ways in which the properties of the system can hang together to form maximal totalities (ultrafilters) of mutually co-existing properties, associated with the different classical states. If the algebra of idempotent magnitudes is not imbeddable into a Boolean algebra, can we still interpret the idempotents as properties of the system?

Firstly, exactly why does non-imbeddability pose a difficulty for a realist interpretation of the theory? What goes wrong if we introduce a classical measure space X, and attempt to represent the quantum

mechanical magmitudes by random variables on X? It turns out that
the functional relations between the magnitudes cannot be preserved.
We can represent all the maximal ('non-degenerate') magnitudes by
distinct random variables, but each non-maximal magnitude will have
to be represented by a family of random variables, one for each
maximal magnitude of which the non-maximal magnitude is a function.
If A, B, C, ... are (non-commuting) maximal magnitudes represented
by the random variables f_A, f_B, f_C, ... and P is a non-maximal
magnitude such that

$$P = g_1(A) = g_2(B) = g_3(C) = ...,$$

then P will be represented by a family of random variables

$$g_1(f_A), g_2(f_B), g_3(f_C), ...$$

which are statistically equivalent for all μ_ψ (where ψ is a quantum
mechanical statistical state). Since no μ_ψ is atomic for all maximal
magnitudes, it is not the case that

$$g_1(f_A(x)) = g_2(f_B(x)) = g_3(f_C(x)) = ...$$

at any point $x \in X$.[8]

The term *contextualistic* has been introduced in the literature to
describe such a hidden variable reconstruction of the quantum statistics.[9] Now, a contextualistic hidden variable theory does not provide a
realist interpretation of quantum mechanics. An idempotent magnitude P can only be regarded as representing a property of the system
relative to some maximal magnitude. If we are considering a system S
which is a subsystem of a larger system $S + S'$, then S only has
properties relative to magnitudes which are maximal in the Hilbert
space of $S + S'$. This is the case even if S and S' are separated and no
longer interacting. So, strictly, no system has any properties, except
relative to a maximal magnitude which is selected (how?) for the whole
universe. The part has no reality except as a facet of the whole.
Alternatively, we might say that the measuring instrument as a mac-
rosystem always selects some maximal magnitude. But then, the properties of a micro-system are created by the measuring instrument in the
act of measurement. At any rate, according to a contextualistic hidden
variable theory, a micro-system does not have all its properties in the
straightforward realist sense ('independent of context'). The algebra of

idempotents of a quantum mechanical system is not interpreted as representing the possibility structure for the properties of a micro-system regarded as an individual entity.

Thus, the question arises whether a realist interpretation of the quantum statistics is at all defensible. Does it make sense to understand the transition from classical to quantum mechanics as the transition from a Boolean to a (strongly) non-Boolean possibility structure? That is, is it possible to maintain an interpretation of the strongly non-Boolean algebra of idempotents of the theory as a non-classical possibility structure? This is the philosophically interesting question posed by the quantum theory.

An affirmative answer to this question has been proposed by Hilary Putnam.[10] The quantum logic of Kochen and Specker suggests a similar thesis.[11] Consider the class of propositional functions $\varphi(x_1, ..., x_n)$ of classical logic. The variables $x_1, ..., x_n$ here range over propositions. An example of such a function is:

$$x_1 \wedge (x_2 \wedge x_3) \equiv (x_1 \wedge x_2) \wedge x_3.$$

It is more convenient to consider the associated class of Boolean functions, where the variables range over equivalence classes of propositions, elements in the Lindenbaum-Tarski algebra of the logic. In this case, $\varphi(x_1, x_2, x_3)$ represents the associated Boolean function

$$(\psi_1 \wedge \psi_2')' \wedge (\psi_2' \wedge \psi_1)',$$

where $\psi_1 = x_1 \wedge (x_2 \wedge x_3)$, $\psi_2 = (x_1 \wedge x_2) \wedge x_3$, and the $'$ denotes Boolean complementation. (Recall the definition of the biconditional in terms of conjunction and negation.) The classical notion of validity selects a set of tautologies which corresponds to the set of Boolean functions taking the value 1 in the 2-element Boolean algebra $\mathscr{L}_2$, for all possible substitutions of elements in $\mathscr{L}_2$ for the variables in φ. In other words, we consider the set of all 2-valued homomorphisms on the Lindenbaum-Tarski algebra of the logic. The set of classical tautologies corresponds to the set of Boolean functions mapped onto 1 in $\mathscr{L}_2$. Equivalently, the classical tautologies are those Boolean functions φ mapped onto the unit in a Boolean algebra $\mathscr{B}$ by all substitutions of elements in $\mathscr{B}$ for the variables in φ. Whether we consider homomorphisms into $\mathscr{B}$ or homomorphisms into $\mathscr{L}_2$, we recover the same set of tautologies. This is a non-trivial property of Boolean algebras, which ultimately depends on Stone's theorem.[12]

Kochen and Specker have generalized the classical notion of validity to allow substitutions for the variables in φ from partial Boolean algebras. The algebra of idempotents of a quantum mechanical system is a partial Boolean algebra not imbeddable into a Boolean algebra (except in the 2-dimensional case). The generalized notion of validity generates a set of tautologies for a quantum mechanical system which is different from the set of classical tautologies. This means that there are Boolean functions φ which are satisfiable by some substitution from the algebra of idempotents of a quantum mechanical system, but are not satisfiable by any substitutions from a Boolean algebra. If we want to interpret the partial Boolean algebra of idempotents of a quantum mechanical system as a possibility structure, then such a φ should be interpreted as representing a possible *property of properties of a quantum mechanical system which is not a possible property of properties of any classical system*. In other words, such a φ represents a possible way in which the properties of a quantum mechanical system can hang together – and this is not a possible way in which the properties of a classical system can hang together.

An example proposed by Kochen and Specker is discussed in detail in this Symposium by William Demopoulos. The function contains 117 variables and is a conjunction over triples of variables, each triple representing a mutually exclusive and collectively exhaustive set of idempotents, i.e. each triple can be taken as representing a different kind of property.[13] The non-satisfiability of this formula classically means that it is not possible to partition a set – any set at all – into 117 subsets, representing properties of a classical system, in such a way as to make the same *kinds of properties* (mutually exclusive and collectively exhaustive triples) that can be constructed out of 117 quantum mechanical properties (represented by elements in a partial Boolean algebra of subspaces of a Hilbert space). In this sense, a quantum mechanical system has different kinds of properties than a classical system; it has a more complex *possibility* structure. This means that the totality of possible ways in which the properties of a quantum mechanical system can be structured – the properties of properties of a quantum mechanical system – is not given by the set of all subsets of the set of possible properties. A realist interpretation of quantum mechanics therefore requires a property theory in which there are different kinds of properties than can be put in one-one correspondence with the power set of possible properties.

One might object that $\varphi(x_1, ..., x_{117})$ can only represent a property of properties if there exists a 2-valued homomorphism which selects a set of properties, such that one and only one property from each kind of property (i.e. each triple) belongs to the set. There is, of course, no such selection function, and hence no such set of properties. Now, the objection suggests that in the classical (i.e. Boolean) case, there are Boolean functions like φ which are satisfiable in a general Boolean algebra $\mathscr{B}$ and which represent possible properties of properties of systems, and also Boolean functions which are satisfiable in some $\mathscr{B}$ but do not represent possible properties of properties, and that the criterion which demarcates between these two cases is the existence or non-existence of a 2-valued homomorphism. But a Boolean function like φ (i.e. representing a conjunction of exclusive disjunctions) is satisfiable in a Boolean algebra $\mathscr{B}$ (i.e. takes the value 1 in $\mathscr{B}$ for some substitution of elements in $\mathscr{B}$ for the variables in φ) *if and only if* φ is mapped onto 1 in $\mathscr{L}_2$ by a 2-valued homomorphism. This is a structural property of Boolean algebras. There are no functions like φ which are satisfiable in some $\mathscr{B}$ and not satisfiable in $\mathscr{L}_2$. Hence, it is not at all clear that the existence of 2-valued homomorphisms is an acceptable criterion in the general case for interpreting functions like φ as representing possible properties of properties.

If the property theory is non-classical–i.e. non-Boolean, or non-set-theoretical–then the probability theory on these properties has to be non-classical too. We require a generalization of the classical Kolmogorov theory of probability appropriate to a strongly non-Boolean possibility structure. Such a generalization has, in effect, been developed by Umegaki,[14] and recently extended by Davies and Lewis.[15]

The theory begins with a suggestion by Moy[16] that classical conditionalization relative to a σ-field can be regarded as a linear map on the algebra of random variables. The conditional expectation of a random variable A relative to a σ-field $\mathscr{F}'$ is, roughly, another random variable, A', measurable with respect to $\mathscr{F}'$, which has the same expectation value as A on elements of $\mathscr{F}'$. More precisely, we begin with a probability space $(X, \mathcal{F}, \mu)$ and fix some sub-field $\mathcal{F}'$ relative to which the conditionalization is performed. The conditional expectation of A relative to $\mathcal{F}'$ is a random variable A' which satisfies the conditions:

(i) A' is measurable with respect to $\mathcal{F}'$

(ii) $\operatorname{Exp}(AI_\Phi) = \operatorname{Exp}(A'I_\Phi)$, for all $\Phi \in \mathcal{F}'$, where I_Φ is the indicator function for the set Φ, or equivalently

(ii)' $\operatorname{Exp}(AZ) = \operatorname{Exp}(A'Z)$, for all Z measurable with respect to $\mathcal{F}'$.[17]

This is a (classical) generalization of the elementary notion of conditionalization relative to an event: we consider a family of events forming a σ-field instead of a single event. If we fix the field $\mathcal{F}'$ relative to which the conditionalization is carried out, every random variable is mapped onto another random variable measurable with respect to $\mathcal{F}'$. The conditional expectation of an idempotent magnitude is the conditional probability of the corresponding event (i.e. the event that the property obtains), so conditional probabilities of events are included in this generalization. Notice that the notion of measurability in (i) and the expectation value in (ii) or (ii)' refer to the given measure μ, i.e. A' is μ-measurable with respect to $\mathcal{F}'$, and

$$\int AI_\Phi \, d\mu = \int A'I_\Phi \, d\mu \quad \text{for all} \quad \Phi \in \mathcal{F}'.$$

Now, the concept of classical conditionalization as a linear map of a certain kind on the commutative algebra of random variables generalizes in a natural way to the non-commutative algebra of magnitudes of a quantum mechanical system. As an example, consider three orthogonal idempotent magnitudes in a 3-dimensional Hilbert space, say P_{α_1}, P_{α_2}, P_{α_3}. These are projection operators onto the three orthogonal lines defined by the orthogonal unit vectors α_1, α_2, α_3, the eigenvectors of a (maximal) magnitude A. The three corresponding atomic properties, a_1, a_2, a_3 (i.e. the properties associated with the values a_1 of A, a_2 of A, a_3 of A) generate a maximal Boolean subalgebra $\mathcal{B}_A$ in the partial Boolean algebra of properties of the system. In the generalized theory, the analogue of summation or integration is the trace function: $\operatorname{Tr}(A)$ yields the sum of the eigenvalues (possible values) of the linear operator representing the magnitude A. The analogue of integration with respect to a probability measure μ, or a weighted sum, is the trace of the product WA, where W is a statistical operator.

It turns out that the conditional expectation of P_β, say, relative to $\mathcal{B}_A$ is the magnitude or random variable

$$P'_\beta = \sum_i P_{\alpha_i} P_\beta P_{\alpha_i} = \sum_i |(\alpha_i, \beta)|^2 P_{\alpha_i}$$

in the sense that

$$\text{Tr}\,(P_\beta P_{\alpha_i}) = \text{Tr}\,(P'_\beta P_{\alpha_i}), \quad \text{for} \quad i = 1, 2, 3$$

or equivalently $\text{Tr}\,(P_\beta Z) = \text{Tr}\,(P'_\beta Z)$, for all Z compatible with A. This means that $P'_\beta = \sum_i P_{\alpha_i} P_\beta P_{\alpha_i}$ is the conditional expectation of P_β relative to $\mathcal{B}_A$ with respect to the statistical operator $I/\text{Tr}(I)$ (where I is the unit operator), i.e. with respect to a uniform initial measure on the atoms of $\mathcal{B}_A$ (each atom being assigned equal a priori probability of $1/3$).

The conditional expectation P'_β yields the conditional probability of the property b (corresponding to the idempotent P_β) relative to a_j as:

$$p(b \mid a_j) = \text{Tr}\,(P'_\beta P_{\alpha_j})$$

$$= |(\alpha_j, \beta)|^2.$$

But this is just the probability assigned to an idempotent P_β by the (pure) statistical state α_j. In other words, the probability assigned to an idempotent P_β by the statistical state α_j can be interpreted as the conditional probability of the property b given the property a_j, with respect to a uniform initial measure on the atoms of $\mathcal{B}_A$.

Von Neumann's projection postulate turns out to be just the quantum mechanical (i.e. appropriate non-Boolean) rule for conditional probabilities, i.e. the transition

$$W \rightarrow W' = \sum_i P_{\alpha_i} W P_{\alpha_i} = \sum_i \text{Tr}\,(W P_{\alpha_i}) P_{\alpha_i}$$

represents the conditionalization of the statistical operator W relative to the maximal Boolean sub-algebra $\mathcal{B}_A$ generated by the atomic idempotents P_{α_i}.[18] It is interesting to note that von Neumann's generalization of this formula for non-maximal measurements is in fact incorrect. Von Neumann proposed the rule:

$$W \to W' = \sum_i \mathrm{Tr}\,(WP_i)P_i/\mathrm{Tr}\,(P_i),$$

where P_i represents the projection operator onto an m-dimensional subspace $(m \neq 1)$. It was pointed out by Lüders[19] that the correct rule is in fact:

$$W \to W' = \sum_i P_i W P_i$$

in the general case (whether or not P_i is an atomic idempotent). Thus the correct rule in the general case is given by the conditionalization of W relative to a Boolean subalgebra (maximal or non-maximal).

The transition from classical to quantum mechanics only poses a philosophically interesting problem because of the difficulties in the way of a realist interpretation of the theory. A proper resolution of these difficulties requires the interpretation of the non-Boolean structure of idempotent magnitudes of the theory as a property structure, or possibility structure of events, i.e. as yielding all combinatorial possibilities for molecular events or properties. I have indicated some of the more revolutionary implications of this interpretation for the notions of truth (via satisfiability) and probability.

University of Western Ontario
Tel Aviv University

NOTES

* Research Supported by Canada Council Grant S74–1032.

[1] The eigenstates corresponding to the eigenvalues a_1, a_2 of the magnitude A are denoted by Greek letters α_1, α_2, and so on.
[2] In the general case, for ranges of possible values of the magnitudes: $a \in R$, $b \in S$, etc.
[3] For details, see J. Bub 'Randomness and Locality in Quantum Mechanics', in P. Suppes (ed.), *Logic and Probability in Quantum Mechanics*, D. Reidel Publishing Co., Dordrecht, 1975, pp. 397–420.
[4] In this case, $p_\psi(a_i - b_j) = p_\psi(b_j - a_i) = \mu_\psi(\Phi_{a_i} \cap \Phi_{b_j})$.
[5] See J. Bub, 'Randomness and Locality in Quantum Mechanics', in P. Suppes (ed.),

Logic and Probability in Quantum Mechanics, D. Reidel Publishing Co., Dordrecht, 1975, pp. 397–420.

[6] In general, the sub-algebra of idempotents of a non-commutative algebra is non-Boolean.

[7] The idempotents of quantum mechanics are mapped onto the indicator functions, i.e. random variables which take the value 1 on some subset of X and zero elsewhere. Note that *atomic* quantum mechanical idempotents (projection operators onto 1-dimensional subspaces) are mapped onto *non-atomic* indicator functions (indicator functions onto subsets of X comprising several classical states).

[8] For details, see J. Bub, 'Randomness and Locality in Quantum Mechanics', in P. Suppes (ed.), *Logic and Probability in Quantum Mechanics*, D. Reidel Publishing Co., Dordrecht, 1975, pp. 397–420. Also J. Bub, *The Interpretation of Quantum Mechanics*, D. Reidel Publishing Co., Dordrecht, 1975, Chapter VII.

[9] Abner Shimony, 'Experimental Test of Local Hidden Variable Theories', in B. d'Espagnat (ed.), *Foundations of Quantum Mechanics*, Academic Press, New York, 1967.

[10] H. Putnam, 'Is Logic Empirical?', in R. Cohen and M. Wartofsky (eds.), *Boston Studies in the Philosophy of Science* **V**, D. Reidel Publishing Co., Dordrecht, 1969, pp. 216–241.

[11] S. Kochen and E. P. Specker, 'The Problem of Hidden Variables in Quantum Mechanics', *J. Math. Mech.* **17**, 59–87 (1967).

[12] See J. Bub, *The Interpretation of Quantum Mechanics*, D. Reidel Publishing Co., Dordrecht, 1975, Chapter IX.

[13] Note that the same variable is repeated in several triples.

[14] H. Umegaki, 'Conditional Expectation in an Operator Algebra', *Tohoku Math. J.* **6**, (1954) 177–181; **8** (1956) 86–100; *Kodai Math. Semi. Rep.* **11** (1959) 51–64; **14** (1962) 59–85.

[15] E. B. Davies and J. T. Lewis, 'An Operational Approach to Quantum Probability', *Commun. Math. Phys.* **17** (1970) 239–260. I am indebted to Paul Benioff for this reference.

[16] S.-T. C. Moy, 'Characterisations of Conditional Expectations as a Transformation on Function Spaces', *Pacific Journal of Mathematics* **4**, 47–64 (1954).

[17] A' is not unique. Two versions of the conditional expectation, A' and A'', take the same value at all points in X, except for a set of points of measure zero.

[18] M. Nakamura and H. Umegaki, 'On von Neumann's Theory of Measurements in Quantum Statistics', *Math. Jap.* **7**, 151–157 (1961–1962).

[19] G. Luders, *Ann. Phys.* **8**, 322 (1951).

WILLIAM DEMOPOULOS

COMPLETENESS AND REALISM IN QUANTUM MECHANICS*

Kochen and Specker ([5], Theorem 1) have shown that a finite partial Boolean algebra $\mathscr{D}$ of quantum mechanics has no two-valued homomorphisms. In my view, this theorem bears on the completeness of quantum mechanics insofar as this is related to the problem of hidden variables and the probabilistic character of the theory: the quantum theory is complete but indeterministic in the sense that the maximal amount of information regarding a system may be significantly probabilistic. It should be noted that this interpretation of the theorem requires a modification of Kochen and Specker's analysis of statistical theories – basically, it requires treating the algebraic structure of physical magnitudes as theoretically primitive. These matters are discussed in some detail in [1].

The present paper concerns the bearing of this theorem on a realist theory of the truth of quantum mechanical propositions. Briefly the question is: Does the indeterminism implied by the absence of two-valued homomorphisms require the indeterminateness of quantum mechanical states of affairs? Somewhat less provocatively: Does the absence of two-valued homomorphisms require abandoning the thesis that the propositions represented by elements of $\mathscr{D}$ have determinate truth values?

One possible approach to this problem may be dismissed immediately: one might ignore the algebraic structure which quantum mechanics postulates for physical magnitudes and treat them as independent random variables. This approach, while preserving realism in general, i.e. while preserving the thesis that truth values are determinate and independent of our knowledge of them, requires a purely conventionalist interpretation of the algebraic structure of the theory. This is therefore more of an evasion than a solution of the problem which quantum mechanics poses for a realist interpretation of physical theories.

Butts and Hintikka (eds.), *Foundational Problems in the Special Sciences*, 81–88.

The view I wish to propose is roughly this: First, to say that a proposition P_i is true is simply to assert that the system has the property which P_i predicates of it. The properties, in this case, are of the kind: 'the value of the magnitude S_i is zero' or 'the value of the magnitude S^2 is two', etc. To say that every proposition is true or false is therefore clearly equivalent to the claim that every magnitude has a definite value.

In both classical and quantum mechanics, physical magnitudes are representable by random variables – real-valued functions – defined on a space of elementary events or states of affairs. (These terms are used interchangeably.) In the case of quantum mechanics events are represented by the partial Boolean algebra of subspaces of Hilbert space, in the classical case, by the subsets of phase space. [This requires a certain generalization of the notion of a random variable for the quantum mechanical case. The generalization is explicitly developed in a paper of Finch [3], and in less detail by Kochen and Specker ([4], Section 2).] Now the inverse image of the singleton subsets of the Real line belonging to the range of possible values of any physical magnitude partitions the Hilbert space of quantum mechanics into mutually exclusive and severally exhaustive sub*spaces*, just as, in classical mechanics each magnitude determines a partition of the phase space into mutually exclusive and severally exhaustive sub*sets*. This suggests the condition: every magnitude always has a definite value if and only if every magnitude determines such a partition of the space of possible elementary events.

In particular, given any idempotent magnitude A of classical or quantum mechanics, if $A^{-1}(\{1\}) = a$, then $A^{-1}(\{0\}) = a'$; a and a' are mutually exclusive and together exhaust the whole 'logical space' of states of affairs, whether this be phase space or Hilbert space. A^{-1} is a σ-homomorphism defined on the Boolean algebra of measurable subsets of the Real line. Thus, in the case of idempotent magnitudes the necessary and sufficient condition for saying that every magnitude has a definite value reduces to the requirement that the states of affairs a, a' representing the properties 'the value of A is one', 'the value of A is zero', satisfy the laws of non-contradiction and excluded middle.

The above view, if adopted, has rather far-reaching consequences for our conception of the properties of physical systems. To see this, let us first review the satisfiability of propositional formulae from the point of view of classical logic.

There are roughly two views regarding the interpretation of propositional formulae: The standard, Fregean, interpretation treats the propositional connectives as operations in the two-element Boolean algebra Z_2. The value of a propositional formula $\varphi(x_1, \ldots, x_n)$ is given by the value of the corresponding operation $\varphi^*: Z_2^n \to Z_2$ for the assignment $a_1, \ldots, a_n \in Z_2$ to the free variables $x_1, \ldots, x_n$. Elements of Z_2 are interpreted as 'truth values'.

Another view – going back to Russell – interprets the connectives appearing in φ as operations in a general Boolean algebra B. For a given assignment $a_1, \ldots, a_n$ of elements of B to its free variables, the formula φ denotes the event or state of affairs $\varphi^*(a_1, \ldots, a_n)$; i.e. the value of the map φ^* corresponding to φ at the point $\langle a_1, \ldots, a_n \rangle \in B^n$. $\varphi^*(a_1, \ldots, a_n)$ cannot be regarded as the truth value of φ in quite the same sense in which this is possible under Frege's interpretation of propositional formulae. For suppose the properties $P_1, \ldots, P_n$ are represented by the events $a_1, \ldots, a_n$: i.e. $P_i \mapsto a_i$ under the correspondence which the inverse image of the magnitude mentioned in P_i establishes. If I know the designation of $\varphi(P_1, \ldots, P_n)$, then I know which state of affairs must obtain in order for $\varphi(P_1, \ldots, P_n)$ to be true. But the designation of $\varphi(P_1, \ldots, P_n)$ is simply an element of B generally distinct from the zero or unit of B. Given merely that $\varphi(P_1, \ldots, P_n)$ designates $\varphi^*(a_1, \ldots, a_n)$, I cannot assert whether $\varphi(P_1, \ldots, P_n)$ is true. On Frege's interpretation, $\varphi(P_1, \ldots, P_n)$ designates either the *always obtaining* state of affairs or the state of affairs which *never* obtains: not only do I know a truth-condition for φ, but in addition, I know whether or not the state of affairs which gives the truth condition, obtains.

Russell's view effectively generalizes Frege's interpretation of propositional formulae to include the possibility of designating contingent states of affairs. Mathematically, the generalization is not significant, since by the semi-simplicity property of Boolean algebras, B is representable by a direct sum of two element Boolean algebras, so that any function $\varphi^*: B^n \to B$ may be replaced by its image $\varphi^* h$ under the (Stone) isomorphism carrying B into the appropriate power of Z_2.

It is well-known that Frege interpreted the truth values of Z_2 as special 'objects' – The True and The False – and maintained that 'truth' is indefinable. (See, e.g. Dummett [2], Chapter 13.) Some departure from this view is required in order to exploit Tarski's analysis of truth. The recursive definition of satisfaction in terms of the truth values 0 and 1 obviously cannot be understood as a definition of this notion in terms of truth and falsity, so that something like the interpretation of the elements of Z_2 as degenerate states of affairs is required. But once this is granted, the more natural procedure would seem to be the one which admits all possible states of affairs in specifying the truth-conditions of propositional formulae. But this is just the proposal of Russell.

[The above characterization of Russell's view is not quite correct: Russell gave a special status to maximal and minimal states of affairs corresponding to atoms and their complements – co-atoms – of B. He called the obtaining of such states of affairs positive and negative *facts*. The holding of a proposition could be true in virtue of a positive or negative fact. The obtaining of other states of affairs was held to be reducible to these. The mathematical justification for singling out positive and negative facts is not generally appreciated: Every complete atomic Boolean algebra is isomorphic to the power set of the set of its atoms or co-atoms. In the general representation theory developed by Stone – where the restriction to atomic Boolean algebras is dropped – the space of atoms is replaced by the space of maximal (proper) filters, co-atoms by the space of maximal (proper) ideals, and the isomorphic field of sets is included in the power set of this space.]

When we attempt to generalize the definition of satisfaction for partial Boolean algebras, the significance of the Russellian point of view is clear: Because of the failure of semi-simplicity, we cannot interpret a propositional formula φ as a map φ^* defined on sequences of elements of Z_2. The following definition of the domain D_φ of a propositional formula φ and its corresponding map φ^* recursively specifies the value of φ in an arbitrary partial Boolean algebra N, just as Tarski's definition recursively specifies the value of φ in Z_2, or more generally, in B. (Cf. Kochen and Specker [5], Section 6, from which this definition has been abstracted.)

Let $a = \langle a_1, ..., a_n \rangle$ be an element in N^n, the n-fold Cartesian product $N \times ... \times N$ of the partial Boolean algebra N.

(1)   If φ is the polynomial 1, then $D_\varphi = N^n$ and $\varphi^*(a) = 1$.

(2)   If φ is the polynomial x_i $(i = 1, ..., n)$, then $D_\varphi = N^n$, and $\varphi^*(a) = a_i$.

(3)   If $\varphi = \psi \otimes \chi$ (where $\otimes$ is either $+$ or $\cdot$), then D_φ consists of those sequences a which belong to the intersection of the domains of ψ and χ (i.e. $a \in D_\psi \cap D_\chi$), and also satisfy the compatibility condition $\psi^*(a) \subsetdot \chi^*(a)$. The map $\varphi^*(a)$ is defined by $\varphi^*(a) = \psi^*(a) \otimes \chi^*(a)$.

The above definition differs from the classical one only by allowing for the possibility that φ designates a function of a partial Boolean algebra which is not semi-simple. All that the recursiveness of the definition of satisfaction requires is that the satisfaction of $\varphi(x_1, ..., x_n)$ should be determined given that certain of its atomic subformulae are satisfied and others are not satisfied by a sequence $a_1, ..., a_n$. In this sense, the definition of satisfaction and hence of truth is recursive. But in general, there are no homomorphisms from N into Z_2 which play the role of classical realizations of a language L whose Lindenbaum-Tarski algebra is isomorphic to N and which determine the truth or falsity of every propositional formula of L.

The exact significance of the above remarks is best seen by considering the propositional formula corresponding to the orthogonality relations of the 1-dimensional subspaces belonging to the partial Boolean algebra $\mathscr{D}$. This formula is constructed as follows: Associate with each line L_i in $\mathscr{D}$ a propositional variable x_i. With each orthogonal triple of lines associate the propositional formula

$$x_i + x_j + x_k + x_i x_j x_k.$$

In terms of the lattice operations $\cup, \cap, '$, this becomes

$$(x_i \cup x_j \cup x_k) \cap (x_i \cap x_j)' \cap (x_j \cap x_k)' \cap (x_j \cap x_r)'.$$

Let φ denote the product

$$\prod (x_i + x_j + x_k + x_i x_j x_k)$$

of all such formulae corresponding to all orthogonal triples.

Now then, because there is no homomorphism $h: \mathscr{D} \to Z_2$ there is no substitution of elements from the two element Boolean algebra for the propositional variables of φ under which the identity $\varphi = 1$ holds. Hence the propositional formula

$$1 - \varphi,$$

where 1 is the truth functional connective which always takes the same constant value 1, is a classical tautology, since $\varphi * h$ is never 1. But substituting the mutually orthogonal and severally exhaustive subspaces L_i used in the construction of φ, each factor of φ takes the value 1 in $\mathscr{D}$ and therefore $\varphi = 1$ holds in $\mathscr{D}$, so that the tautology $1 - \varphi$ does not hold in $\mathscr{D}$. Thus under this substitution of elements of $\mathscr{D}$ the classical tautology just constructed is false.

Our present interest however is with the formula φ rather than the tautology $1 - \varphi$, for it represents a property which holds for some quantum mechanical system although no classical system has the property represented by φ. Classically, the property φ is inconsistent. Discussions of the logical structure of quantum mechanics have emphasized the failure of classical identities. Much less attention has been paid to the dual fact: the satisfiability of a formula which, from the point of view of classical logic, is a logical contradiction. Every substitution of elements from Z_2 for the variables appearing in φ yields the zero of Z_2. Yet quantum mechanically, φ describes a possible state of affairs. There is therefore a sense in which there are more nontrivial quantum mechanical properties – e.g. the property represented by φ – than there are classical properties. Classically it is logically impossible that any system should have the property represented by φ. Quantum mechanically the formula $\varphi(x_1, ..., x_n)$ is not a contradiction and is, in fact, satisfiable by the propositions $P_1, ..., P_n$ corresponding to the lines $L_1, ..., L_n$.

Recall that the necessary and sufficient condition for a magnitude to have a definite value is that the inverse image of the singleton subsets of possible values of the magnitude should partition the phase space or Hilbert space into mutually exclusive and severally exhaustive subsets or subspaces. With this explication, each factor of φ asserts that exactly one of the properties P_i, P_j, P_k holds or that the magnitude determining the partition of H_3 – i.e. the magnitude whose associated properties include P_i, P_j, P_k – has a definite value. Classically there is always associated with this fact the possibility of forming a selection subset D_t of the set D of all possible properties: D_t consists of the set of all properties holding at a particular instant t, and may be defined by $D_t = \{h_t^{-1}\{1\}$: for some $h_t: D \to Z_2\}$. But this association breaks down with the failure of the homomorphism theorem.

Suppose $\varphi(P_1, ..., P_n)$ is interpreted as asserting a property of the system in the way I have indicated: $\varphi(P_1, ..., P_n)$ asserts the property that every magnitude associated with the orthogonal triples of $\mathscr{D}$ has a definite value. Then $\varphi(P_1, ..., P_n)$ is an example of a propositional function of properties $P_1, ..., P_n$ belonging to D which is not associated with any selection subset of D. The peculiarity here, from a classical point of view, is that there are more properties of the properties belonging to D than there are subsets of D – or else there are more subsets of D than are generated by the classical power set operation.

There is no way of avoiding this consequence if we wish to satisfy the following two demands: (i) the magnitudes of quantum mechanical systems have the algebraic structure which the theory ascribes to them; (ii) the magnitudes always have definite values, and the values they have are independent of our knowledge; equivalently, the truth or falsity of every proposition which the theory allows us to formulate regarding such systems is determined by objective states of affairs which are independent of us.

The logical interpretation of quantum mechanics holds that the satisfiability of a propositional formula such as $\varphi(x_1, ..., x_n)$ by a sequence of propositions $P_1, ..., P_n$ has its classical meaning – each factor holds if and only if exactly one of the P_i holds. This condition is satisfied when φ is interpreted by a function defined on a partial Boolean algebra such as $\mathscr{D}$. This interpretation resolves one important component of the problem of understanding the theory realistically. The principal modification of our customary thinking about physical systems and their properties is this: Both classically and quantum mechanically $\varphi(P_1, ..., P_n)$ asserts a property of $P_1, ..., P_n$, viz. the property that exactly one of the properties P_i, P_j, P_k of each factor of the product holds. Both classically and quantum mechanically, this higher order property is itself a property of the system. The difference is that under any classical interpretation of the holding of φ it denotes the empty property, while φ has a non-zero quantum mechanical interpretation.

NOTE

* Research partially supported by Canada Council Grant S74 1032.

BIBLIOGRAPHY

[1] W. Demopoulos: 1975, 'Fundamental Statistical Theories', in *Logic, Probability, and Quantum Mechanics* (ed. by P. Suppes), D. Reidel, Dordrecht.
[2] Michael Dummett: 1973, *Frege: Philosophy of Language*, Duckworth.
[3] P. D. Finch: 1975, 'Quantum Mechanical Physical Quantities as Random Variables', in *Statistical Theories of Physics* (ed. by W. Harper and C. A. Hooker), D. Reidel, Dordrecht.
[4] Simon Kochen and E. P. Specker: 1965, 'Logical Structures Arising in Quantum Theory', in *Theory of Models* (ed. by J. Addison *et al.*), North-Holland.
[5] Simon Kochen and E. P. Specker: 1967, 'The Problem of Hidden Variables in Quantum Mechanics', *Journal of Mathematics and Mechanics* **17**, 59–87.

The University of Western Ontario

III

FOUNDATIONS OF BIOLOGY

THE ONTOLOGICAL STATUS OF SPECIES AS EVOLUTIONARY UNITS

1. THE NATURE OF THE SPECIES PROBLEM

Reference to the species problem today might sound quaint and vaguely anachronistic. Perhaps the species problem was of some importance ages ago in the philosophical dispute between nominalists and essentialists or a century ago in biology when Darwin introduced his theory of organic evolution, but it certainly is of no contemporary interest. But 'species', like the terms 'gene', 'electron', 'non-local simultaneity', and 'element', is a theoretical term embedded in a significant scientific theory. At one time the nature of the physical elements was an important issue in physics. The transition from the elements being defined in terms of gross traits, to specific density, to molecular weight, to atomic number was important in the development of atomic theory. The transition in biology from genes being defined in terms of unit characters, to production of enzymes, to coding for specific polypeptides, to structurally defined segments of nucleic acid was equally important in the growth of modern genetics. A comparable transition is taking place with respect to the species concept and is equally important.

It is easy enough to say that species evolve; it is not so easy to explain in detail the exact nature of these evolutionary units. Evolutionary theory is currently undergoing a period of rapid and fundamental reformulation, and our conception of biological species as evolutionary units is being modified accordingly. There is nothing unusual about a theoretical term changing its meaning as the theory in which it is embedded changes, but sometimes such development has an added dimension. Not only is the meaning of the theoretical term altered, but the ontological status of the entities to which it refers is also modified. For centuries, philosophers and scientists alike have treated species as

Butts and Hintikka (eds.), Foundational Problems in the Special Sciences, 91–102.
Copyright © 1977 by D. Reidel Publishing Company, Dordrecht-Holland. All Rights Reserved.

secondary substances, universals, classes, etc. Nothing has seemed clearer than the relation between particular organisms and their species. In contemporary terminology, this relation is termed class membership. Species are classes defined by means of the covariation of the traits which their members possess. On this same interpretation, organisms are individuals and the species category itself is a class of classes.

However, species have proved to be very peculiar classes. Their membership is constantly undergoing change as new organisms are born and old ones die. At any one time, one can rarely discover a set of traits which is possessed by all the members of a species and by no members of some other species. In addition, the members of successive generations of the same species are usually characterized by slightly different sets of traits. These facts of life have forced philosophers to view the names of particular biological species (as well as all taxa) as cluster concepts. However, in this paper, I would like to argue for a radically different solution to the species problem. Just as relativity theory necessitated shifting such notions as space and time from one ontological category to another, evolutionary theory necessitates a similar shift in the ontological status of biological species. If species are units of evolution, then they cannot be interpreted as classes; they are individuals. The purpose of this paper is to show why evolutionary theory requires such a change in the ontological status of species.

There are three basic premises to the argument which I will present. Two are quite familiar; one is a variation on a familiar theme; none is totally uncontroversial. In this paper, however, I will not attempt to defend these premises. Instead I will show the consequences which follow for the nature of biological species *if* they are accepted.

2. THREE BASIC PREMISES

My first premise is that the ontological status of theoretical entities is theory-dependent.[1] Does time really flow like a river independent of the existence and distribution of material bodies? Are species really individuals? I don't think that such questions can be answered without reference to a particular scientific theory. A particular atom of gold is an individual and not a property, universal, process, relation, or what have you, because atomic theory requires atoms to be viewed in this way. The ontological status of electrons is equivocal for similar reasons, the theory in this case being quantum mechanics. A segment of DNA bounded by initiation and termination codons is an individual because molecular genetics requires genes to be interpreted in this way. If species are to be viewed as individuals, it will be because current evolutionary theory necessitates such a conceptualization.

The concept of an individual in philosophy is very broad, including such entities as sense data and bare particulars. In this paper, I will limit myself to a narrower use of the term to refer just to those entities which are characterized by unity and continuity, specifically spatiotemporal unity and continuity (see Hull, 1975). This is the sense in which organisms, planets, houses, and atoms are individuals. Such individuals have unique beginnings and endings in time, reasonably discrete boundaries, internal coherence at any one time, and continuity through time. There are individuals for which such spatiotemporal unity and continuity are not required; e.g. nations, political parties, ideas. The United States did not become any the less an individual when Alaska and Hawaii became states. In addition, there are situations in which a nation ceases to exist for a time and then the 'same' nation comes into being again later. However, species as units of evolution fulfill the stricter requirements. They are 'individuals' in the same sense that organisms are individuals. Both are localized in space and time.

On the surface, the distinction between a class and an individual could not seem sharper. A class, on the one hand, is the sort of thing that can have members. The name of a class is a class term defined intensionally by means of the properties which its members possess. The relation between a member and its class is the intransitive class-membership relation. An individual, on the other hand, is the sort of thing that can have parts. The name of an individual is a proper name possessing ideally no intension at all. The relation between a part and the individual of which it is part is the transitive part-whole relation. However, classes can be construed in ways which make them all but indistinguishable from individuals. Classes are often defined in terms of simple one-place predicates, like the class of atoms with atomic number 79. Classes can also be defined in terms of relations, sometimes a relation between two classes, sometimes a relation which the members of a class have to each other, sometimes a relation which each member of the class has to a specified focus. For example, 'planet' can be defined as any relatively large non-luminous body revolving around a star. Although the relation mentioned in this definition is spatiotemporal, the class itself is spatiotemporally unrestricted because planets can revolve around any star whatsoever.

The problematic cases are those 'classes' defined in terms of a spatiotemporal relation which the 'members' have to each other or to a specified individual. For example, 'forest' could be defined in terms of a sufficiently large number of trees no further apart than a certain distance from at least one other tree in the complex. On this definition, the term 'forest' would be a spatiotemporally unrestricted class, but each particular forest would not be. Similarly, 'tributary system' could be defined in terms of those rivers which flow into one main river. On this definition, 'tributary system' would be a spatiotemporally unrestricted class, but each particular tributary system would not be. Such complexes as the Black Forest and the Mississippi River tributary system can be treated as classes only at the expense of collapsing the distinction between classes and individuals, an excellent reason for not doing so.[2] The extension of this analysis to 'species' defined in terms of descent and the names of particular species should be obvious.

The final consideration central to my argument concerns the nature of scientific laws. According to the traditional conception, natural laws must be spatiotemporally unrestricted. To the extent that a scientific law is true, it must be true for all entities falling within its domain anywhere and at any time. The regularities in nature which scientific laws are designed to capture cannot vary from place to place or from time to time. I must hasten to add, however, that my acceptance of this traditional notion of a scientific law does not commit me to a raft of other beliefs commonly associated with it. Laws are not all there is to science. Descriptions, for example, are also important. Nor do I think that recourse to scientific laws is the only way in which an event can be explained. The issue of the nature of scientific laws must be raised, however, because evolutionary theory is often cited as evidence against the view that laws must be spatiotemporally unrestricted. But properly construed, the laws which go to make up evolutionary theory are as spatiotemporally unrestricted as any other laws of nature. Species evolve, languages evolve, our understanding of the empirical world evolves, but the laws of nature are eternal and immutable.[3]

3. EVOLUTIONARY THEORY – IS IT DIFFERENT?

The vague feeling which philosophers have had that evolutionary theory is somehow different from other process theories stems from three sources, two of them clearly mistaken. One source of the belief that the laws of evolutionary theory are spatiotemporally restricted stems from the process-product confusion, from confusing evolutionary processes such as mutation, selection, and evolution with their product – phylogeny. The statement that mammals arose from several species of reptile is clearly historical in the sense that it describes a temporal succession of events, but such descriptions or phylogenetic sequences are no more part of evolutionary theory than a description of the successive stages in an eclipse is part of celestial mechanics.

A second source of the vague feeling which philosophers have had that evolutionary theory is peculiar is their conceiving of species as classes. In the 19th century, philosophers such as John Stuart Mill distinguished between laws of succession and laws of co-existence. They viewed the apparent universal distribution of traits among the organisms which make up biological species as the best example of such laws of co-existence. Thus, for 19th century philosophers, the evolution of species meant the evolution of laws. This line of reasoning is still common today, especially among social scientists.[4] But if species are interpreted as individuals, the evolution of species poses no problem for the traditional conception of a scientific law. A description of an individual as it develops through time is hardly a candidate for the status of a scientific law. If individuals are localized in space and time and scientific laws must be spatiotemporally unrestricted, then no law of nature can contain essential reference to a particular individual. If species (as well as all taxa) are interpreted as individuals developing through time, then any statement which contains essential reference to such individuals can no more count as a scientific law than Kepler's laws – if Kepler's laws had been true only for the sun and its planets.[5] In point of fact, evolutionary theory contains no reference to particular taxa, just what one would expect if taxa are actually individuals and not classes. On this view, "All swans are white' could not count as a scientific law even if it were true.

However, there is a sense in which evolutionary theory might turn out to be peculiar. Biological evolution is a selection process. If selection processes are different in kind from other processes, then selection theories such as evolutionary theory might turn out to be different in form from ordinary process theories. To the extent that there is a difference, it seems to be this: selection processes require the existence of at least two partially independent processes operating at different levels of organization on different time scales. Hence, one finds biologists contrasting 'ecological time' with 'geological time', a strange manner of speaking to say the least. In a selection process, variation and selective retention result in the evolution of some unit of much larger scope. The entities which vary and which are selected come into being, reproduce themselves, and pass away in such a fashion that these larger units gradually change. However, the resolution of this particular problem has no special relevance to the thesis of this paper. Whether or not selection processes turn out to be reducible to ordinary processes, the consequences for the ontological status of biological species remains the same: they must be interpreted as individuals.

4. Species as individuals

According to the traditional formula, three levels of organization are involved in organic evolution. Genes mutate, organisms compete and are selected, and species evolve. We now know that this formula is too simple. Mutation can be as slight as the change of a single base pair or as major as the gain or loss of entire chromosomes. Competition and selection can take place at a variety of levels from macromolecules (including genes) and cells (including gametes) to organisms and kin groups. In addition, an important level of organization exists between kin groups and entire species – the population. Populations are the effective units of evolution. There is no question that genes, cells, organisms and kin groups are related by the part-whole relation. Genes are part of cells, cells are part of organisms, and organisms are part of kin groups. The point of the dispute is whether the relation changes abruptly from part-whole to class-membership above the level of organisms and possibly kin groups.

Several issues are involved in this dispute, including the existence of group selection. Organisms and kin groups form units of selection. Can populations and possibly even entire species form units of selection? Populations and species evolve. Can entities at lower levels of organization like colonies also form units of evolution? What kind of organization is required for something to function as a unit of selection? A unit of evolution? How do units of selection differ from units of evolution? Must evolution always occur at levels of organization higher than that at which selection is taking place? What does it mean to say that something is a 'unit' of selection or a 'unit' of evolution? Can such units be classes as well as individuals?

Considerable disagreement exists among biologists over the answers to the preceding questions, increased to some extent by certain unfortunate terminological conventions like using the terms 'organism' and 'individual' interchangeably. For example, E. O. Wilson (1974, p. 184) argues that the preceding "are not trivial questions. They address a theoretical issue seldom made explicit in biology":

In zoology the very word colony implies that the members of the society are physically united, or differentiated into reproductive and sterile castes, or both. When both conditions exist to an advanced degree, as they do in many of these animals, the society can equally well be viewed as a superorganism or even an organism. The dilemma can therefore be expressed as follows: At what point does a society become so well integrated that it is no longer a society? On what basis do we distinguish the extremely modified members of an invertebrate colony from the organs of a metazoan animal?

Theodosius Dobzhansky (1970, p. 23) has long argued that:

A species, like a race or a genus or a family, is a group concept and a category of classification. A species is, however, also something else: a superindividual biological system, the perpetuation of which from generation to generation depends on the reproductive bonds between its members.

Michael Ghiselin (1974) objects to terming species 'organisms' lest the term imply that species can function as units of selection the way that organisms do. He prefers the generic term 'individual'. However, the message for our purposes is the same. There is something about evolutionary processes that requires that the units of mutation, selection, and evolution be treated as individuals integrated by the part-whole relation.

All versions of evolutionary theory from Darwin to the present have included a strong principle of heredity. It is not enough for a gene to be able to mutate; it must be able to replicate and pass this change on. It is not enough for an organism to cope more successfully in its environment than its competitors; it must also be able to reproduce itself and pass on these variations to its progeny. But exactly the same observations can be made about populations and species, possibly not as units of selection but certainly as units of evolution.

The role of spatiotemporal unity and continuity in the evolutionary process is easily overlooked, especially when it is being described in terms of populations. The term 'population' is systematically ambiguous in the biological literature. In its broadest sense, a population is merely a collection of individuals of any sort characterized by the distribution of one or more traits of these individuals. Although the members of populations in this broad sense could be chosen at random, usually they are selected on the basis of some criterion; e.g. people on welfare, herbivorous quadrupeds, stars increasing in brightness. As biologists such as Ernst Mayr (1963) have repeatedly emphasized, the populations which function in the evolutionary process are 'populations' in a much more restricted sense of the term. Descent is required. But descent presupposes replication and reproduction, and these processes in turn presuppose spatiotemporal proximity and continuity. When a single gene undergoes replication to produce two new genes or a single cell undergoes mitotic division to produce two new cells, the end products are spatiotemporally continuous with the parent entity. In sexual reproduction, the propagules, if not the parent organisms themselves, must come into contact. The end result is the successive modification of the 'same' population.

Populations are made up of successive generations of organisms. These generations may be temporally disjoint or largely overlapping, but a certain degree of genetic continuity is required for a population to function as a population in the evolutionary process. To be sure, new organisms can migrate into a population and others leave, changing the genetic composition of the population. New genes can be introduced by means of mutation. But such changes cannot be too massive or too sudden without disrupting the evolutionary process. Given the differences in time spans required for mutation, selection and evolution, sufficient continuity is required to allow for the cumulation of the adaptive changes necessary for evolution. Channeling is required, sufficient channeling to warrant conceptualizing such lineages as individuals.

Identity in populations is determined by the same considerations which determine identity in organisms. Constancy of neither substance nor essence is required in either case; spatiotemporal continuity is. Just as all the cells which comprise an organism can be changed while that organism remains the 'same' organism, all the organisms which comprise a population can be changed while that population remains the 'same' population, just so long as such changes are gradual. Just as all the traits which characterize an organism at one stage of its development can change without that organism ceasing to be the 'same' organism, all the distributions of traits which characterize a population at one stage in its evolution can change without that population ceasing to be the 'same' population, just so long as such changes are gradual.[6]

The factors responsible for spatiotemporal continuity in evolution are fairly straightforward; the factors which promote evolutionary unity are not. In order for selection to result in evolutionary change, it must act on successive generations of the 'same' population. But the units of evolution must also be sufficiently cohesive at any one time to evolve as units. Two issues are at stake: the mechanisms which tend to promote evolutionary unity and the degree of unity necessary for an entity to count as an individual. Certain biologists have argued that in order for organisms to form populations or species, they must reproduce sexually. Gene exchange is the only mechanism capable of producing evolutionary unity (see Mayr, 1963; Dobzhansky, 1970; Ghiselin, 1974). Others have argued that gene exchange is not all that powerful a force in promoting evolutionary unity. The same selection pressures which produce unity in asexual species produce it in sexual species as well. Gene exchange is only of minor significance (see Meglitsch, 1954; Ehrlich and Raven, 1969).

From the discussion so far, it might seem that populations and species are far from paradigm individuals. In most cases, there is no one instant at which a species comes into existence or becomes extinct. Both speciation and phyletic evolution usually take hundreds, if not thousands, of generations. One mechanism which has been suggested for narrowing the borderline between a newly emerging species and its parent species is Mayr's Founder Principle. According to this principle, speciation in sexual species occurs always or usually by means of the isolation of one or a few organisms. Such isolates rarely succeed in forming a population, but when they do, the resulting population tends to be quite different from its parent species and can undergo additional rapid change. The end result is that the transition between two species is reduced both in time and with respect to the number or organisms involved.

Nor are the organisms which go to make up a species always in close proximity, let alone spatially contiguous. However, the exigencies of reproduction require at least periodic proximity of the relevant entities, whether organisms or merely their propagules. But before the notion that species are spatiotemporal individuals is dismissed, it should be noted that exactly the same observations can be made with respect to organisms as individuals, albeit on a reduced scale; and organisms are supposed to be paradigm individuals. Anyone who thinks that the spatiotemporal boundaries of organisms are that much sharper than those of species should read up on slime molds and grasses. Any problem which can be found in the spatiotemporal unity and continuity of species also exists for particular organisms. If organisms can count as individuals in the face of such difficulties, so can species.[7]

5. Conclusion

I think that the preceding considerations give ample support for the conclusion that species are not spatiotemporally unrestricted classes. However, doubt may remain whether or not they should be interpreted as individuals. Perhaps they belong in some hybrid category like 'individualistic classes', as Leigh Van Valen has suggested, or 'complex particulars', as Fred Suppe (1974) has proposed, but the evaluation of such suggestions must wait for further elucidation of these notions. For now, species fit as naturally into the idealized category of a spatiotemporally localized individual as do particular organisms. Once again, if organisms are individuals, so are species.

University of Wisconsin-Milwaukee

NOTES

[1] The claim that ontological status of theoretical entities is theory-dependent is not the same thing as the well-known claim that they are theory-laden, though the rationale behind the two claims is much the same.

[2] Another reason for not considering complexes defined in terms of a spatiotemporal relation between spatiotemporally localized individuals as genuine classes is the role which class terms play in science. The chief use of class terms in science is to function in scientific laws. A scientific classification is important to the extent that it produces theoretically significant classes. In fact, I think that the class-individual distinction is best made in terms of the contrasting roles played by classes and individuals in scientific laws.

[3] If it was discovered that the laws of nature, as we currently conceive them, actually changed through time, our conception of a scientific law would not have to be modified if these changes were themselves regular. Hence, statements characterizing these regular changes would become the new basic laws of nature.

[4] The highly touted book by M. D. Sahlen and E. R. Service (1960) relies throughout on confusing evolutionary processes with the products of such processes. They consistently argue that political laws evolve because political systems evolve. See also T. L. Thorson (1970) for the same crude mistake.

[5] The distinction assumed in this discussion is between laws of nature and true accidental generalizations. It is only fair to acknowledge, however, that this distinction is currently one of the most problematic in philosophy of science. At the risk of some circularity, I suspect that the only way in which this distinction can be made is by reference to the actual or eventual inclusion in a scientific theory. Any generalization, true though it may be, cannot count as a scientific law if it remains isolated from all other such generalizations.

[6] The subject matter of this paragraph deserves a paper of its own. Neither organisms nor species require constancy of substance or essence for individual identity, but as individuals, organisms and species differ in some important respects. Organisms possess a program which directs and circumscribes their development to some extent. The development of species is much more open-ended. One might wish to argue that an organism's genome is its essence. If so, then this essence is an individual essence and an essence of a very peculiar sort, since genomes are neither eternal nor immutable and two organisms can have the same individual essence without becoming the same individual. Finally, rapid change can occur in a single generation in a population without that

population ceasing to be the same population under special circumstances; e.g. if gene frequencies differ markedly in males and females.
[7] Comparable difficulties exist for the philosophical notion of the 'self'. Unlike organisms and species, most of the problem cases for the 'self' as an individual are hypothetical; see Derek Parfit (1971).

BIBLIOGRAPHY

Dobzhansky, T.: 1970, *Genetics of the Evolutionary Process*, Columbia Univ. Press, New York.

Ehrlich, P. R. and Raven, P. H.: 1969, 'Differentiation of Populations', *Science* **165**, 1228–1231.

Ghiselin, M. T.: 1974, 'A Radical Solution to the Species Problem', *Systematic Zoology*, 536–544.

Hull, D. L.: 1975, 'Central Subjects and Historical Narratives', *History and Theory* **14**, 253–274.

Lewontin, R. C.: 1970, 'The Units of Selection', *Annual Review of Ecology and Systematics* **1**, 1–18.

Mayr, E.: 1963, *Animal Species and Evolution*, Harvard Univ. Press, Cambridge, Mass.

Meglitsch, P. A.: 1954, 'On the Nature of Species', *Systematic Zoology* **3**, 49–65.

Parfit, D.: 1971, 'Personal Identity', *The Philosophical Review* **80**, 3–27.

Sahlin, M. D. and Service, E. R.: 1960, *Evolution and Culture*, Univ. of Michigan Press, Ann Arbor, Mich.

Suppe, F.: 1974, 'Some Philosophical Problems in Biological Speciation and Taxonomy', in J. A. Wojciechowski (ed.), *Conceptual Basis of the Classification of Knowledge*, Verlag Dokumentation, Pullach/Muchen.

Thorson, T. L.: 1970, *Biopolitics*, Holt, Rinehart & Winston, Inc., New York.

Wilson, E. O.: 1974, 'The Perfect Societies', *Science* **184**, 54.

THEORIES AND OBSERVATIONS OF DEVELOPMENTAL BIOLOGY

1. INTRODUCTION

Developmental biology is one of the most active and fastest expanding areas of biology if not of all natural sciences. The development of many species is intensively studied at many different levels of organization, from the molecular to that of the integral organism. Yet there is no underlying or unifying theory of development, no theory with its own formalism and laws, and there are not even any informal generalizations applicable to unrelated organisms. This lack of a theory is certainly not due to lack of activity or imagination on the part of theoretical biologists, but rather to the basic intractability of the problem. In the present paper I attempt to indicate the major directions of effort and the connections in the presently pursued theoretical work. I cannot, however, begin to descuss the biological consequences and supporting data of these theories, in other words their empirical status, it would require an entire treatise to do so. An overview of some of the abstractions may still give a useful perspective of this field to the workers in developmental theory construction as well as to philosophers of biology who have not covered this area up to now.

2. DEVELOPMENTAL OBSERVATIONS

Developmental biology consists in the first place of an incredible accumulation of observational reports concerning the structures of a given group of organisms at various stages of their development. These developmental descriptions (which is called 'descriptive embryology' or 'Entwicklungsgeschichte' in German) are usually obtained by preserving and sectioning a number of individual organisms at various ages and then arranging the observed structures in an assumed sequence.

Butts and Hintikka (eds.), Foundational Problems in the Special Sciences, 103–118.
Copyright © 1977 by D. Reidel Publishing Company, Dordrecht-Holland. All Rights Reserved.

This method has the disadvantage that the actual 'development', i.e., the change in time, of a given organ is conjectural only, since the observations refer to disconnected stages of what are assumed to be homologous structures.

Sometimes there are time extended observations possible on the same organism, concerning particular features which can be followed without destroying the organism. These observations, representing measurements of certain parameters, can then be plotted against time. The literature is full of empirical curves of this kind.

An entirely different approach to development, called 'experimental embryology', consists of interfering in the normal process through chemical, physical or surgical treatments. We may think of all the studies which have been carried out with metabolic inhibitors, centrifugation or transplantation on embryos or other growing tissues. The interpretations of results obtained by these techniques while very fruitful and valuable, are also open to criticism because each agent or treatment may have manifold effects besides their specific influence on certain components of the developing structure.

Still another approach is through the study of mutant forms, a large and expanding field which is now called 'developmental genetics'. Here deviations from normal development are produced not by external agents but by mutations which arise in a clone of organisms. The mutations may originally have been produced by physical or chemical agents, or they may have occurred spontaneously. The drawing of developmental conclusions from known genetic changes brings with it, however, the same problems as those associated with external agents: each gene can have many different and non-specific effects on development.

Finally, there is a 'biochemistry of development' following the presence and fate of various compounds in a developing embryo, particularly looking for patterns in time and space in which certain components appear or disappear. The last, more-or-less irreversible, steps of development which are usually designated 'differentiation', can usually be attacked very decisively by biochemical methods, while the early phases of development, ('determination' and 'pre-pattern formation') not so well. This is because differentiation involves the accumulation of chemical components of which the concentration can be followed and even the synthesis can be elucidated, while the chemical nature of determination is still mostly unknown.

3. DEVELOPMENTAL THEORIES

One might distinguish two basic tendencies in past attempts at constructing theories of development. One tendency is to stay as close to observations as possible, thus in the formal expressions using only terms which are observationally interpretable. The other tendency is to consciously incorporate theoretical terms in the expressions, terms, that is, which cannot be interpreted immediately, but the use of which lead to testable consequences.

There is, of course, room for philosophical argument whether any term in any scientific theory can be considered purely observational or purely theoretical (cf. Hanson, 1958; Hempel, 1961; Nagel, 1961; Putnam, 1962; Spector, 1966; Ruse, 1973) but there is no question that in any field one can find terms which are primarily and obviously of one or of the other kind.

Thus, the statements and mathematical expressions which constitute a scientific theory can be roughly classified into those which contain only observational terms, those which also contain theoretical terms, and those which have only theoretical terms. The computational power of the theory depends mainly on statements of the latter kind, while experimental testing of the theory is primarily aided by statements of the first kind.

One might ask which are the observational and which are the theoretical terms in present day developmental biology, and what kinds of observational, mixed, or theoretical statements one finds in the literature. To go into this question in detail would be impossible in this short review, we can only attempt to provide brief descriptions of the currently used theories with indications of the theoretical-observational status of their most important concepts.

4. DEVELOPMENTAL THEORIES WITHOUT THEORETICAL TERMS

The most obvious way for constructing a developmental theory is by following the example of physics. When one is faced with a phenomenon taking place in space and time, and undoubtedly development is such a phenomenon, the first thing a physicist will do is to try to find functions describing the parameters of the phenomenon in terms of spatial and temporal coordinates. In case of a developing organism a parameter which can be measured most simply is the overall size of the organism, expressed in terms of weight or volume. Thus one tries to describe this parameter as a function of time, let us say $W(t)$. Once a statistically smoothed curve is found for a given genetically homogeneous population of organisms developing in a homogeneous environment, then one can try to express this empirical curve in a mathematical form. It is true, of course, that to any curve drawn through n points one can always fit a power series of the form:

$$W(t) = A_0 + A_1 t + A_2 t^2 + ... + A_{n-1} t^{n-1}.$$

But such an expression is of very little use as far as any underlying growth mechanism is concerned: the constants $A_0, ..., A_{n-1}$ one cannot assign any physiological meaning.

Various mathematical functions have been proposed for growth descriptions which are better interpretable (cf. Medawar, 1945; Richards, 1968): exponential and power functions, of which the logistic function may be the best known for S-shaped growth curves:

$$W(t) = \frac{B}{1 + ke^{-\lambda B t}},$$

where k, λ and B are positive constants, B being the upper bound for W. This function is the integral of the differential equation:

$$\frac{dW}{dt} = \lambda B W - \lambda W^2.$$

The first term on the right-hand side of this equation yields exponential growth at low values of W, while the second term restricts further growth at higher values of W. A more general form is (von Bertalanffy, 1948)

$$\frac{dW}{dt} = \alpha_1 W^\beta - \alpha_2 W^\gamma, \quad \text{where} \quad \beta \le \gamma.$$

Some time ago there was hope that by compiling growth data for many different organisms and by finding for them constants α_1, α_2, β, and γ, one will arrive at a growth theory which will yield the possible combinations of constants and which will connect these with other observations. This hope has not been fulfilled, although large tables are available today for growth constants.

That no general theory has emerged on this basis has possibly been due to the fact that when an organism is considered in its entirety, the resulting growth is a composite of many part-processes and therefore the essential features of the processes are obliterated. This shortcoming has been corrected by measuring different parts separately and trying to find correlations among them (cf. Medawar, 1945; Harte, 1976). In this way 'allometric' growth formulas were obtained, of the form

$$y(t) = b \cdot x(t)^a,$$

where x and y are time functions of sizes of various body parts, and a and b are allometric constants. Again, it has not been possible to find unifying principles yielding these constants.

A still more sophisticated approach of the same kind is that of subdividing the growing organism into unboundedly many parts and expressing the growth of each of these by a general formula (Richards and Kavanaugh, 1945; Erickson, 1976). This can be done by observing markers on the surface of a growing organism (e.g. a root or leaf), finding the velocity vectors for each of the markers, and then by interpolation between the markers obtain relationships between these vectors and the positions at which they are measured. In other words, in an (x, y, z) coordinate system one finds the component vectors v_x, v_y, v_z, where

$$v_x = \frac{\partial x}{\partial t},$$

and then calculate the *divergence* which is defined as

$$\text{div } \mathbf{v} = \frac{\partial v_x}{\partial x} + \frac{\partial v_y}{\partial y} + \frac{\partial v_z}{\partial z}.$$

The mathematical concept of divergence has its origin in hydro- and aero-dynamics. A region of space where divergence is positive is called a 'source', where it is negative, a 'sink', and in an incompressible fluid divergence must be everywhere zero.

Erickson (1956, 1966) has pointed out that the one dimensional divergence $(\partial v_x/\partial x)$ is the same as the relative elemental rate of elongation and the two-dimensional divergence $[(\partial v_x/\partial x)+(\partial v_y/\partial y)]$ the same as the relative elemental rate of increase of area. In other words, the latter is the rate of increase of an (infinitesimal) elemental area per unit area.

Provided that growth takes place in a time independent way, by plotting the divergence value against a spatial coordinate one obtains a characteristic curve describing the growth of the organism with a far greater precision and generality than by any of the methods mentioned above. This work is now being pursued in various laboratories, and has already contributed very valuable new insight into growth processes.

From a methodological point of view, these approaches can be summed up by pointing out that growth is considered in all of them as a description of an object purely in time and space, without recognizing any other properties of the object. The same methods could be used to describe the growth of crystals, clouds or galaxies, the specific nature of living organisms is not being considered.

An entirely different approach to spatial and temporal organization of development is that where the part – whole relation is the basic concept. Development consists, of course, of the creation of new parts, and sloughing off of old parts of the same whole organisms. Experimental studies involve cutting and/or transferring parts from one organism to another, creating new wholes as it were. Sometimes one is interested in the further development of an isolated part. The observation that several complete sea urchins develop from a divided embryo (Driesch, 1921), in other words, the 'equipotency' of embryonic parts was a constant source of wonder and speculation in the early part of this century (now explained by the repeated occurrences of the genetic material in each of the cells of the embryo). Regeneration of lost parts belongs also among these phenomena.

It was for this reason that theoretical biologists have concerned themselves with a formalization of development based on the part–whole relation. Woodger (1937, 1939) and Tarski (1937) have constructed an axiom system (with connections to Lesniewski's 'mereology') with the following two primitive notions:

'x is part of y' (written also as 'xPy') and 'x is before y in time' (written as 'xTy'). Furthermore, one needs the defined notion 'x is a (composite) sum of set α', or conversely, 'α is a dissection of x' (written as 'xSα'). This concept is defined as follows: α is a dissection of x if and only if

(1)   every element of the set α is a part of x, and
(2)   every part of x (let us say y) is such that there is an element of α (let us say z) such that y and z have some part in common (let us say u).

The second part of this definition is illustrated below where the object x is dissected into several objects one of which is z, and where y is part of x, and u (the shaded area) is the common part of y and z.

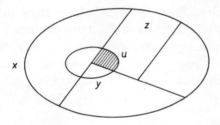

The use of the terms P, T, and S are governed by the following axioms:

1.   'Being part of' is a transitive relation (if xPy and yPz then xPz).
2.   Every non-empty set (of objects) is the dissection of one and only one object (*the* composite sum of α exists for every non-empty set α)
3.   'Being before in time' is an asymmetric relation (if xTy then it is not the case that yTx).
4.   For every non-empty set α and β, the composite sum of α is before the composite sum of β in time if and only if for every element x of α and y of β it is the case that x is before y in time.
5.   For any pair of objects x and y, if no part of x is after y in time, then everything which is after y in time is after x in time.

6. Similarly, for any pair of objects x and y, if no part of x is before y in time, then everything that is before y in time is before x in time.

    A definition of 'momentary things' is provided in the following way: An object x is momentary if and only if there are no two parts u and v of x such that u is before v in time.
7. Every object x has some momentary parts.

Then 'slices', 'beginning and end slices' are defined, and finally 'division' and 'fusion', as applicable to cells or other objects. A large number of formal results can be obtained from these definitions and axioms. They can describe and classify known developmental situations, and they lend elegance and conciseness to otherwise very messy descriptions of developmental events and life cycles. Nevertheless, these descriptions lack a certain theoretical power. In other words, it is difficult to see how new, previously unrecognized, regularities and generalizations could arise from them. Possibly this is because these formalisms share the lack of theoretical terms with the ones expressed as differential equations In both cases, the theories stay very close to the observations, yield perfectly respectable, uncontroversial, but as far as scope and generality is concerned, rather weak results.

5. DEVELOPMENTAL THEORIES WITH THEORETICAL TERMS

There have been, in the last 20 years or so, a number of attempts to introduce theoretical, non-empirical, content into statements of developmental biology. In each case, a certain mathematical concept was found useful to illuminate a particular developmental problem, this mathematical formalism was then elaborated and used as a 'paradigm' for further examples and problems.

One such paradigm is that of 'positional information' specified with the help of 'gradients' in developing 'fields' (Wolpert, 1969; Wolpert et al., 1975; originally Child, 1941). According to this assumption, special morphogenetic substances are present in developing organisms, which are produced at certain centers, gradients are formed between their sources and their sinks, and these gradients govern various developmental processes. Eventually, at specified places new gradients arise, and so one gets more and more complex structures. The cells which are in a field governed by a gradient acquire a 'positional information'; in a way they 'know' where they are and therefore what to do. Actually, it would be better to call this a 'boundary specification' mechanism, because in fact the cells do not need to know their absolute positions, only their relative ones with respect to boundaries specified with the help of gradients and thresholds values for concentrations. The morphogenetic substance or substances responsible for these gradients remain in most cases hypothetical, except for instance for growth hormones in plants and aggregation substances in slime molds and fungi. The mathematics of diffusion and transport processes by which such gradients can be set up (cf. Crick, 1970) is, on the other hand, well established.

Another, entirely different paradigm for development is that of Goodwin (1963), Goodwin and Cohen (1969), built on the existence of coupled oscillators. Under certain conditions, two oscillators with different frequencies and phases may interact and get locked in some frequency and phase. In this way temporal regulation of development may be taken care of.

Interference patterns can also be generated by two or more oscillators, giving rise to spatial morphogenesis. Although a number of biochemical oscillating systems have been found (in yeast and slime molds) the widespread existence of oscillators with the right frequencies in developing organisms remains hypothetical. The mathematics of oscillatory processes is of course well known and is used extensively in electronics to mention only one field of application. So a connection between biology and physics on this basis could become very fruitful.

A third paradigm is Turing's (1952) reaction-diffusion mechanism for morphogenesis, which has recently been further elaborated by Cohen (1971), Robertson and Cohen (1972), and Gierer and Meinhardt (1972). In this case spatial and temporal pattern generation is brought about by the diffusion of several chemical substances, and their reactions with each other, in an otherwise homogeneous medium. It has been known for some time (Liesegang rings, Zhabotinsky patterns) that one can obtain banded patterns in liquids or gels by reaction-diffusion mechanisms. These mechanisms could also be responsible for morphogenesis in living organisms. One needs to find suitable trimolecular reactions coupled with diffusion in order to substantiate this claim (the mechanism does not work with bimolecular reactions because one needs non-linear differential equations for the reactions in order to get the mathematical results).

The fourth developmental paradigm we may mention is that of Thom (1975). He uses the topological theory of differentialable manifolds as applied to stable dynamic systems.The behaviour of such a system may exhibit abrupt changes (called 'catastrophes') from one stable area of parameters to another.

In systems with at least four parameters one can distinguish among seven qualitatively different 'catastrophes', which should correspond to forms generated by stable dynamic systems. One of the assumptions necessary for this mathematical result is that 'potential functions' must exist to describe the time dependence of the parameters, just as there are potential functions for motion in gravitational and electromagnetic fields. Until now no potential functions have been found for biochemical systems, i.e. for the diffusion and reaction processes, which could account for the metabolism and morphogenetic mechanisms of organisms. According to statements by Thom potential functions may yet be found for these processes. Thus, here again we are dealing with certain hypothetical factors which, should their existence be proved, could provide a theoretical foundation for development. It might be added that the 'epigenetic landscape' idea of Waddington (1957), where he likens development to stones rolling down mountains is conceptually sympathetic to Thom's views, in as much as the motion of stones is known to be describable in terms of potential functions.

A different approach to development from those listed above is based on the concept of an 'algorithm'. An algorithm is a set of 'instructions' which are applied to a certain structure and by which the structure changes into a new one. If this procedure is performed repeatedly using the same set of instructions, one has an algorithmic procedure. The best know algorithms at the present time are computer programs. By these 'computations' are performed on some initial 'data' yielding 'results', i.e. the numeric or alphabetic inputs are stepwise changed into expressions which are considered as the outputs of the procedure. The result must be obtainable in finitely many computation steps for the procedure to be considered an algorithm.

Computational algorithms can be applied to development, either as convenient methods to derive consequences of some of the previously mentioned assumptions, or as paradigms for the developmental process itself. (cf. Apter, 1966; Arbib, 1972; Baer and Martinez, 1974). The development of a multicellular organism may be considered as the execution of a 'program' which is present in the fertilized egg, as well as in all the cells of the developing organism. The fact that all higher plants and animals have a cellular organization, and that all their cells possess the same complement of DNA make this assumption rather attractive.

Several recent developmental models are algorithmic in the sense that they consist of computer programs in which sets of instructions are repeatedly applied to consecutive stages in the development of an organism (Cohen, 1967; Ede and Wilby, 1975, 1976; Bezem and Raven, 1975, 1976; Ransom, 1975). Each of these programs is *ad hoc* developed, and while they are useful for a specific developmental process within a limited group of organisms, they would be difficult to generalize over larger classes. The difficulty lies in the fact that computer programs are formally not sufficiently understood in such a way that would allow us to make comparisons among programs as to their power or complexity.

An attempt has recently been made to provide a uniform formalism within which developmental programs can be expressed. We have defined (Lindenmayer, 1968, 1971, 1975; Herman and Rozenberg, 1975) generating systems for strings of symbols, where each symbol represents the state of a cell or of a group of cells at some moment, and where a new string of symbols is generated from a previous one by application of a set of 'production' rules to each symbol simultaneously. Thus the cells of a filamentous organism would correspond to the strings of symbols, and its development corresponds to a sequence of strings generated by the rules beginning with an initial string which may consist of a single symbol. By having productions which result in the substitution of two new symbols for one previous one, we can program cell divisions, and by productions which substitute nothing for a symbol, we can have cell death taking place.

Formally two kinds of string-generating systems have been defined, those without interactions among the neighbouring cells (OL-systems), and those with interactions (IL-systems). An OL-system is an ordered triple $\langle \Sigma, P, \omega \rangle$, where Σ is a finite, non-empty set (alphabet of symbols); P is a set of ordered pairs (productions) of the form $a \rightarrow x$ such that a is in Σ and x is in Σ^* (the set of all finite strings over the alphabet Σ), and for each a in Σ there is at least one string x in Σ^* such that $(a \rightarrow x)$ is in P; and ω is a non-empty string over Σ (the initial string). If for each a in Σ there exists only one x in Σ^* such that $(a \rightarrow x)$ is in P, then the OL-system is deterministic, otherwise it is non-deterministic. If P contains no pairs such as $(a \rightarrow \varepsilon)$, where ε is the empty string, then the OL-system is propagating, otherwise it is non-propagating.

A string x directly derives string y in the OL-system S if and only if for some $n \geq 1$ there are symbols a_i and strings p_i, for all i between one and n, such that x is the string $a_1 a_2 \cdots a_n$ and y is the string $p_1 p_2 \cdots p_n$, and all $(a_i \rightarrow p_i)$ are productions of the system S. Furthermore, a string x derives string y (in some finite number of steps) if and only if there are strings $q_1, q_2, ..., q_{m-1}$ for some $m \geq 0$ such that x directly derives q_1, q_1 directly derives q_2, etc., and q_{m-1} directly derives y.

The language of an OL-system S is defined as the set of all strings which are derived from the initial string of S by the productions of S. A developmental language is thus an (unordered) set of all developmental stages of an organism, while a developmental sequence is the ordered succession of stages. A developmental sequence (also a language) may consist of infinitely many strings, in which case their lengths cannot be bounded, or of finitely many strings. In this latter case, either the sequence is terminating, due to the disappearance of all symbols in the last string (producing the empty string), or the sequence may mainly consist of cyclically repeating strings.

A string-generating system with interactions (an IL-system) is defined as follows. An IL system is an ordered quadruple $S = \langle \Sigma, P, \#, \omega \rangle$, where Σ is a finite, non-empty set (of symbols); P is a finite subset of

$$\{\#, \varepsilon\} \Sigma^* \times \Sigma \times \Sigma^* \{\#, \varepsilon\} \times \Sigma^*;$$

$\#$ is a symbol not in Σ (the endmarker); and ω is a non-empty string over Σ (the initial string). The meaning of the above Cartesian product is that each production is of the form $\langle \alpha, a, \beta \rangle \to \gamma$, where a is a symbol in Σ, and α, β, γ are (possibly empty) strings of symbols over Σ such that α may begin with an endmarker and β may end with an endmarker.

String x directly derives string y in the IL-system S if and only if for some $k \geq 0$ there are symbols a_1, a_2, ..., a_k in Σ and strings p_1, p_2, ..., p_k over Σ such that x is the string $a_1 a_2 \ldots a_k$, y is the string $p_1 p_2 \ldots p_k$, and for every i, $1 \leq i \leq k$, there exist m, $n \geq 0$ such that $\langle \text{suffix}_m (\# a_1 a_2 \ldots a_{i-1}), a_i, \text{prefix}_n (a_{i+1} a_{i+2} \ldots a_k \#) \rangle \to p_i$ is a production of S. An IL-system S must be strongly complete, i.e. for any non-empty string x over its alphabet Σ there must be a string y over Σ such that x directly derives y in S. Provided that the endmarker $\#$ is present at each step on both ends of the strings, the definition for x deriving y (in a finite number of steps) in IL-system S is the same as that for OL-systems. The language generated by IL-system S is defined as the set of strings (without their endmarkers) derived from $\# \omega \#$ by the productions of S, where ω is the initial string of S.

There is a considerable body of mathematical results on OL-systems, IL-systems and related concepts such as their table versions, and canonical extensions (see bibliography of L-systems in Lindenmayer and Rozenberg, 1976).

In the past year or two graph-generating L-systems have been introduced (Culik and Lindenmayer, 1976; collection of papers in Lindenmayer and Rozenberg, 1976), which are applicable to non-filamentous organisms.

The mathematical theory of these algorithmic models with discrete states and discrete spatial and temporal units is more tractable than those with continuous parameters. Particularly in cases where in a developing structure many different subregions exist, each with its own growth rate or differentiating tendency, such a control-theoretical cellular generating system can be of use. The biological interpretability is also easier since the various hypothetical 'states' can be directly identified with the accumulation of chemical substances or with physical characteristics of cells or cell groups which are of importance for development. Similarly, the inputs received by a cell can be identified with inducers or hormones entering it.

This mathematical formalism is, on the other hand, not as well suited for interactive developmental mechanisms where only one or two substances are diffusing and/or reacting with each other. Partial differential equations can clearly provide better results.

In general, the usefulness of this particular algorithmic approach is apparent at the present time for describing complex development controlled by cell ancestries (cell lineages). The results up to now are only indicative of possible lineage mechanisms for given organisms, they remain hypothetical as long as the cellular 'states' and 'inputs' are not founded on cytological or biochemical observations.

6. CONCLUSIONS

We have surveyed the present status of theory construction in developmental biology. Numerous attempts have been made and are being pursued in order to provide theoretical foundations for the immense amount of observations available in this field. While possibly none of the approaches listed is by itself sufficient, it may be hoped that some combination of these ideas will eventually be successful. There is clearly need for concepts both close to the observational level and higher on the theoretical level before a coherent and powerful theory will be realized.

University of Utrecht

BIBLIOGRAPHY

Apter, M. J.: 1966, *Cybernetics and Development*, Pergamon Press, Oxford.
Arbib, M. A.: 1972, 'Automata Theory in the Context of Theoretical Embryology', in *Foundations of Mathematical Biology*, R. Rosen (ed.), Academic Press, New York, Vol. II, pp. 141–215.
Baer, R. M. and Martinez, H. M.: 1974, 'Automata and Biology', *Annual Review of Biophysics and Bioengineering* **3**, 255–291.
Bertalanffy, L. von; 1948, 'Das organische Wachstum und seine Gesetzmässigkeiten', *Experientia* Vol. 4.
Bezem, J. J. and Raven, C. P.: 1975, 'Computer Simulation of Early Embryonic Development', *Journal of Theoretical Biology* **54**, 47–61.

Bezem, J. J. and Raven, C. P.: 1976, 'Some Algorithms used in the Simulation of Embryonic Development', in A. Lindenmayer and G. Rozenberg (eds.), *Automata, Languages, Development*, North-Holland Publ. Co., Amsterdam, pp. 147–158.

Child, C.: 1941, *Patterns and Problems of Development*, Chicago University Press, Chicago.

Cohen, D.: 1967, 'Computer Simulation of Biological Pattern Generation Processes', *Nature* **216**, 246–248.

Cohen, M. H.: 1971, 'Models for the Control of Development', *Symposia of the Society for Experimental Biology*, No. 25, pp. 455–476, Cambridge University Press.

Crick, F.: 1970, 'Diffusion in Embryogenesis', *Nature* **225**, 420–422.

Culik, K. II and Lindenmayer, A.: 1976, 'Parallel Graph Generating and Graph Recurrence Systems for Multicellular Development', *International Journal of General Systems* **3**, 53–66.

Driesch, H.: 1921, *Philosophie des Organischen*, Verlag W. Engelmann, Leipzig.

Ede, D. A. and Law, J. T.: 1969, 'Computer Simulation of Vertebrate Limb Morphogeneses', *Nature* **221**, 244–248.

Ede, D. A. and Wilby, O. K.: 1976, 'Analysis of Cellular Activities in the Developing Limb Bud System', in A. Lindenmayer and G. Rozenberg (eds.), *Automata, Languages, Development*, North-Holland Publ. Co., Amsterdam, pp. 15–24.

Erickson, R. O. and Sax, K. B.: 1956, 'Elemental Growth Rate of the Primary Root of *Zea Mays*', *Proc. American Philosophical Society* **100**, 487–498.

Erickson, R. O.: 1966, 'Relative Elemental Rates and Anisotropy of Growth in Area', *Journal of Experimental Botany* **17**, 390–403.

Erickson, R. O.: 1976, 'Growth in Two Dimensions, Descriptive and Theoretical Studies', in A. Lindenmayer and G. Rozenberg (eds.), *Automata, Languages, Development*, North-Holland Publ. Co., Amsterdam, pp. 39–56.

Gierer, A. and Meinhardt, H.: 1972, 'A Theory of Biological Pattern Formation', *Kybernetik* **12**, 30–39.

Goodwin, B. C.: 1963, *Temporal Organization in Cells*, Academic Press, London.

Goodwin, B. C. and Cohen, M. H.: 1969, 'A Phase-Shift Model for the Spatial and Temporal Organization of Developing Systems', *Journal of Theoretical Biology* **25**, 49–107.

Hanson, N. R.: 1958, *Patterns of Discovery*, Cambridge University Press.

Harte, C.: 1976, 'Measurement of Growth Correlation and the Genetic Control of Growth', in A. Lindenmayer and G. Rozenberg (eds.), *Automata, Languages, Development*, North-Holland Publ. Co., Amsterdam, pp. 75–87.

Hempel, C. G.: 1961, 'The Theoretician's Dilemma', *Minnesota Studies in the Philosopy of Science*, Vol. 3, pp. 42–62.

Herman, G. T. and Rozenberg, G., with a contribution by Lindenmayer, A.: 1975, *Developmental Systems and Languages*, North-Holland Publ. Co., Amsterdam.

Lindenmayer, A.: 1968, 'Mathematical Models for Cellular Interactions in Development, Parts I and II', *Journal of Theoretical Biology* **18**, 280–299, 300–315.

Lindenmayer, A.: 1971, 'Developmental Systems Without Cellular Interactions, their Languages and Grammars', *Journal of Theoretical Biology* **30**, 455–484.

Lindenmayer, A.: 1975, 'Developmental Algorithms for Multi-Cellular Organisms: A Survey of L-Systems', *Journal of Theoretical Biology* **54**, 3–22.

Lindenmayer, A. and Rozenberg, G. (eds.): 1976, *Automata, Languages, Development*, North-Holland Publ. Co., Amsterdam.

Medawar, P. B.: 1945, 'Size, Shape and Age', in W. E. LeGros Clark and P. B. Medawar (eds.), *Essays on Growth and Form*, Oxford University Press, pp. 157–187.

Nagel, E.: 1961, *The Structure of Science*, Routledge and Kegan Paul, London.

Putnam, H.: 1962, 'What Theories are Not', in E. Nagel *et al.* (eds.), *Logic, Methodology and Philosophy of Science*, Stanford University Press, pp. 240–251.

Ransom, R.: 1975, 'Computer Analysis of Division Patterns in the *Drosophila* Head Disc', *Journal of Theoretical Biology* **53**, 445–462.

Richards, F. J.: 1969, 'The Quantitative Analysis of Growth', in F. C. Steward (ed.), *Plant Physiology*, Vol. VA, Academic Press, New York.

Richards, O. W. and Kavanagh, A. T.: 1945, 'The Analysis of Growing Form', in W. E. LeGros Clark and P. B. Medawar (eds.), *Essays on Growth and Form*, Oxford University Press, pp. 188–230.

Robertson, A. and Cohen, M. H.: 1972, 'Control of Developing Fields', *Annual Review of Biophysics and Bioengineering* **1**, 409–464.

Ruse, M.: 1973, *The Philosophy of Biology*, Hutchinson University Library, London.

Spector, M.: 1966, 'Theory and Observation', *British Journal of the Philosophy of Science* **17**, 1–20, 89–104.

Tarski, A.: 1937, 'Appendix' in J. H. Woodger, *The Axiomatic Method in Biology*, Cambridge University Press, pp. 161–172.

Thom, R.: 1975, *Structural Stability and Morphogenesis*, W. A. Benjamin Inc., Reading, Massachusetts.

Turing, A. M.: 1952, 'The Chemical Basis of Morphogenesis', *Philosophical Transactions, Royal Society, London*, Series B, **237**, 37–72.

Waddington, C. H.: 1957, *The Strategy of the Genes*, George Allen and Unwin, London.

Wolpert, L.: 1969, 'Positional Information and the Spatial Pattern of Cellular Differentiation', *Journal of Theoretical Biology*, **25**, 1–47.

Wolpert, L., Lewis J., and Summerbell, D.: 1975, 'Morphogenesis of the Vertebrate Limb', in *Cell Patterning*, Ciba Foundation Symposium, No. 29, pp. 95–130, Assoc. Scientific Publishers, Amsterdam.

Woodger, J. H.: 1937, *The Axiomatic Method in Biology*, Cambridge University Press.

Woodger, J. H.: 1939, *The Technique of Theory Construction*, Int. Encyclop. of Unified Science, O. Neurath (ed.), Univ. of Chicago Press, Vol. II, No. 5.

ORGANIC DETERMINISM AND TELEOLOGY
IN BIOLOGICAL RESEARCH

The share of theoretical research in modern scientific knowledge tends sharply to grow. Now that we have entered the 'age of biology' this process is revealed perhaps in the most characteristic form in the attempts to construct a generalizing theory describing the basic regularities of living systems, with the inevitable resort by the scientists to the fundamental principles of the theory of knowledge and an analysis of the heuristic potentialities of the traditional and new methods of research in their interconnection and interdependence. Among the principles most intensively discussed over the past few years are the principles of causality, determinism in their specific expression, notably, in biological knowledge in connection with the problem of organic purposefulness and its teleological interpretation.

1. SPECIFICS OF ORGANIC DETERMINATION IN THE LIGHT OF CYBERNETICS

In modern science, this general problem, connected with the need to bring out the specifics of organic determination of living systems is tackled both within empirical research and within the framework of theoretical generalizations of an extremely extensive class. This applies, in particular, to the general theory of systems on the basis of whose propositions attempts have already been made to produce a symbolic presentation of the changes within a system which is some distance away from the state of equilibrium but, in a sense, 'striving' to attain that state in the future. In this context, L. von Bertalanffy characterises the living systems as 'equifinal', that is, as being capable of achieving a similar final result practically regardless of initial conditions. This property of 'equifinality', which H. Driesch brought out

Butts and Hintikka (eds.), Foundational Problems in the Special Sciences, 119–129.
Copyright © 1977 by D. Reidel Publishing Company, Dordrecht-Holland. All Rights Reserved.

vitalistically and interpreted teleologically, is associated in modern science with the special character of the interaction of living systems, their activity, the specifics of the processes of regulation and control, which are being intensively studied above all by biocybernetics.

Without going in detail into the concepts established by cybernetics and connected with the control of self-organising systems, coding, information transfer, etc., let us recall the most essential characteristics of self-regulating and self-governing systems, which also include living systems. These relate above all to the capability of modifying their state under the impact of information signals, that is, the capability of selective response. Complex systems of this type also have the capability of memorising the best effect of earlier responses, which is why they are characterised as self-regulating and self-teaching systems. Such systems can receive information signals from other systems and the environment and transmit them after an indefinitely long interval of time. They are capable of modifying their working algorithms and their own organisation depending on changing information signals, which ensures not only the survival of these systems, the self-reproduction of the organisation achieved, but also their improvement and development.

It was also highly essential that self-regulating and self-governing systems realise these characteristic properties with the aid of feed-back mechanisms (that is, a continuous exchange of information between the governing device and the executive organ). Self-regulation is effected in the form of a cyclical process which runs along a closed circle. The feed-back can also exist in the form of a fixed device with the character of secondary regulation built up over the primary dynamic interactions between processes in complex systems. But these primary interactions also ensure a feed-back effect, even if indirectly. This means, for instance, that living beings with a developed nervous system effect feed-back ensuring the processes of self-regulation in the form of signals which are sent back to the central nervous system. But there is also a feed-back effect in intracellular processes, biochemical interactions between individual structures of the genotype and the phenotype, etc. This is also effected in populations in their relations with their environment and within the biosphere as a kind of extensive class integrated system.

These characteristics of complex systems, brought out by cybernetics, are of universal importance. They will be found in any directed process of active adaptation connected with the selection of the optimal variant for modifying the structural or functional properties of a system. It goes without saying that these changes themselves have many a value and their overall trend, which is realised integrally has a statistic, probabilistic nature.

In this sense, these cybernetic characteristics are brought out in the analysis of biochemical processes of living systems, in their molecular-genetic interactions. This is most clearly revealed in the mechanism of protein synthesis, where an exceptional role belongs to DNA molecules, the carriers of the genetic (heredity) code, in accordance with which molecular synthesis is effected and the cells and the living system as a whole are ultimately self-reproduced. The information coded in DNA whose loci are formed by genes, is a programme of genetic processes and their realisation in the development of organisms as individuals and as a species. The concept of the genotype of the individual as a peculiar 'programming device' has helped to gain a deeper understanding of the biological importance of heredity information in organisms as a concentrated and duly coded flow of the influence of the environment throughout the individual life of organisms and the historical development of their species.

This makes it clear why organisms which have travelled a more complicated evolutionary path and whose heredity has taken shape under the impact of the most diverse factors of the environment have the most diverse information. That is why, in their responses to its changes they display the maximum activity and are capable, with the aid of fixed secondary regulation devices, of working out oriented adaptive restructuring of their organisation and behavioral acts as is the case, in particular, among the higher animals with a developed nervous system.

This growing activity of living systems is simultaneously connected with the increasing switch of self-regulation mechanisms to individuals, whereas this kind of 'individualisation' is much less frequent, say, among plants, where the basic self-regulation (and evolving) unit is the population, the species. As for the individual, while information processes are effected through it, it itself exists as a peculiar 'variant' in the course of 'selection' by the population of optimal values of adaptation in new environmental conditions. This is what is fixed in the concept of natural selection, in its statistical action, which is only possible when it comes to an ensemble, a definite discrete set, which is the more effective the greater the diversity achieved in the types of structuring living systems and their changes.

Still, the initial primary mechanisms, which results not only in self-preservation but also in the self-improvement of living systems, and in morphophysiological and adaptive development, are formed on the individual level. On this level are created the prerequisites for adaptive responses, their 'thematic orientation', which is realised in the population by means of selection.

Adaptive actions, whose forms may be either genetically coded (programmed) in accordance with a definite scheme or may be worked out in the course of individual life, are characterised as being oriented precisely because they are determined by a definite programme with an unusually large information capacity. Indeed, one may even say that this programme is excessively complex. This becomes clear if we consider the exceptional complexity and the inexhaustible 'inventiveness' of the environment in which organisms carry on their vital and adaptive behavioral activity, and which determines, in vigorous interaction with this activity, the formation of the complexity of the programme itself.

That is why the programme coded both in the genetic heredity structures and in the physiological systems with a fixed acceptor of the result of action (P. Anokhin) may appear as the 'anticipatory model' of an action not yet performed, as its result. Cybernetics must be credited above all with having shown the possibility of such models existing in nature by giving new facets to the understanding of the problem of purpose and purposefulness in the overall anti-teleological framework of their scientific explanation. Where teleology saw the idealistically interpreted action of 'final causes', 'rational purposes', etc., cybernetics established the material causal relationship, having shown in strict scientific terms and in complete accord with Darwin's theory, the more general grounds for treating purposefulness in nature as a material relationship. In this way it has preserved and rationalised the objective meaning of this relationship, having carried on Darwin's effort in research into the material causes of purposefulness in nature. At the same time, expelling teleology from what might be called the 'inside' of purposefulness itself, cybernetics has more than mechanically drawn a line, for instance, between organic purposefulness and man's purposeful activity. By analysing their common principles as mechanisms for realising oriented processes in self-governing systems, it has clarified the 'rational' meaning of the ancient analogy between the adaptive functioning and development of living systems and purposeful human activity. Consequently, it has 'won' a vast empirically existing bridgehead which science has traditionally by-passed, believing it to be firmly held by teleology, its age-old adversary.

Summing up what has been said, we have reason to state that cybernetics does not introduce the concept of purpose into the science of living systems and does not invest it with a boundlessly extensive meaning by stripping it of anthropomorphic and biomorphic elements, but merely finds material *analogues* and purposes in the objective characteristics of self-regulating systems, designating these by means of information and feed-back terms, that is, producing semantic invariants of purpose. Is it at all useful to convert these invariants into the initial ones which had served as the starting points for analogy? Is it not better to retain the term 'purpose' only in its immediate, specific sense, which is connected with the *comprehension* with an ideal notion of the final results of activity, leaving the *relative* use of the term to characterise natural processes, as relative, for instance, is the concept of organic purposefulness?

Consideration of this question could also be continued in application to the concept of 'purposeful causality' in nature. Cybernetics has provided a materialist explanation for the specific links and interactions, for instance, in living nature which teleology had presented by direct analogy with human activity in the spirit of finalism, implying that material processes in living systems were determined by an ideal 'final purpose', etc. There again it also made use to the end of the 'rational meaning' contained in the analogy, taking it to mean only the *beginning, the starting point* of cognition, and *not its completion*, as in finalism.

The results of this cognition, expressed in information and feed-back terms, are in complete accord with the conception of organic determinism, helping to enrich and develop it. It is, accordingly, established that in the functioning and development of self-regulating self-governing systems there appears a new type of link, which is characterised, in particular, as being cyclical. From the information standpoint the cyclical connection effected in the form of interaction between differently oriented processes, can be designated either as direct or feedback connection. In it we observe processes of a peculiar predetermination, fixed in the programme, in the form of a code model of subsequent actions and determining the statistically realised orientation of these actions. The mechanism of such connections itself appears in the form of their 'duplication', a superimposition on the objective material process either of its ideal scheme, and epistemological sample–purpose–or of a material programme, a code model, which, it is known, may or may not have a figurative value.

Accordingly, the interactions observed between the various states of a self-regulating system and expressed in the form of causality may be characterised by the concept of cyclical connection between cause and effect (direct and feed-back) of which the so-called purposeful causality is in effect a type. In this way its existence is not denied, but the limits of its applicability, which may be extended, again only in a relative sense, are specified. Is it necessary, in this instance, to pose this question once again: what do we really gain by 'transcending the limits' of the meaning of the term? What new knowledge do we gain if we impose–even if only terminologically–for instance, on a plant 'selection of purpose', 'striving towards an aim', 'purposeful activity', and so on?

This is not such a simple question as it may appear to those who are not abreast of the history of the philosophical struggle over the problem of determinism and teleology and who, for this reason may very easily 'revise' its lessons. Nor does the question here at all boil down to purely semantic aspects, or to a dispute about words, although that is also important. We find that somewhere close to these disputes and sometimes from them stem such forms of 'mythology on the cybernetic level' which coalesce directly with teleology and finalism. After all, the discussions on the cognitive importance of teleology have not been taking place in philosophical vacuum, and the recognition (however formal) of the scientific effectiveness of the 'teleological principle'–even if only as a definite *methodological instrument*–may be easily transformed into the assertion about a 'revival' of teleology and finalism as a *philosophical, theoretico-cognitive* conception confronting the principles of determinism.

Let us note that this kind of situation which results at any rate, in formal assertions about a 'revival' of teleology and finalism exists in biological science, in particular. Many leading natural scientists, philosophers and methodologists have been trying to escape from the situation and to dissociate themselves, even in formal terms, or, you might say, semantically, from teleology and finalism.

Thus, E. Mayr said in a paper at a symposium on theoretical biology in 1966 that biologists have long since come to feel the ambiguity of designating the purposeful behaviour of individuals programmed by the properties of its genetic code as 'teleological'. He held that

scientific biology has not found any evidence that would support teleology in the sense of various vitalistic or finalistic theories ... The complexities of biological causality do not justify embracing non-scientific ideologies, such as vitalism or finalism, but should encourage all those who have been trying to give a broader basis to the concept of causality (Waddington, 1968, pp. 49–50, 54).

In an effort to give adequate expression to this 'broader basis' in his familiar concepts and terms, Mayr accepts Pittendrigh's designation of the behaviour of systems, 'not committed to Aristotelian teleology', as being *'teleonomic'*, confining its use to systems operating on the basis of some programme or a code of information, and taking this to mean 'the *apparent* purposefulness of organisms and their characteristics', as Julian Huxley put it.

Commenting on these considerations of Mayr's, C. Waddington agreed with him on the point that the teleological or vitalistic type of explanation was not acceptable and that there was need for 'tele-nomic', or 'quasi-finalistic' explanations, to use the term he introduced. However, he did not agree with Mayr that

natural selection is not purposive. In itself it is, of course, no more purposive than is the process of formation of interatomic chemical bonds. But just as the latter process is the basic mechanism underlying the protein syntheses which are integrated into the quasi-finalistic mechanism of embryonic development, so natural selection is the basic mechanism of another type of quasi-finalistic mechanism, that of evolution. The need at the present time is to use our newly won insights into the nature of quasi-finalistic mechanisms to deepen our understanding of evolutionary processes (Waddington, 1970, p. 56).

Rejecting teleology, Waddington seeks to interpret the quasi-finalistic explanation of biological processes within the framework of materialism (mechanicism, to use his own term), even assuming that 'the general system of concepts which is beginning to take shape ... is in a sense close to Marxist dialectical philosophy'. (Astaurov, 1970, p. 8).

The terms 'telenomy' and 'quasi-teleology', or 'quasi-finalism', are being presented as alternatives to the teleological interpretation of causality in the spirit of finalism. In effect, they describe the causal relations expressed in the language of cybernetics by means of the concepts of programme and feed-back, that is, they describe cyclical, reciprocal causality, including predetermination of the results of action and a corresponding orientation of the latter. One may, of course, argue about the aptness of these terms, which on the whole still revolve round the initial concept of teleology. But we can well understand the urge to dissociate oneself from it, while retaining the *method of research* into complex systems through an analysis of the relation of purposefulness which is only nominally qualified as 'teleological' (Frolov, 1958, 1965).

2.   Purposive approach in research

The question of this method of research is of importance in itself. It is connected with the heuristic use of the concepts of purpose and purposefulness in the study not only of the processes which could be designated as purposeful in the immediate sense of the word, but even in relative forms, to express their objective orientation. Here the relative nature of the concept of purposefulness of processes, empirically accepted and resulting from the development of the new content in the old semantic form (as we find, for instance, in the case of organic purposefulness), is allowed deliberately as a definite method of research. This implies the so-called purposive approach which is now and again interpreted as being part of a general functional analysis of complex systems of the organic-integrated type.

The 'teleological' approach is usually taken to mean the functional approach in its broad sense, which involve the study of processes, the dynamics of the elements of the system characterised as a specially stable type of behaviour of these elements (or subsystems), that is, as their derivative function. In biology, from which these concepts historically take root, later to be more broadly applied, function is always the result of vital activity, of the concerted operation of a definite organ (or system of organs), each of which also has a systemic quality that is recorded structurally and dynamically.

Consideration of organically integrated systems and their components in the light of the results of their functioning means bringing out one of the specific properties of these systems. But only in some instances can such consideration assume the form of purposive analysis (and, consequently, be designated as 'teleological'), because the concept of function does not always assume the meaning of a definite orientation of processes, to say nothing of their purposefulness.

That is why it is possible to separate the functional approach in the narrow sense of the term (analysis of the behaviour of systems which is unconnected with notions of orientation), the functional purposive approach (analysis of the behaviour of systems characterised as oriented or relatively oriented), and the purposive approach proper, under which the researcher turns to the final stage, to the result of the process as its aim, starting from which the cause is analytically established from its effect.

Consequently, the final stage of the process tends to be regarded as its purpose, that is, the functional-purposeful or simply the purposeful approach is realised regardless of the meaning with which we invest the concept of purpose, because the latter appears as the *regulatory principle* whose potentialities were analysed by Kant. In the purposive approach, the concept of purpose may reflect a real phenomenon (as in the analysis of forms of human activity) and may designate it adequately; it may represent directed (fixing the relation between the initial and the final stages) processes in the form of a material or an ideal model, in the form of theoretical constructions, a purposive hypothesis, etc.

Thus, the purposive approach may be used not only in the study of 'equifinal' systems but also in the spheres where one deals with cyclical and oriented interactions, where processes of progressive development are studied. The purposive approach may also be used in situations where the final result of a process may be established empirically. In such instances, it is constructed ideally, hypothetically. Here again the analysis is based on the assumption that the result of the process is *ostensibly* present in reality in the form of a peculiar purpose. In such an approach, the resort to this 'purpose' appears as a special method of hypothetical anticipation, description of a process subject to subsequent scientific analysis.

On the strength of this, the purposive approach in general cannot, as a matter of principle, be contrasted to the traditional and the new methods of causal analysis of living systems (historical, experimental, etc.), as we find in the event of its teleological interpretation when this is assumed to be 'the most characteristic' method for biology. It has a definite cognitive value only in connection with other methods, and within their system, reflecting the overall dynamics and strategy of scientific research, establishing and dividing up the forms of objects and helping to clarify their functional role and origin.

The purposive approach is extensively used for the purposes of description, evaluation and explanation in sociological and economic research, in aesthetics and other fields of knowledge and culture. Its terms can help to interpret many methods of research in mathematics, statistical physics, technology (one need merely mention the construction of idealised objects, extreme principles, etc.).

However, in all these instances, when this or that postulated or actually existing result of a process assumes the form or purpose in subsequent retrospect of the process, there is no question of the 'teleological principle' as an alternative to determinism. In contrast to teleology and finalism, scientific research and explanation is achieved here in the light of the many values and diverse orientations of objective interactions in nature and society, of their 'thematical orientation', programmed *action of instruments but not of results as a whole*, the direction towards which as a real or conditional and apparent purpose is brought out only integrally, in the form of a general tendency. The basis of such research and explanation is provided by the dialectico-materialist view of the conception determinism–organic determinism.

Journal 'Voprosy filosofii', Moscow

BIBLIOGRAPHY

Astaurov, B. L. (ed.): 1970, *Towards a Theoretical Biology*, Russian Translation of Waddington, 1968, Moscow.

Frolov, I. T.: 1958, 'Determinism and Teleology', *Voprosy filosofii*, No. 2 (in Russian).

Frolov, I. T.: 1965, *Essays in Methodology of Biological Research*, Moscow (in Russian).

Waddington, C. H. (ed.): 1968, *Towards a Theoretical Biology. I. Prolegomena*, Birmingham.

GUNTHER S. STENT

EXPLICIT AND IMPLICIT SEMANTIC CONTENT OF THE GENETIC INFORMATION

1. GENETIC INFORMATION

The latter-day success of molecular biology in providing a physical account of the structure and function of the hereditary material was in large part due to the focus of its founders on the concept of 'genetic information'. As was spelled out thirty years ago by Erwin Schrödinger (1945) in his seminal book *What Is Life*, the gene, as a hereditary determinant, can be viewed as an information carrier whose physical structure corresponds to an aperiodic succession of a small number of isomeric elements of a hereditary codescript. Eventually, the gene was identified as a segment of a double helical DNA molecule residing in the chromosomes of the cell nucleus. The isomeric elements of the hereditary codescript turned out to be the four nucleotide bases adenine, guanine, thymine, and cytosine, which embody the genetic information by their aperiodic linear sequence along the DNA molecule. And as far as the meaning of the information contained in an individual gene is concerned, it was first put forward as an *a priori* dogmatic postulate and later established as an empirical fact, that the linear sequence of DNA nucleotide bases stands for the linear sequence of amino acids of a particular protein molecule. The two sequences are related to each other via a genetic code, under which each nucleotide base triplet stands for one of the twenty kinds of amino acids of which protein molecules are built. However, the chromosomal DNA also contains some nucleotide base sequence segments which do not stand for protein molecules, and, strictly speaking, do not constitute genes. Some of these segments serve as templates for the assembly of nucleic acid components of the cellular apparatus for protein synthesis, such as transfer and ribosomal RNA molecules, and other segments serve as control sites at which the expression of the genes is

Butts and Hintikka (eds.), Foundational Problems in the Special Sciences, 131–149.

regulated. Thus it is fair to say that there exists at present a highly satisfactory state of understanding of the nature of the informational content and meaning of the structural elements of the genetic material. Indeed, efforts are now underway to subject the semantic character of DNA nucleotide sequences to a detailed linguistic analysis and to formulate a grammar of a "genetic language" (Ratner, 1974).

2. GENOME AND PHENOME

However, the present status of the problem of the overall meaning of the entire DNA complement of an organism, or its *genome*, is not really all that satisfactory. What does the genome, in fact, represent? A not uncommon answer to this question is that the genome is obviously a one-dimensional representation of the organism. For instance, it was suggested at a Conference on Communication With Extraterrestrial Intelligence (Sagan, 1973) that to transmit via radio signals the DNA nucleotide sequence of a cat to a Distant Alien Civilization is equivalent to sending the Aliens the cat itself. This suggestion, though made partly in jest, allows us to recognize that the correct answer to the question into the meaning of the genetic information is not so obvious. On the contrary, what *is* obvious is that the Alien Intelligence, even if it possessed a table of the terrestrial genetic code, would not be able to reconstruct the cat from its DNA nucleotide base sequence. To make this reconstruction, the Aliens would have to know a good deal more about terrestrial life than the formal relations between DNA nucleotide base sequences and protein amino acid sequences.

What they would have to know, above all, is that the actual cat, or the feline *phenome*, arises from a fertilized egg containing the feline DNA nucleotide sequences by an *epigenetic* process of embryonic development.[1] Moreover, the Aliens would have to understand the epigenetic relation between phenome and genome, an understanding which we unfortunately still lack. As I shall try to show here, the Aliens could not, regardless of their level of intelligence and technical sophistication, work out this relation by a semantic analysis of the feline DNA, for the reason that the epigenetic relations cannot be found in the DNA. And what goes for the Aliens goes for us too.

3. THE EPIGENETIC LANDSCAPE

As C. H. Waddington (1957) set forth long before the informational theories of molecular biology had even received their experimental validation, the genetic information does not represent an organism but merely some functional components of an *epigenetic landscape.* What Waddington meant by this poetic term is a plot of multivariant functional relations in multidimensional space. In this space, ontogenetic time is the independent variable and the properties that describe both the organism and its environment are the dependent variables. The functional relations from which this landscape is constructed are the chemical and physical processes which relate the changes in these properties with the flow of ontogenetic time. The topography of this landscape, therefore, represents the developmental pathways along which the embryo moves from fertilized egg to adult. Waddington's main reason for using the landscape metaphor was to point out that in this space the developmental pathways are bound to form a system of interconnected valleys that slope 'downward' from the summit of the egg in the direction of ontogenetic time towards the 'sea level' of the adult organism. This feature would assure that the pathways are relatively resistant to perturbations of the functional relations and to fluctuations in the dependent variables, and hence guarantee a reasonably invariant relation between genome and phenome. The role of the genes in shaping this landscape derives from their control of critical chemical processes in the developmental sequence, or (as we now know) from their governance of the production of protein molecules capable of catalyzing specific chemical reactions.

The idea of the epigenetic landscape certainly brings us closer to an understanding of the relation between the genetic information and the organism to which it gives rise. And it can fairly be said that the discovery of the functional relations of that landscape, or as Francois Jacob (1973) has called them, "the algorithms of the living world", is one of the main goals, if not *the* main goal, of contemporary genetic biology. But I give it as my opinion that, thus far at least, this goal has not generally been brought into sufficiently sharp conceptual focus, because of semantic difficulties inherent in the notion of 'phenome', on the one hand, and in the notion of 'meaning', as applied to the genetic information, on the other hand.

4. OSTENSION AND CONNOTATION

It is not without irony that the problem of meaning is no less trouble-some in the domain of human communication, for which the scientific concept of information was developed in the first place, than in the study of the genome-phenome relation to which semantic notions have been extended by molecular biologists. Since the question of how meaning arises from language, or even what it is we are saying about a word when we say what it means, still awaits its answer, it is hardly surprising that we encounter also conceptual difficulties with the metaphorical use of semantic terminology in genetic biology. It so happens, however, that some of the philosophical contributions to the problem of linguistic meaning can assist us also in the troublesome matter of the meaning of the genetic information.

One such philosophical contribution to semantics that helps in the clarification of the meaning of 'phenome' is the distinction between an *ostensive* and a *connotative* definition. An ostensive definition of 'phenome' is produced by pointing to an organism and saying "this is what I mean by 'cat' ", whereas a connotative definition is produced by drawing up a list of necessary and sufficient properties which an organism must have to be a cat.

Now, although our ordinary ontological notion of a cat has an ostensive basis, this concept cannot serve as the phenome of whose relation to the feline genome we can expect to be able to give a scientific account. The reason for this is that the ostension-based Gestalt by which we recognize any cat as a cat is so highly abstract that it lacks most of the biological properties which are essential for even beginning any attempt to explain its gene-directed embryological de-velopment. If, in order to fill in these conceptual lacunae, we proceed to examine an actual cat, we find that the animal presents us with such a wealth of concrete detail that a full description of the phenome would produce a virtually limitless catalogue of properties in want of explanation. Thus the conceptual endpoint, or sea-level, of the feline epigenetic landscape must be the list of necessary and sufficient prop-erties of a connotatively defined cat phenome, rather than an actual cat. It follows, therefore, that the selection of a connotative phenome for study is the first, and most crucial step in any embryological analysis. If the set of properties whose epigenetic development one sets out to explain is too simple, the explanation when eventually found, may be of only trivial significance. If the chosen set is too complex, the explanation, though it might be highly significant if found, would probably be too difficult to fathom. Thus the art of practicing developmental biology consists of defining an optimal connotative phenome.

5. Explicit and Implicit Meaning

A second philosophical insight into semantics that can help us to understand how meaning arises from the genetic information is the recognition that the meaning of semantic structures may depend on the *context* in which they are produced. For instance, linguists draw attention to the fact that although the *literal* meaning of the sentence 'I want you to shut the door' is a declaration of the speaker's state of mind, in the context of polite social intercourse the usually intended meaning of that sentence is, in fact, a command. A related, context-dependent distinction between two different kinds of meaning, which I believe is useful for our present purpose, is that between *explicit* and *implicit* meaning. The explicit meaning is that which a semantic structure has by virtue of the syntactic relation of its elements. Hence, the explicit meaning can be extracted from the structure by subjecting it to a linguistic analysis. The implicit meaning, by contrast, is not really contained in the structure itself and arises secondarily from the explicit meaning by virtue of its context. For instance, the explicit meaning of the sentence 'John Smith is traveling to New York', is that a particular individual is on his way to a particular geographical location. However, depending on the context in which the sentence is produced, it can also have a large number and variety of implicit meanings. For instance, if it is produced at San Francisco airport, it would imply that Mr. Smith is about to board a particular flight, that his suitcase is waiting to be loaded on a particular baggage truck, that he cannot be reached by telephone for the next six hours, and so on. The same sentence with the same explicit meaning would carry a different set of implicit meanings if it were produced at a roadside service station in Colorado.

Actually, it is difficult to draw a sharp line of demarcation between explicit and implicit meanings, since the extraction of the explicit meaning is itself rarely context-free. For instance, in the example just given, the intrinsic ambiguity in the explicit meaning of whether 'New York' refers to the state or the city is resolved by the San Francisco airport context under which the term can be safely assumed to refer to the city. Thus the distinction between explicit and implicit meaning is relative rather than absolute, with a meaning being the less explicit and the more implicit the more dependent it is on the context.

Furthermore, because of its high degree of dependence on the context, the implicit meaning is open-ended, in that it can become ever more remote from the explicit meaning as the context is widened.

When we apply this distinction to the semantic relation between genome and phenome it becomes evident that the explicit meaning of the genetic information consists of the protein amino acid sequences encoded in the genes, and of the nucleotide sequences of the ribosomal and transfer RNA molecules encoded in other non-genic DNA sectors. The explicit meaning would include also those physico-chemical properties of the non-genic control DNA segments that produce regulatory functions. These meanings are explicit in the sense that they can be extracted from an analysis of the DNA nucleotide base sequence itself, provided one knows that the DNA base sequence is transcribed into a complementary RNA base sequence, and one has access to the table of the genetic code. But these explicit meanings form only the basic skeleton of the functional relations that shape the epigenetic landscape. The bulk of these relations are merely implicit in the genetic information.

By way of an example of such an implicit meaning we may consider the three-dimensional conformation of protein molecules. Although it is in some sense true that the spatial conformation of a protein molecule is 'genetically determined', this 'determination' devolves from the DNA-encoded specification of the one-dimensional amino acid sequence of the molecule. Once assembled from its constituent amino acids, the protein molecule automatically folds to assume its functional three-dimensional structure. The physico-chemical principles which govern this folding are in part understood, although it is not yet possible (but soon may be) to predict the three-dimensional structure which a protein with a given amino acid sequence will assume. But it is important to note that these folding rules are nowhere represented in the DNA nucleotide base sequences, being part of the context rather than of the genetic informational structures. The enzymatic function of the protein molecule, which can also be said to be 'genetically determined', is an even more purely implicit meaning of the genetic information than the spatial conformation. Once a protein molecule *has* been assembled specifically from its constituent amino acids and *has* assumed its specific three-dimensional structure, certain parts of that structure turn out to possess the power to catalyze some particular chemical reaction. The stereochemical principles which govern that catalysis are also partly understood, although (to my knowledge, at least) it is not yet possible to predict on the basis of the known three-dimensional structure of a protein molecule the kind of reaction which that molecule can catalyze. It goes without saying that the principles of chemical catalysis are not represented in the DNA nucleotide base sequences either; they enter into the meaning of the genetic information at a second order level of a contextual hierarchy, at whose first order level we found the protein folding process.

This procedure of identifying implicit meanings of the genetic information can be continued almost indefinitely to higher and higher levels of the contextual hierarchy. For instance, the physiological function of a chemical substance whose formation is catalyzed by a particular protein molecule can likewise be said to be 'genetically determined', as can be the overt behavioral feature to which that physiological function gives rise. The nearly unlimited horizon of implicit meanings of the genetic information shows that, as has long been recognized (cf. Woodger, 1953), the notion of the 'inborn nature', or genetic determination of characters is so all inclusive as to be nearly devoid of meaning. After all, there is no aspect of the phenome to whose determination the genes cannot be said to have made their contribution. Thus it transpires that the concept of genetic information, which in the heyday of molecular biology was of such great heuristic value for unraveling the structure and function of the genes, i.e. the explicit meaning of that information, is no longer so useful in this later period when the epigenetic relations which remain in want of explanation represent mainly the implicit meaning of that information.

6. Development of the Nervous System

In order to demonstrate the pertinence of these abstract semantic discussions for present-day biology, we may consider one presently particularly active area of research, namely the study of the development of the metazoan nervous system. The nervous system is an especially suitable object for epigenetic investigations, because, as neuroanatomical and neurophysiological studies have shown, it is the precise manner of the interconnection of the cellular components of that system to which the organism's behavior is attributable. Thus here a connotative definition of a phenome in want of explanation can be provided in terms of a circuit diagram of specified cellular elements. Although we cannot be sure as yet that it *will* be possible to give an account of the epigenetic landscape that produces the neural circuitry, we can be reasonably confident that the explanation, if found, will not be trivial.

The general problem of the development of the nervous system has been formulated by Seymour Benzer (1973) in the following terms:

When the individual organism develops from a fertilized egg, the one-dimensional information arrayed in the linear sequence of the genes on the chromosomes controls the formation of a two-dimensional cell layer that folds to give rise to a precise three-dimensional arrangement of sense organs, central nervous system and muscles. These elements interact to produce the organism's behavior, a phenomenon whose description requires four dimensions at least. Surely the genes, which so largely determine anatomical and biochemical characteristics, must also interact with the environment to determine behavior. But how?

7. An Information-Theoretical Fallacy

One possible answer to the question of how the genes determine behavior is that they, in fact, contain the information for the circuit diagram of the nervous system (Benzer, 1971). However, it has been argued that the circuitry of the nervous system cannot, in fact, be genetically determined because the total amount of genetic information does not suffice to specify the neuronal connections that need to be made (Horridge, 1968). According to this argument, the linear sequence of roughly 10^{10} DNA nucleotide bases in the genome of a higher vertebrate animal contains an upper limit of 2×10^{10} bits of information (since each base, being one of four possible types, embodies 2 bits). On the other hand, if each one of the roughly 10^{10} cells in the nervous system of such an animal were connected to just two other cells, then it would require of the order of $10^{10} \log_2 10^{10}$, or 3×10^{11} bits to specify this network. Thus even under the most absurdly oversimplified view of the complexity of the nervous system, the total information content of the genetic material, even if it had no non-nervous determinative role, would be too low by an order of magnitude to allow the specification of nerve cell connections.

Although this anti-genetic argument has little merit, it is useful to examine it because it exemplifies two not uncommon errors of thought which must be corrected before the relation of the genome to the development of the nervous system phenome can be profitably considered. The first of these errors derives from a spurious application of information theory to biological problems. That is to say, it derives from the failure to recognize that the quantitative concept of information applies only to processes in which the probabilities of realization of alternative outcomes are known or clearly defined. To illustrate this point, we may consider a biological example which bears some formal analogies to the problem of the circuitry of the nervous system but which is presently much better understood, namely the determination of the structure of protein molecules. A typical species of protein molecule consists of about 300 amino acid building blocks, or of about 4000 atoms held to each other in specific chemical linkage, with each atom, on the average, being connected to about two other atoms. We may ask how many bits of information are needed to specify the chemical structure of that protein molecule. If we were to proceed by the same calculation as that which was just applied to the nervous system, we would reckon that about $4 \times 10^3 \log_2 (4 \times 10^3)$ or 5×10^4 bits are needed. But here we encounter an apparent paradox, because the gene that encodes the chemical structure of a 300-amino acid protein molecule consists only of a sequence of about 900 nucleotide bases, and hence contains a maximum of $2 \times 900 = 1800$ bits of information. Thus the information content of the gene would be too low by more than an order of magnitude to encode the chemical structure it is known to determine. The insights of molecular biology readily resolve

the development of the nervous system can be governed only by a rough program under which there arises an overproduction of cells and connections, from among which an appropriate subset is selected by various testing procedures.

By way of example of this principle, we may consider the development of the visual pathway of the cat that is to be sent to the Alien Civilization. Its visual pathway begins in the retina at the back of the two eyes. There a two-dimensional array of about a hundred million primary light receptors converts the radiant energy of the image projected via the lens on the retina into a pattern of electrical signals. Since the electrical response of each light receptor cell depends on the intensity of light that happens to fall on it, the overall activity pattern of the light receptor cell array represents the light intensity existing at a hundred million different points in the visual space. These light receptor cells are connected through three or four intermediate stages to nerve cells in a particular area of the cerebral cortex. The vision of the cat is such that its optical system allows both eyes to see the same visual field (in contrast to wall-eyed mammals, such as cows, whose pair of eyes look at different sectors of the visual space). In order to provide for fusion of the binocular visual input into a single visual percept, each cortical cell receives electrical signals from a few thousand light receptor cells in the retinas of both eyes. For such binocular fusion to occur, however, it is necessary that the two sets of retinal light receptor cells which connect to the same cortical cell receive their light from exactly the same points in the visual space.

How do these retino-cortical connections arise during embryonic development of the cat? We might envisage that there is some, as yet unknown, gene-determined process which directs the formation of nerve cell outgrowths and contacts in such a manner that light receptor cells from the corresponding areas of right and left retinas connect to the same cortical nerve cell. But here we must take into account that for binocular vision the 'correspondence' of retinal areas depends not only on the topography of the retina but also on the physical optics of the eye. That is to say, which pair of retinal light receptors in the right and left eye happens to see the same point in the visual space is governed by the exact structure and positioning of right and left lenses. Although the epigenetic realization of the physical optics might also occur by gene-determined processes, it is well-nigh inconceivable that the independent formation of the retinas and the lenses could be genetically pre-programmed to occur with such a high degree of precision that the image of a given point in the visual field always falls exactly on that pair of light receptor cells which the genes have managed to connect to the same cortical cell.

As neurological studies of the development of the retino-cortical connections of the cat have revealed, this perplexing developmental problem is solved by providing for a (possibly genetically determined) *overconnection* of light receptor and cortical nerve cells. That is to say, at birth prior to visual experience, each cortical cell is connected to light receptor cells from a much larger retinal area than is actually compatible with sharp vision. This imprecise congenital visual system is then refined by early post-natal visual experience of the kitten, by a neurophysiological process that identifies those corresponding retinal areas of the two eyes which, given the actual physical optics which the young animal happens to have, do receive light from the same point in the visual space. Thanks to this identification, the developing nervous system selects among the excess of existing retino-cortical connections just those which bring to each binocular cortical cell a coherent visual input.

9. NEUROLOGICAL MUTANTS

Although it thus appears that the genes cannot precisely pre-program the connections of the nervous system, it is nevertheless obvious that they must play some considerable role in the genesis of its structure. And, hence the genes would also make an important contribution to the determination of an animal's behavior. The recognition of this fact has given rise to a neurobiological specialty which aims at discovering just how the genes perform this determinative epigenetic function, and which is a sub-field of the more general discipline known as 'developmental genetics' (Markert and Ursprung, 1971).

The principal approach used thus far in the effort to establish the role which the genes play in the determination of nerve cell connections is to isolate gene mutations affecting the behavior of an animal and noting changes in the structure of the nervous system responsible for the altered behavior (Brenner, 1973). This approach is evidently based on the belief that the procedure of isolating mutants and analyzing the resulting abnormalities, which molecular geneticists used with such brilliant success in the elucidation of the explicit meaning of the genetic information, will also be of service in unraveling its implicit meaning. This belief is undoubtedly correct, since what Waddington called the 'remodeling of the epigenetic landscape' by a gene mutation and its attendant changes can help to identify the functional relations that produce the normal developmental pathways. But it must be borne in mind that though a mutated gene may help to identify a particular epigenetic function, the connection between that function and the mutant gene can be very indirect and involve many other members of the functional network. In view of their general remoteness from the primary action of the genes, the great majority of epigenetic algorithms are unlikely to refer to any gene at all.[2]

10. THE SIAMESE CAT

In order to appreciate the kind of insights into the development of the nervous system that this genetic approach is likely to provide we may now consider, as a paradigmatic case, the Siamese cat, which happens to carry a mutation that affects its behavior and produces anatomically and physiologically identifiable changes in the nervous system.

The visual pathway of the cat is arranged such that the cerebral cortex on the right side of the animal receives visual input only from the left half of the visual space and the left cerebral cortex only from the right half of the visual space. To produce this right-left crossover of the visual input, the optic nerve fibers reporting from light receptor cells located in the *nasal* half of the retina (i.e. the half which is next to the nose and which receives light from the same side of the visual space as the side of the body on which the eye is located) cross over to the cerebral cortex on the other side of the body, whereas the optic nerve fibers reporting from light receptors located in the *temporal* half of the retina (i.e. the half which is next to the temples and which receives light from the opposite side of the visual space) do not cross over and connect to the cortex on the same side of the body. In normal felines, i.e. in the ordinary domestic cat, the line of demarcation for crossing over of optic nerve fibers is exactly midway between the nasal and temporal edges of the retina. However, as was discovered by R. W. Guillery, in Siamese cats the normal line of demarcation is displaced from the midline towards the temporal edge of the retina (Guillery, 1974; Guillery *et al.*, 1974). As a result of this displacement, some optic nerve fibers reach the cerebral cortex on the 'wrong' side of the brain.

It is beyond the intended scope of this discussion to consider in detail the extremely interesting changes which this faulty cerebral projection of the optic nerve fibers produces in the nervous system and behavior of the Siamese cat. Suffice it to say that in response to the aberrant visual input both the cerebral cortex and the behavior of these animals is reorganized in specific and functionally obviously adaptive ways, so as to minimize the pernicious effect of this genetically determined developmental malformation. Moreover, the nature of this cerebral reorganization lends strong support to our previous conclusion that in the course of development of the nervous system the final connections depend partly on a selective process based on the functional testing of a tentative, imprecise circuitry. What I do want to consider here, however, is the way in which the mutation of a gene carried by the cat helps us discern the genetic component of behavior.

11. A SAMPLE ALGORITHM SET

Although the informational content of the genome of the Siamese cat certainly differs from that of the ordinary cat in more than one gene, Guillery has been able to identify the mutant gene whose change is responsible for the aberrant crossover of the optic nerve fibers. It is the 'tyrosinase' gene in which the amino acid sequence of a protein is inscribed that catalyzes a reaction step in the biosynthesis of the dark pigment melanin. In the Siamese cat this gene carries a mutation which renders the mutant protein unable to carry out its catalytic function at 37°C and thus prevents the synthesis of the dark pigment at body temperature. It is this mutation which is responsible for the characteristic Siamese coat color, namely the lightly pigmented body fur framed by black hair patches on the tips of the ears, the paws and the snout.

But what is the possible connection between the formation of melanin and the directed outgrowth of the optic nerve fibers from the retina to the right or left cerebral cortex? And why does the absence of pigment produce an aberrant crossing over, particularly in view of the fact that the retinal cells that provide the optic nerve fibers do not normally contain significant amounts of melanin in any case? The answers to these questions are not yet available, but it is not difficult to invent a set of hypothetical epigenetic algorithms that would provide a plausible formal explanation. For this purpose, we envisage that the embryonic retinal nerve cells have some property whose measure x increases monotonically across the retina from the temporal to the nasal edge. We envisage, furthermore, that all those cells for which $x > x_0$ send their optic nerve fibers to the opposite side of the brain, with the remainder of the cells sending their fibers to the same side. The normal developmental system (i.e. the retinal region of the epigenetic landscape) is so poised that $x = x_0$ at the retinal midline. Now although the retinal nerve cells do not themselves contain melanin, they are, in fact, the direct developmental descendants of the layer of melanin-containing cells that form the pigmented epithelium which lies behind the retina and shields it from stray light. To complete our sample algorithms we need merely envisage that if the epithelial precursor cells are not pigmented, as they are not in the Siamese cat embryo, the normal developmental schedule of formation of their retinal nerve cell descendants is slightly perturbed. And, as a consequence of this slight remodeling of the epigenetic landscape, the temporal-nasal retinal gradient of the postulated property is also slightly perturbed in such a manner that x already reaches the value x_0 on the temporal side of the retinal midline. Thus the absence of pigment in the epithelial precursor cells would cause those optic nerve fibers reporting from light receptors in the temporal half of the retina for which $x > x_0$ to make an 'incorrect' outgrowth to the cerebral cortex on the opposite side of the brain.

12. An object lesson

We are now in a position to understand the sense in which the 'tyrosinase' gene in the feline genome takes part in the 'determination' of the visual pathway and behavior: it directs the assembly of a particular protein molecule from its constituent amino acids in the precursor cells of the optic nerve fibers. The presence of the pigment whose synthesis is catalyzed by that protein is a necessary condition for the 'normal' development of the pathway, in that the absence of the pigment from the precursor cells sets off a cascade of dysfunctional, albeit specific aberrations, which eventually leads to a profound reorganization of a part of the brain of the animal. The visual pathway of the Siamese cat is, therefore, an ideal case for the genetic approach: a mutation in a single known gene that determines the amino acid sequence of a known protein that catalyzes a known chemical reaction whose end product has a known physiological function causes a known specific and striking structural change of the nervous system. But, alas, it tells us rather little about the *explicit* genetic component of behavior that we did not know already. Rather the tremendous interest of the Siamese cat lies in the help it is likely to provide in the effort to discover the epigenetic algorithms that govern the contextual realization of the *implicit* meaning of the genetic information. For instance, the probable connection between the absence of retinal epithelial pigment and the misdirection of the optic nerve fibers suggests some testable hypotheses about the rules that determine whether an optic nerve fiber grows to the same or to the opposite side of the brain as its retina of origin. The eventual statement of these hypothetical rules may contain such terms as enzymes, gradients, growth rates, threshold concentrations, preferential adhesion and nerve impulse frequencies, but the word 'gene' is unlikely to find frequent mention.

It is possible, of course, that the genetic approach may yet lead to the discovery of DNA nucleotide sequences whose explicit meaning is, in fact, *directly* concerned with the determination of the structure of the nervous system. And if such genes do turn up, the information they could provide would undoubtedly be of tremendous help in our effort to discover the developmental algorithms. But, all the same, it seems most likely that the great majority of mutants isolated for abnormal behavior will manifest changes in their nervous system because the mutation has occurred in a gene whose meaning bears implicitly rather than explicitly on the functional relations that form the epigenetic landscape.

13. NEUROLOGICAL EPIGENETICS

In the light of these considerations we can try to appreciate the nature of the contribution which the genetic approach can make to the understanding of the nervous system. There can be no doubt that the genetic approach is of great practical and technical significance. First, within the context of human psychology and. medicine it is of the utmost importance to understand the hereditary component of normal or pathological behavior. For instance, if it could be shown that schizophrenia is 'determined' by a particular mutant gene, then the value of this knowledge would be in no way diminished by the realization that a vast contextual hierarchy separates the explicit molecular biological meaning of that gene from its implicit epigenetic meaning for behavior. Second, within the context of neurophysiology the method of 'genetic dissection' of behavior (Benzer, 1973) is likely to be of great assistance for functional analysis of known nerve cell networks. For instance, an abnormal behavior and a concomitant abnormal structure of the nervous system of a mutant genotype can obviously provide insights into how the normal circuitry generates the normal behavior. Third, within the context of developmental biology the remodeling of the epigenetic landscape by mutant genes can, as we saw in the case of the Siamese cat, help us to recognize the functional relations that create the normal pathways that lead to the endpoint of the connotatively defined phenome. But as far as the discovery of how the genes interact with the environment to determine behavior is concerned, we can see that, thanks to the past achievements of molecular biology, that discovery has already been made: the genes determine behavior, just as they determine any other aspect of the phenome, by directing the synthesis of specific macromolecules.

Thus the deep biological problem in want of a solution is not how genes determine behavior but to find the algorithms of the living world that produce the nervous system. The horizon of that discipline, which might be called 'Neurological Epigenetics', lies far beyond the genes, and encompasses the context under which the explicit meaning of the genetic information gives rise to the implicit meaning that is the organism.

University of California, Berkeley

NOTES

[1] The term 'epigenesis' was coined by Caspar F. Wolff in 1759 (Theorie der Generation. Ostwald's Klassiker der Exakten Wissenschaften No. 84, 1896), to designate the view that embryonic development consists of the diversification and differentiation of an initially undifferentiated protoplasm. This view was advanced in opposition to the then dominant theory of *preformation*, which envisaged that the fertilized egg already contains a miniaturized version of the adult organism and that embryonic development consists of a mere increase in size from microscopic to macroscopic dimensions. The term 'epigenesis' was, therefore, already in use for about 150 years when, early in the present century, W. Johannsen proposed the terms 'gene' and 'genetics' to designate the hereditary determinants discovered by Mendel and the science of heredity.

[2] It is to be noted that many students of embryonic development do not share this view. For instance, F. H. C. Crick and P. A. Lawrence (*Science* **189**, p. 345, 1975) are of the opinion that the fact that *en* mutants of *Drosophila* show a specific abnormality in wing pattern 'strongly supports the idea that the role of the [normal] *en*⁺ allele is ⋯ to instruct the cells so that they do not intermingle with cells of the neighboring anterior type'.

BIBLIOGRAPHY

Benzer, S.: 1971, 'From Gene to Behavior', *Journal of the American Medical Association* **218**, 1015–1022.

Benzer, S.: 1973, 'Genetic Dissection of Behavior', *Scientific American* **229**, No. 6, 24–37.

Brenner, S.: 1973, 'The Genetics of Behavior', *British Medical Bulletin* **29**, No. 3, 269–271.

Guillery, R. W., Casagrande, V. A., and Oberdorfer, M. D.: 1974, 'Congenitally Abnormal Vision in Siamese Cats', *Nature* **252**, 195–199.

Guillery, Ray: 1974, 'Visual Pathways in Albinos', *Scientific American* **230**, No. 5, 44–54.

Horridge, G. A.: 1968, *Interneurons*, W. H. Freeman & Co., San Francisco, p. 321.

Jacob, F.: 1973, *The Logic of Life*, Pantheon, New York.

Markert, C. L. and Ursprung, H.: 1971, *Developmental Genetics*, Prentice-Hall, Inc. Englewood Cliffs, N.J.

Ratner, V. A.: 1974, 'The Genetic Language', *Progress in Theoretical Biology* **13**, 143–227.

Sagan, C.: 1973, *Communication with Extraterrestrial Intelligence*. M.I.T. Press, Cambridge, p. 331.

Schrödinger, E.: 1945, *What Is Life?* Cambridge University Press, New York.

Waddington, C. H.: 1957, *The Strategy of the Genes*, Allen and Unwin, London.

Woodger, J. H.: 1953, 'What do we mean by Inborn?', *British Journal of the Philosophy of Science* **3**, 319.

IV

FOUNDATIONS OF PSYCHOLOGY

O. L. ZANGWILL

CONSCIOUSNESS AND THE BRAIN

I. INTRODUCTION

How is one to approach the problem of consciousness in relation to the brain? "Had it not been for the anatomists", wrote G. F. Stout, "conscious individuals might have thought, felt, willed and known about themselves and each other, without ever suspecting that they had brains at all." (Stout, 1924, p. 17.) In this, if only in this, Freud might well have agreed with him. Although everyone today knows that he has a brain and few would deny that it has something to do with how he thinks, feels, wills and knows about himself, psychologists still disagree as to whether attempts to understand the workings of the brain have relevance to psychology.[1] In consequence, it is perhaps worth while having a look at the contemporary psychological scene to determine whether, and if so to what extent, advances in brain research are shedding light on psychological issues.

While psychology was defined in Stout's day as the systematic study of conscious experience, since the advent of Behaviourism there has been a strong tendency to restrict its scope largely, if not exclusively, to the study of observable behaviour. J. B. Watson himself was anxious wherever possible to explain behaviour in terms of the nervous and glandular mechanisms which he believed to govern it and to repudiate consciousness as a valid concept in psychological inquiry (cf. Watson, 1929). To this end, he relied heavily on the reflex concept, used to such good effect by Sherrington (1906) in his analysis of the integrative action of the nervous system, together with the concept of the conditioned reflex, as formulated by Pavlov (1910) in his early studies of digestion. Pavlov's younger colleague, K. S. Lashley, came to be much more critical of both these concepts but nonetheless devoted most of his life to elucidating the neural basis of behaviour (cf. Lashley, 1923). Indeed he came to see as virtually inevitable the eventual coalescence of neurology and psychology.

Butts and Hintikka (eds.), Foundational Problems in the Special Sciences, 153–165.

Thus Behaviourism, in its earlier days at least, might justifiably be represented as a 'takeover' of psychology by the neurosciences. But Lashley's anticipated coalescence has certainly not come to pass; indeed in the past thirty five years the gap between neurology and psychology has, if anything, grown wider. This may be traced in part to the somewhat negative aspect of Lashley's work and in part to the overwhelming preoccupation on the part of Watson's successors with the nature and explanation of learning.[2] As a result, simple models of learning in terms of Pavlovian conditioned reflexes soon became unfashionable and many attempts were made to evolve theories of learning which paid much more attention to the realities of behaviour and which for the most part eschewed explanation in physiological terms. Of these, by far the most thoroughgoing was that of Clark Hull (1943) whose work at first appeared to give promise of a genuinely deductive approach to the explanation of behaviour. At the same time, in spite of retaining a few neurophysiological variables the Hullian system had only the most tenuous links with the advancing neurosciences. Lashley's physiologically based behavourism was in eclipse.

This eclipse is still in evidence. In the writings of B. F. Skinner (e.g. 1972, 1974), for example, even the tenuous links with neurophysiology admitted by Hull seem to have been virtually extinguished. Although Skinner's rejection of 'mentalism' is no less absolute than Watson's, he replaces it not by central nervous activities but by what can only be called a behavioural technology. Skinner sees behaviour as governed by only two principles, contingencies of survival, as embodied in genetical constitution, and contingencies of reinforcement, as manifested in operant conditioning. There is little, if anything, in his work to suggest that explanation in psychology might be furthered by a much more central concern for the nervous system envisaged as the instrument of behaviour.[3]

In spite of this retreat from physiology, it is difficult not to see much virtue in the behaviourist revolution. It virtually put an end to the sterile structuralist psychology which E. B. Titchener and others brought to America from Germany and greatly invigorated the study of comparative psychology and of human development. But need it have been quite so radical? A biological basis for psychology had been established in America well before Watson through the teaching of such men as William James, John Dewey, James Rowland Angell and Robert Yerkes. A somewhat similar functionalist outlook had arisen in Britain with Sir Francis Galton and others, such as Lloyd Morgan, working directly in the Darwinian tradition. Even Stout, remote as he was in biology, conceded the case for a science of physiological psychology, "··· in which physiology and psychology co-operate, each receiving light and guidance in its own special domain." (Stout, 1924, p. 17.) Had psychology in America remained content with its Darwinian background and functionalist outlook and resisted the facile charms of Behaviourism, we might be much further along the road to understanding the physiological basis of mind.

II. FECHNER AND PSYCHOPHYSICS

In seeking a starting point for physiological psychology, one might do worse than begin with the work of a man often described as the 'Father of Experimental Psychology'. I refer to Gustav Theodor Fechner, whose famous two-volume *Elemente der Psychophysik* appeared in 1860. Although Fechner is remembered by philosophers as a pantheist and a mystic, it is important to bear in mind that he contributed an important principle to experimental psychology. This is the Weber-Fechner law, which specifies the relationship between the intensity of physical stimulation and differential sensitivity. But Fechner's interests were by no means limited to psychophysics in the sense in which this term is ordinarily understood. Ultimately, he supposed, psychophysical studies would be largely replaced by psychophysiological studies, whereby correlations would be established between subjective magnitudes and the intensity of the underlying brain processes.[4] This, Fechner hoped, would allow the creation of a quantitative science of brain-mind relationships. Although techniques appropriate to such a science were not available in Fechner's time and are indeed only in their infancy today, we might seem to see in the techniques of modern electrophysiology, neurochemistry and psychopharmacology the first fruits of Fechner's remarkable foresight.

It would be inappropriate to dwell at length on Fechner's ideas of mind and consciousness, in spite of their often surprising relevance to modern inquiry. For example, he made much play with the notion of a limen of consciousness, i.e. the absolute threshold for awareness generated by central nervous excitation. He supposed that this threshold could be influenced both by variation in level of consciousness, as in the daily fluctuation of sleep and wakefulness, and by shifts in the level of attention, whether voluntary or involuntary. We see in this an anticipation of the modern notion of cortical arousal. Fechner also devoted careful consideration to the effects of brain lesions on psychological capacity, and indeed devoted an important section of the *Elemente* (Vol. 2, pp. 381–427) to this topic. He was also well aware that the loss of large amounts of brain tissue, extending on occasion even to an entire cerebral hemisphere, does not necessarily depress the level of consciousness and that not all psychological functions are to be strictly localized in the brain cortex. These views stand close to what is believed today. While Fechner's fame may continue to be linked primarily with his psychophysical law, he can nonetheless be truthfully regarded as the founder of what has come to be called neuropsychology.

III. THE DUPLEX BRAIN

An important principle adduced by Fechner (1860), and later restated more formally by the philosopher Eduard von Hartmann (1868) is that continuity of anatomical structure is an essential condition of the unity of consciousness. Believing, as he did, that consciousness bears a special relation to the cerebral hemispheres, Fechner was led to consider in some detail the issues raised by the brain's duplex structure. More specifically, he argued that if one could divide the brain longitudinally by section of the corpus callosum, something like the doubling of a human being would be brought about. "The two hemispheres", he wrote, "while beginning with the same moods, predispositions, knowledge and memories, indeed the same consciousness generally, will thereafter develop differently according to the external relations into which each will enter". (1860, **2**, p. 537.) In short, splitting the brain will divide the stream of consciousness. Although Fechner explicitly considers that such an operation, at all events in man, is beyond the bounds of possibility, it has in fact been undertaken on a number of occasions in recent years and is indeed now a recognised form of surgical treatment for severe and intractable epilepsy.

It is important to appreciate that Fechner was writing before the demonstration by Broca (1861) and others in France that disorders of speech result almost without exception from lesions of one hemisphere, as a rule the left thought occasionally the right. As is well known, Broca contended that only a limited portion of the left hemisphere viz. the third frontal convolution, governs articulate speech – a contention, which although still widely accepted has nonetheless been frequently challenged.[5] It is also uncertain whether this functional difference between the hemispheres does or does not depend on an asymmetry of structure.[6] As to the functional difference itself, however, this is no longer theory but fact.

Asymmetry of cerebral hemisphere function is usually considered under the rubric of *cerebral dominance.* This admittedly ill-defined concept embodies the belief that there is an inborn difference in the respective potentialities of the two hemispheres to subserve speech and that this difference is in all probability linked to the issue of handedness. Certainly, the left hemisphere subserves speech and related linguistic functions in the vast majority of human beings, most of whom are of course right-handed. The right hemisphere would seem to subserve speech in only a small minority of human beings, in which left-handers outnumber right-handers. Although there are exceptions to the rule linking aphasia with injuries of the hemisphere opposite to the preferred hand, the existence of some connection between laterality and language would seem well-established. Indeed evidence of this link is perhaps strengthened by the fact that the left hemisphere appears to be much more essential than the right for the initiation of motor activity. At all events, disorders of purposive action (apraxia) involving both sides of the body commonly result from lesions of the left hemisphere but never from lesions of the right hemisphere alone. Cerebral dominance, therefore, appears to embrace purposive action no less than the control of speech.

At the same time, it would be wrong to conclude that the right hemisphere is without importance for mental life. Indeed injury or disease of this hemisphere, more especially of its posterior portions, not infrequently gives rise to striking disabilities in spatial perception and in performances such as drawing, copying and the use of tools, which demand fine co-ordination and spatial judgement. In some cases, visual recognition, in particular of persons and places, is strikingly impaired. These defects may on occasion persist for many years and constitute an appreciable handicap in everyday life. There is also evidence that certain disorders in auditory perception, in particular those involving appreciation of musical qualities, are likewise selectively impaired by lesions of the right cerebral hemisphere (Milner, 1962).

We thus seem to arrive at a concept of cerebral hemisphere function that entails a very real 'division of labour' between the two halves of the brain. While no one denies that there is an important sense in which the brain acts as a whole in the control of behaviour, this unity conceals a diversity of component mechanisms, differing in their degree of dependence upon the activity of one or other cerebral hemisphere. This relative asymmetry of cerebral hemisphere function is brought out even more strikingly by studies of individuals who have undergone section of the cerebral commissures and to these we must now turn.

IV. Effects of brain bisection

We owe to R. W. Sperry the initiative in taking advantage of the operation of commissurotomy for investigating hemispheric asymmetry and disconnection in the human subject. His work, much of it in association with the neurosurgeon, Joseph Bogen, has opened an entirely new chapter in our understanding of interhemispheric relationships and has made a contribution of the first order to modern experimental psychology.

When it was first introduced, the commissurotomy (split-brain) operation was not considered to produce any very noteworthy changes in the psychological sphere. Behaviour, in its general and social aspects, was substantially unchanged and performance on intelligence and certain other tests differed little, if at all, from the level established before operation. At first sight, therefore, Fechner's claim that such an operation would produce the "doubling of a human being" would seem devoid of foundation. But such a conclusion has now been shown to be distinctly misleading. Sperry's work has demonstrated, that if the right questions are asked and appropriate experiments designed to answer them, very substantial disconnection between the two halves of the brain can be brought out. For example, an object explored by one hand can be recognised by that hand, though not by the other hand. If the hand is the left, the object, although it can be recognised, cannot be named. In the same way, a picture of an object flashed to the left half of the visual field can, provided that there is no movement of the eyes, be recognised only if it is re-exposed in that field. The object cannot however be named. If the picture is flashed to the right visual field, however, it can be both recognised and named. That is to say, if conditions are so arranged that sensory input reaches only one hemisphere of the brain, it can be appreciated and reacted to only by that hemisphere and the opposite hemisphere is unaware of its existence. Naming or other verbal specification of the input, moreover, can take place only if that input is received by the hemisphere controlling speech.[7]

Many of the findings regarding hemispheric asymmetry of function have been confirmed by the results of split-brain studies. For example, the patients, even though right-handed, draw and copy patterns more accurately with the left hand than with the right, and have been shown to exhibit better spatial judgement in dealing with visual or tactile inputs to the right hemisphere than with those to the left. Moreover, strategies in visual perception appear to differ appreciably according to which hemisphere is prepotent (Levy-Agresti and Sperry, 1968). In general, the right hemisphere proceeds by global impression and direct matching, somewhat in the way Mach (1886) describes when one directly perceives a complex visual configuration, such as a tree.[8] The left hemisphere, on the other hand, proceeds by means of verbal analysis and sequential ordering of key features in the display. Similar differences probably characterise the functions of each hemisphere in recognition and recall, the right being predominantly involved in visualization and the left in vocalization (cf. Zangwill, 1972). Thus

individual differences in the manner of recall may find their explanation in differential hemisphere activation according to the demands of
the task.

V. Consciousness and the cerebral hemispheres

It is findings of this nature that have led Sperry and his colleagues to
conclude that "each hemisphere seems to have its own conscious
sphere for sensation, perception and other mental activities, and the
whole realm of gnostic activity of the one is cut off from the corresponding experience of the other hemisphere". (Sperry *et al.*, 1969.)
Although the person remains a unified person, there is no question
that, under appropriate conditions of unilateral input, concurrent conscious events, one in each hemisphere, can be convincingly demonstrated.[9] Such findings raise knotty problems for orthodox theories of
the mind-body relation. What, if any, significance do they possess for
our understanding of mind and consciousness when the brain is intact?

One view, which has been advocated by Joseph Bogen (1969) is that
even in the intact brain one must postulate a duplex representation of
conscious experience. This, as Bogen himself indicates, is a view that
goes back at least to Wigan, an early Victorian psychiatrist, who
advocated it in a curious book called *The Duality of the Mind* (1844).
Sperry, too, admits, that under certain circumstances sensory experience may be duplex as, for example, when a point on the face (which is
bilaterally represented in the brain) is stimulated. That we experience a
single event in such cases may be due either to suppression of the
sensory information in one hemisphere or to some form of fusion,
analagous perhaps to what takes place in normal binocular vision.
Nonetheless, the hypothesis of a literal duplication of experience
generally lacks parsimony, and certainly is not imposed upon us by the
evidence of split-brain studies.

An alternative view has been put forward by Eccles. In his Eddington Lecture delivered at Cambridge in October, 1965, Eccles argued
that only nervous activities in the dominant (i.e. left) cerebral hemisphere are, in Sherrington's term, "conjoined with consciousness".
Those in the right hemisphere, on the other hand, can be understood
on mechanistic principles of a kind applicable to the workings of a

computer. Although the potentialities of the right hemisphere, especially in relation to language and voluntary activity, were not clearly appreciated at the time Eccles first put forward his view, he has nonetheless repeatedly restated it without essential change in later publications (Eccles, 1970, 1973).

Eccles bases his argument on two main considerations. First, that it is impossible to establish the existence of consciousness in the absence of some mode of symbolic communication; and secondly, that the exercise of volitional control in the production of movement by the disconnected right hemisphere has not been established. These arguments, to say the least, smack of special pleading (Zangwill, 1973). In the first place, the right hemisphere, albeit mute or almost so, has been shown to possess intellectual and memory capacity entirely comparable with that of its fellow and to be capable of emotional response at a level ordinarily regarded as wholly human (Sperry et al., 1969). While it cannot be interrogated directly as to its conscious state any more than can an animal, it is impossible to deny it consciousness in the sense in which the term is used in ordinary discourse. And secondly, there is now good evidence (Gazzaniga et al., 1967), that under certain circumstances the right hemisphere can directly generate and control movements of the right hand, thus disproving the contention that it cannot initiate voluntary movement. Furthermore, it might seem odd to accept the computer analogy in relation to one hemisphere of the brain while denying that it has relevance to the workings of the other hemisphere or indeed to the brain as a whole.

It should also be said that patients who have undergone removal of the entire left hemisphere in cases of malignant and rapidly growing cerebral neoplasm give every indication of normal consciousness. For example, a patient described by Smith (1966), although grossly handicapped in speech and motor capacity, was alert, socially responsive and capable of attaining a low average score on a diagrammatic intelligence test. Unless one rejects the concept of consciousness altogether in the manner of the radical behaviorists, there would seem no reason to deny that this unfortunate individual possessed a stream of consciousness.

A further objection to Eccles' position is the demonstration by Levy et al. (1972) that concurrent perceptual events can take place in the two separated hemispheres. This is perhaps the most convincing illustration of Fechner's principle that the unity of consciousness is dependent upon structural continuity within the nervous system. In the split-brain patient, independent sensory input to each hemisphere may result in concurrent, one might perhaps say co-conscious, perceptual events.[10] Eccles seems nowhere to have considered the implications of such observations for his views about consciousness and the brain.

Although Sperry sometimes writes as though the split-brain patient possesses two quite independent streams of consciousness, or even two minds, it does not necessarily follow that the systems associated with the activity of each hemisphere respectively are of equivalent scope, complexity and continuity. Indeed Sperry (1973) has himself described the role of the dominant hemisphere as the more aggressive and dominant in control of the motor system. It effectively controls access to the channels of communication (and in this respect appears largely to inhibit the right hemisphere) and undoubtedly initiates the greater part of volitional activity involving actions of either hand alone or of the body as a whole. As Sperry (1973) puts it, the split-brain patient runs primarily on the left hemisphere. "It is the hemisphere we see in action and the one with which we regularly communicate." Right hemisphere control is seen only episodically, and for the most part when conditions are so arranged as to limit visual input to this hemisphere and to limit response to the left hand. While it would be unwise to generalise from the split-brain patient to the individual whose brain is intact, it is nonetheless likely that the behaviour of the healthy person is predominantly under the control of the left cerebral hemisphere. In this respect, at least, one may agree with Eccles (1965). It must be left to the future to determine the extent to which this control is modified by factors such as age, sex and handedness, as well as by the nature of the environment and the relative importance of skills which do not depend primarily on linguistic capacity.

VI. SOME PHILOSOPHICAL IMPLICATIONS

The issues posed by split-brain patients undoubtedly raise conceptual and linguistic difficulties with which neurologists and psychologists are in general ill-equipped to deal. Ordinary language, with its implicit assumption that the individual's mind is undivided, cannot easily be used to describe a situation in which perception and action are governed by one cerebral hemisphere acting largely or wholly independently of the other. Thus in describing the findings in experiments with split-brain patients, Sperry and his colleagues often use "the right (or left) hemisphere" as the subject of a sentence when we should ordinarily speak of "the person". Yet this is surely unsatisfactory. It is the person who experiences or acts, even if his brain has been split and his experiences or actions are governed by one or other hemisphere acting in isolation. But to say that "the person perceives the picture by virtue of his right hemisphere" is clumsy and suggests only that the right hemisphere is necessary for this process, not that the state of consciousness itself is generated by this hemisphere alone and is inaccessible to its fellow. In my view linguistic philosophers could perform a most useful service in teaching neurologists and others how they should think and talk about these paradoxical phenomena.

University of Cambridge

NOTES

[1] For a recent discussion of the relations between physiology and psychology, see Joynson (1974).

[2] Jean Piaget once dubbed this obsession with learning theory as the 'American disease'.

[3] In his most recent book it is true, Skinner (1975) tells us that some day the physiologist will "give us all the details of how an organism behaves". But no guidance whatsoever is given to any physiologist who might seek to attain this objective.

[4] This Fechner christened 'inner psychophysics' in contrast with 'outer psychophysics' which specifies quantitative relationships between stimulus and sensation.

[5] See, e.g. Zangwill (1975) for a critical discussion of Broca's area in relation to aphasia.

[6] Evidence of a structural basis for functional asymmetry has been adduced by Geschwind and Levitsky (1968). This is in all probability inborn (see Wada *et al.*, 1975).

[7] For an account of Sperry's methods and findings, see Sperry, Gazzaniga and Bogen (1969). More popular accounts of much of this work have been given by Gazzaniga (1970) and Dimond (1972).

[8] "The tree with its hard, rough, grey trunk, its many branches swayed by the wind, its smooth, soft shining leaves, appears to us at first a single, indivisible whole." (Mach, 1959, p. 102.)

[9] The brilliant experiments of Levy *et al.* (1972), in which two pictures are flashed simultaneously to the patient, one to each hemisphere, demonstrate most convincingly the existence in the separated hemispheres of concurrent perceptual events.

[10] In the split-brain patient, duplex experience of this kind is almost certainly rare under ordinary conditions of daily life, as constant movements of the eyes ensure that visual input reaches both hemispheres. Manipulation of objects, too, as a rule involves the coordinated activity of both hands, in addition to visual control of the pattern of action. It, too, is therefore likely to involve the activity of both hemispheres. Nonetheless, the occurrence of duplex experience under conditions in which input to each half of the brain is independent of the other clearly indicates that each hemisphere can, under certain conditions, independently subserve awareness of events in the external world.

BIBLIOGRAPHY

Bogen, J. E.: 1969, 'The other side of the brain', *Bull. Los Angeles Neurol. Soc.* **34**, 73–105, 135–162.

Broca, P.: 1861, *Bull. de la Soc. anat.*, 2ᵐᵉ series **6**, 330–357.

Dimond, S. J.: 1972, *The Double Brain*, Churchill/Livingstone, London.

Eccles, J. C.: 1965, *The Brain and the Unity of Conscious Experience*, 19th Arthur Stanley Eddington Memorial Lecture, Cambridge University Press.

Eccles, J. C.: 1970, *Facing Reality*, New York, Heidelberg, Berlin/Springer-Verlag, Longman, London.

Eccles, J. C.: 1973, *The Understanding of the Brain*, McGraw-Hill, New York.

Fechner, G. T.: 1860, *Elemente der Psychophysik*, 2 Vols., Von Breitkopf and Härtel, Leipzig.

Gazzaniga, M. S.: 1970, *The Bisected Brain*, Appleton-Century-Crofts, New York.

Gazzaniga, M. S., Bogen, J. E., and Sperry, R. W.: 1969 'Dyspraxia Following Division of the Cerebral Commissures', *A.M.A. Arch. Neurol.* **12**, 606–612.

Geschwind, N. and Levitsky, W.: 1968, 'Human Brain: Left-Right Asymmetries in Temporal Speech Region', *Science* **161**, 186–187.

Hartman, E. von: 1868, Eng. Trans. 1931, *Philosophy of the Unconscious: Speculative Results According to the Inductive Method of Physical Science*, Kegan Paul, Trench and Trubner, London.

Joynson, R. B.: 1974, *Psychology and Common Sense*, Routledge and Kegan Paul, London.

Lashley, K. S.: 1923, 'The Behavioristic Interpretation of Consciousness', *Psychol. Rev.* **30**, 237–272, 329–353.

Levy-Agresti, J. and Sperry, R. W.: 1968, 'Differential Perceptual Capacities in Major and Minor Hemispheres', *Proc. U.S. Nat. Acad. Sci.* **61**, 1151.

Levy, J., Trevarthen, C., and Sperry, R. W.: 1972, 'Perception of Bilateral Chimeric Figures Following Hemispheric Deconnection', *Brain* **95**, 61–78.

Mach, E.: 1886, Dover Edition 1959, *The Analysis of Sensations and the Relation of the Physical to the Psychical*, Dover Publications Ltd., New York.

Milner, B.: 1962, 'Laterality Effects in Audition', in: *Interhemispheric Relations and Cerebral Dominance* (ed. by V. B. Mountcastle), pp. 177–195, John Hopkins Press, Baltimore.

Pavlov, I. P.: 1910, *The Work of the Digestive Glands*, 2nd English edition, Charles Griffin & Co., London.

Sherrington, C. S.: 1906, *The Integrative Action of the Nervous System*, Scribner's, New York; reprinted 1947 by Cambridge University Press.

Skinner, B. F.: 1972, *Beyond Freedom and Dignity*, Cape, London.

Skinner, B. F.: 1974, *About Behaviourism*, Cape, London.

Smith, A.: 1966, 'Speech and Other Functions After Left (Dominant) Hemispherectomy', *J. Neurol. Neurosurg. Psychiat.* **29**, 467–471.

Sperry, R. W.: 1973, *Changing Concepts of Mind*, Gerstein Lecture delivered at York University, Toronto, November 1973 (mimeographed).

Sperry, R. W., Gazzaniga, M. S., and Bogen, J. E.: 1969, 'Interhemispheric Relationships: The Neocortical Commissures: Syndromes of Hemispheric Disconnection', in: *Handbook of Clinical Neurology*, Vol. 4, North Holland Publ. Co., Amsterdam.

Stout, G. F.: 1924, *A Manual of Psychology*, 3rd Edition, University Tutorial Press, London.

Wada, J. A., Clarke, P., and Hamm, A.: 1975, 'Cerebral Hemispheric Asymmetry in Humans', *A.M.A. Arch. Neurol.* **32**, 239–246.

Watson, J. B.: 1929, *Psychology from the Standpoint of a Behaviourist*, 3rd edition, Lippincott, Philadelphia and London.

Wigan, A. L.: 1844, *The Duality of the Mind: A New View of Insanity*, Longman, London.

Zangwill, O. L.: 1972, ' 'Remembering' Revisited', *Quart. J. Exp. Psychol.* **24**, 123–138.

Zangwill, O. L.: 1973, 'Consciousness and the Cerebral Hemispheres', in: *Hemisphere Function in the Human Brain* (ed. by S. J. Dimond and J. G. Beaumont) Elek Science, London, pp. 264–278.

Zangwill, O. L.: 1975, 'Excision of Broca's Area Without Persistent Aphasia', in: *Cerebral Localization* (ed. by K. J. Zülch, O. Creutzfeldt, and G. C. Galbraith) Springer-Verlag, Berlin, Heidelberg, New York, pp. 258–263.

RAIMO TUOMELA

CAUSALITY AND ACTION

I. Causal Theories of Human Action

According to causal accounts of human action an adequate analysis of action cannot be given without reference to the fact that in acting the agent is exercising his causal powers. The main purpose of the present paper is to extend a certain version of a causal theory, previously developed by the author (see Tuomela, 1974a, 1975), to cover any kind of *complex actions*.

We are thus going to study what behavior is to be classified as (simple or compound) action and what an agent can do. In the course of doing this we will investigate such topics as (simple and compound) basic and bodily actions, the scope of an agent's causal powers, and the special nature of intentional actions. A basic thesis to be investigated, clarified and defended is: all an agent ever does is to move his body; the rest is up to nature.

The present investigation is to be regarded as a work within a certain causalist research programme; the causal theory will not really be directly defended here against rivalling intentionalist, coherentist, and other accounts (cf. Sellars, 1973 and Tuomela, 1975, for such a defense).

This introductory section will be primarily devoted to a simplified summary exposition of some central features of the causal theory of action I have tried to develop elsewhere, as mentioned.

Let us start with a preliminary comment on actions. To give an adequate account of the notion of action one has to refer both to the antecedents and the consequents of action. The antecedents will be activated propositional attitudes like wanting, believing, and intending. As will be seen, the notion of intentional action is *conceptually* linked to intending. As to consequents, only behavior with certain achievement or *result* aspects will classify as action. Actions can be regarded

Butts and Hintikka (eds.), Foundational Problems in the Special Sciences, 167–203.
Copyright © 1977 *by D. Reidel Publishing Company, Dordrecht-Holland. All Rights Reserved.*

as a kind of 'responses' to *tasks* so that the correctness of the action as a task-solution can be *publicly* assessed in terms of result-events.

Ontologically viewed a singular action is a complex event brought about by an agent. It is complex in the sense of being process-like: a singular action consists of a sequence of events. (We are going technically to call actions events though in many cases it would be more according to ordinary usage to call them processes.)

Let us now briefly consider the antecedents of action. According to my view intendings, believings and other related propositional attitudes are to be *functionally* characterized as (realistically conceived) dispositional states with a certain propositional structure. Conceptually (or semantically) these states are introduced in terms of intelligent linguistic and non-linguistic behavior (i.e. actions and other intelligent behavior). We can then say that they are introduced by reference to social conventions and social 'practice', as such behavior is conceptualized in terms of a social and public conceptual framework.

This introduction of mental states (and non-dispositional episodes as well) as a kind of 'theoretico-reportive' entities is functional (and hence indirect), and it is given causalistically in terms of the 'input-output' behavior of the person, especially in terms of these inner states *causing* his relevant behavior in different circumstances. Thus, to intend to do *p* means, roughly, being in a dispositional state with the structure *p* (or with the structure *p* has) such that this state, given suitable internal and external circumstances, will cause the bodily behavior needed for the satisfaction of the intention. (In a finer analysis we also need here an *epistemic* criterion to account for the agent's privileged access to his mental states. A mere 'topic neutral' analysis by means of e.g. Ramsey-sentences is insufficient.)

To clarify the issues involved we shall introduce some formalism. We describe the agent's intendings, wantings, believings, etc., in standard first-order logic by means of three-place predicates. Accordingly we think of intendings, etc., as states (or, occasionally, events) and employ a special variable for singular states cum events:

'$(Ex)I(A, p, x)$' translates '*A* intends that *p*', and

'$(Ex)B(A, p, x)$' translates '*A* believes that *p*'.

More literally, we may read these formulas, respectively, as '*x* is a singular state of intending of A with the structure p' and 'x is a singular state of believing of A with the structure p'. These singular states are to be regarded as 'substrata' or 'quasi-bare' states. The same is the case with singular events (e.g. actions), which will be the ontological entities that most concern us later.

Action statements are formalized in the manner of Davidson (1966) by means of an explicit event-ontology. Space does not here permit a more detailed treatment of the problems of the formalization of action statements nor statements about propositional attitudes (for details, see Tuomela, 1974a).

By means of this formalism we seem to be able to do several things simultaneously. First, the above formalization represents our proposal for the *logical form* or, equivalently, the *deep structure* of action statements and statements about propositional attitudes. Secondly, if one accepts an 'analogy account', of thought processes (cf. e.g. Sellars, 1956), one may use the above formalization to analogically represent 'the language of thought'. Thirdly, our formalization can in principle be proposed as a tool for a psychologist or at least for a philosopher analyzing theories in cognitive and motivation psychology.

Put briefly, this kind of *conceptual functionalism* semantically introduces propositional attitudes as states with a certain causal power. It is essential that these states are realistically construed. One can then, for instance, conjoin with this some version of materialism and claim that these states are material states, which future neurophysiology will tell us more about.

A related essential matter is that these dispositional states can be mentally 'manifested' and actualized, i.e. singular mental events or episodes manifest them. This 'manifestation' need not be conscious, and in the 'final' scientific analysis these mental events will presumably be given a *non*-functional (categorical) description.

Now some, though not perhaps all, of these manifesting inner singular events can be said to *activate* the disposition. This I take to mean that these disposition-activating manifestations *cause* behavior (or at least that they occur in a suitable constellation of states and events causing behavior).

The agent's wants and beliefs are the most important dispositional determinants of his actions (there are even conceptual connections between these elements, as indicated). From time to time these wants (cum the relevant beliefs) become activated (due to environmental factors and 'self-stimulation'). Then we say that the agent forms intentions to act on these wants (cum beliefs). In other words, we may say that the wants generate intentions (i.e. states of intending). This little studied type of generation is presumably causal, but still we cannot rule out e.g. indeterministic non-causal generation *a priori*.[1]

A want (in our *broad* sense) is either *intrinsic* (when something is wanted for its own sake) or *extrinsic* (a duty, an obligation, etc. ultimately serving some intrinsic want). Intendings thus can be formed on the basis of either intrinsic or extrinsic wants (cum the relevant beliefs).

Intendings may generate other, usually more specific intendings. For instance, the practical syllogism represents one such intention transferral mechanism (cf. von Wright, 1971; Kim, 1976, and, for arguments concerning its causal character Tuomela, 1975).

Let us consider an agent who has formed an intention to illuminate the room. His deliberation makes him – through intention transferral – to form an *intention to flip the switch*. When the time for flipping the switch arrives, this intention becomes a causally effective *trying* (as we technically call it) to flip the switch (see Tuomela, 1974a, for this notion).

The mental event of trying (= an effective intending or willing), an event participating in the activation of the agent's 'action plan', is to be characterized functionally in terms of what it causes. Thus a trying to raise one's arm is a singular occurrent event which, the world suitably cooperating, causes the bodily events necessary for the satisfaction of the intention. Here it causes the arm's going up which constitutes the overt embodiment of the arm-raising action.

A trying is an intentional mental act, an executive act which is under the agent's control. But yet is is not an action caused by another trying; trying to try is, if anything, only trying. Also notice that I have not required a trying to be a fully conscious event.

It is important to see that each singular claim concerning the causation of arm raising by a trying is a *synthetic* claim. For in the case of any such singular causal claim the trying and the movement (or action, if you like) are ontically *distinct* events and they can be redescribed so that every trace of intentionality as well as of the generic logical connection holding between the intention and the corresponding action disappears from the description.[2]

I have here been making a plea for a *dynamic* account of action, suitable as a basis for scientific psychology. The 'forces' I have been speaking about are of course the causal powers – 'active' and 'passive' – of the agent as exercised in e.g. want-intention generation, intention transferral, and, most centrally, in his tryings, which cause his body to move in the way required for the purpose expressed in his intentions to become satisfied.

It should be emphasized that my account is not in any essential sense reductionistic. I consider it important that the conceptual framework of agency (wants, hopes, beliefs, norms, etc.) be employed. An attempt to reduce away this framework would be an attempt to make psychology do without the thinking, sensing and acting man as its subject matter. (Accordingly, the 'definitions' to be given in the following sections of various action concepts are not reductionistic and in that sense real definitions, although my characterizations are not directly circular.)

II. SINGULAR CAUSATION

One central thing to be clarified in discussing causal accounts of action is naturally causation itself. It is rather unfortunate that most causal theorists have paid little, if any, attention to this problem (cf. Goldman, 1970). I have elsewhere in Tuomela (1974b) developed an explanatory backing law account of singular event causation. In Tuomela (1975) it was applied to problems of action-causation to yield a concept of directed or purposive causation.

In this section I shall summarily sketch some of the features of my account of singular causation, as we are later going to need them in developing a theory of complex actions.

To discuss singular causation we introduce a two-place predicate '$C(a, b)$' which reads 'singular event a is a (direct) deterministic cause of singular event b'. Now we can give truth conditions (in a wide 'assertability' – sense of truth) for '$C(a, b)$' basically in terms of the DE-model of deductive explanation (developed in Tuomela (1972) and elaborated in Tuomela (1974b)) and two other conditions, assuming that the event a and b have occurred:

DEFINITION II.1. The statement '$C(a, b)$' is true if and only if there are singular statements D_1 and D_2 and a causal theory S (possibly including also singular context-describing statements) such that D_1 and D_2 describe a and b, respectively, and the following conditions are satisfied:

(1) S jointly with D_1 gives an actual explanation of D_2 in the sense of the DE-model, viz. there is a Tc for $T = S \& D_1$ such that $\varepsilon(D_2, Tc)$ in the sense of the DE-model;

(2) a is causally prior to b.

(3) a is spatiotemporally contiguous with b.

In this definition '$\varepsilon(D_2, Tc)$' reads 'the conjunction of the members in the set Tc of ultimate sentential components of T deductively explains D_2'. Due to lack of space we cannot here really clarify this notion nor the conditions (2) and (3), but we refer the reader to the above mentioned works.

What kind of formal properties does '$C(a, b)$' have? Merely on the basis of the logical properties of ε it is easy to list a number of them. For instance, it is asymmetric and irreflexive, and it fails to be transitive. One feature to be especially noticed is that the truth of '$C(a, b)$' is invariant with respect to the substitution of identicals in its argument places.

On the basis of the definition of '$C(a, b)$' we immediately get a conditional $\triangleright \rightarrow$:

DEFINITION II.2. '$D_1 \triangleright \rightarrow D_2$' is true if and only if there is a theory S such that $\varepsilon(D_2, S \& D_1)$.

If S is a causal theory and if the conditions (2) and (3) of Def. II. 1. are satisfied, then $\triangleright \rightarrow$ expresses a *causal* conditional: 'if a (under D_1), then b (under D_2)'; or 'a (under D_1) causes b (under D_2)'.

Truth in this definition is to be considered relative to a 'model' or 'world' constituted by the explanatory situation in question.

We immediately notice that if '$C(a, b)$' is true then so is, for some D_1, D_2, '$D_1 \triangleright \rightarrow D_2$' (but not necessarily '$\sim D_1 \triangleright \rightarrow \sim D_2$'). (For other properties of $\triangleright \rightarrow$ see Tuomela (1974b).)

Our notion of event causation discussed above is not yet the notion of *purposive* causation which we need to account for the idea of intendings (cum beliefs) causing the action within practical inference and other relevant contexts. The basic idea of purposive causation is that it is something that (in a sense) *preserves intentionality* or purpose. Thus if an agent acts on a certain conduct plan (e.g. premises of a practical syllogism) the actions caused by his intendings have to satisfy the conduct plan, given that 'normal conditions' obtain.[3] If an agent forms an intention to go to the theater at 8 p.m. tonight this intention should cause him to satisfy the intention (purpose) by causing whatever bodily behavior is needed to satisfy that purpose. Hence it should not e.g. cause him to go to a restaurant at that time. The element of control needed here is provided by the mental act of trying (active intending) by means of which the agent exercises his causal power.

Assume now that we are in the possession of a finitary language of thought (cf. Section I). This language will employ all the concepts in the 'framework of agency', and in order to be learnable it must be finitary in character (as Davidson has forcefully argued). Now we interpret all the extralogical constants of this (first-order) language into a domain of the singular actual events, states, etc., and objects present in *this* behavior situation. We call the resulting Tarskian set theoretical structure $\mathcal{M}$. Let us assume that in principle we can, for all cases, determine $\mathcal{M}$. (Difficult epistemological and methodological problems will normally be involved in that.) Given $\mathcal{M}$, we can discuss the *theory of $\mathcal{M}$* (i.e. $Th(\mathcal{M})$ with respect to the full language of the framework of agency (plus whatever else is required). $Th(\mathcal{M})$ is, as usually, defined as the set of all statements true of $\mathcal{M}$. Now, as is well known, any theory T which is satisfied by $\mathcal{M}$ and which is complete (in the logical sense) is equivalent to $Th(\mathcal{M})$.

Given this framework, we can now go to the heart of the matter and formulate a characterization of purposive causation. We consider an action token u performed by an agent A. (In our later terminology we are considering an action token$_1$.) We assume that u has the internal structure of a sequence of events $\langle t, ..., b, ..., r \rangle$. Each event in this sequence is assumed to be spatiotemporally contiguous with at least its predecessor and its successor event. Presystematically speaking, t is going to represent (part of) the event activating A's conduct plan. In the case of intentional action it will be a trying (= effective intending) cum the relevant believing. The event b will represent the maximal overt bodily component in the action u, and r will be the agent-external result of the action. A result of an action is roughly the overt event intrinsically connected to the action in the way a window's becoming open is connected to the action of opening the window. What (if anything) is needed between t and b in u is up to future scientists to find out. From the philosopher's conceptual point of view it then suffices to represent u by $\langle t, b, r \rangle$. Furthermore, we shall below often simplify matters and speak of the composite event u_o obtained by composing b and r.

Let thus $u = \langle t, u_o \rangle$. Then we introduce $C^*(t, u_o)$ as a two-place predicate to be read 't is a direct purposive cause of u_o'. We 'define', relative to an agent A:

DEFINITION II.3. '$C^*(t, u_o)$' is true (with respect to the behavior situation conceptualized by $\mathcal{M}$) if and only if

(a) '$C(t, u_o)$' is true (with respect to $\mathcal{M}$); and

(b) there is a complete conduct plan description K satisfied$_2$ by $\mathcal{M}$ such that in K t is (correctly) described as A's trying (by means of his bodily behavior) to exemplify that action which u ($= \langle t, u_o \rangle$) is describable as tokening according to K, and u_o is represented in K as a maximal overt component event of u; t and u_o belong to the domain of $\mathcal{M}$.

Of the notions that occur in Def. II. 3. we shall later try to clarify better what a maximal overt part of an action is. Thus we shall discuss both the nature of b and of r. Notice that at least u_o need not belong to the domain of the model as an event separate from $u = \langle t, u_o \rangle$, though it may.

The notion of satisfaction$_2$ means, roughly, ordinary model theoretic satisfaction *and* a requirement that A has a correct wh-belief of the overt or 'external' entities in the domain of the conduct plan structure satisfying K in the model theoretic sense. Here wh-belief means belief about *what* (or *which* or *who*, as the case may be) the entity in question is. Speaking in other terms, the conduct plan K must in a stronger sense than the ordinary model-theoretic one be *about* the objects in the domain of its model. As I have elsewhere (Tuomela (1975)) discussed this notion of satisfaction$_2$ (or belief about an entity or wh-belief) in more detail I will not here discuss further this important problem.

What our above definition comes to saying is that the trying t purposively causes u_o only if t under the description D_1 it has in K causes (in the ordinary sense of event-causation) the event u_o as described by a statement (D_2) in K. Notice that the backing law justifying the truth of '$C(t, u_o)$' of course can employ quite different predicates from those appearing in D_1 and D_2. The essential content of the above definition – over and above the causation requirement – is that t can indeed by truly described as a trying (activated intending) to bring about an action of a certain type such that u_o is describable as an overt part of a token of that action type.

To exclude some factually possible cases of 'underdetermination' we have required in Def. II. 3. that K be complete. (See Tuomela, 1975, for a motivation of this.)

Several derivative causal notions can now be defined on the basis of $C^*(-, -)$. We shall below define the notion of indirect extra-linguistic purposive causation and two linguistic notions corresponding to the extra-linguistic notions.

We start by defining indirect ordinary event-causation. Let e_1, e_2, ..., e_n be singular events. Now we say that e_1 is an indirect (partial) deterministic cause of e_n (we write $IC(e_1, e_n)$) just in case e_1 is causally linked to e_n through a chain of direct causal relationships:

$IC(e_1, e_n)$ if and only if $C(e_1, e_2)$ & $C(e_2, e_3)$ & $\cdots$ & $C(e_{n-1}, e_n)$ for some singular events $e_2, e_3, ..., e_{n-1}$.

We now go on to characterize indirect purposive causation. In principle we can define such a notion for any two mental events rather than only for a mental event and an action. (An analogous remark of course can be made for direct purposive causation.)

We let a singular action u consist of a sequence $\langle t, b, r \rangle$ (short for $\langle t, ..., b, ..., r \rangle$) where t represents a trying event. b represents a bodily event which in the present case will be describable as an action (cf. Section III). The event r is the event of the coming about of the result state of the full action u. The event b will below be assumed to indirectly generate the event r. Thus the composition of b and r (which we call u_o) is describable as an overt part of a generated action.

Now we get in the obvious way a notion of indirect purposive causation:

DEFINITION II.4.   '$IC^*(t, r)$' is true (with respect to $\mathcal{M}$) if and only if

(a)   '$IC(t, r)$' is true (with respect to $\mathcal{M}$) such that t corresponds to e_1, b to e_2, and r to e_n;

(b)   there is a complete conduct plan description K satisfied$_2$ by $\mathcal{M}$ such that in K t is (correctly) described as A's trying to do by his bodily behavior what u tokens according to K, and u_o (composed of b and r) is described as a maximal overt part of an action which is (directly or indirectly) factually, but not (merely) conceptually, generated by some of A's actions in this behavior situation; t and u_o are assumed to belong to the domain of $\mathcal{M}$.

Our Definitions II.3. and II.4. do not yet cover all actions, for there are other more complex actions which will be discussed later. The notions of factual and conceptual generation will be defined in exact terms in Section III; until that they can be understood as meaning generation due to factual and to semantical connections, respectively.

It should be noticed that in II.4. we in effect require that u_o occurs according to the agent's beliefs or other relevant cognitions), that is, we now compare u_o with the contents of A's beliefs. Thus, technically speaking, we are requiring that u_o satisfies the proposition named by the singular term in the second argument of one of A's beliefs (cf. Section I).

We next define the notions of linguistic purposive causation, both for the cases of direct and indirect causation. They are defined as follows:

DEFINITION II.5.   '$D_1 \triangleright\!\!\rightarrow D_2$' is true (with respect to $\mathcal{M}$) if and only if '$C^*(t, u_o)$' is true (with respect to $\mathcal{M}$) such that D_1 describes t and D_2 describes u_o as required by clause (b) of Definition II.3.

DEFINITION II.6.   '$D_1 \triangleright\!\!\rightarrow D_n$' is true (with respect to $\mathcal{M}$) if and only if '$IC^*(t, u_o)$' is true (with respect to $\mathcal{M}$) such that D_1 describes t and D_n describes u_o as required by clause (b) of Definition II.4.

Our conditional $\triangleright\!\!\rightarrow$ could also have been defined with respect to the conditional $\triangleright\rightarrow$. For $\triangleright\rightarrow$ is essentially $\triangleright\rightarrow$ together with the requirements for employing a *causal* backing theory and relativization to a conduct plan as required by clause (b) of our Definition II.3. In addition the truth is explicitly relativized to a given model $\mathcal{M}$ assumed to represent an actual behavior situation. (Notice that in Tuomela, 1974b, the relativization lies implicit in the notion of an explanatory situation.) All the same, $\triangleright\rightarrow$ entails $\triangleright\rightarrow$, which again (jointly) with an almost automatically satisfied cotenability assumption entails, and is indeed much stronger than, Lewis' variably strict conditional $\square\rightarrow$, as shown in Tuomela (1974b).

It should be emphasized here again that the backing theory underlying our technical notions $C(-,-)$, $C^*(-,-)$, $\triangleright\rightarrow$, $\triangleright\rightarrow$, and $\triangleright\rightarrow$ need not be in the vocabulary of the conduct plan K, nor need it be a psychological theory at all. Our account thus does not require even that there be any strict psychological laws at all.

III. BASIC ACTIONS

In recent literature often a dichotomy of actions (action types) into basic and non-basic ones has been made. Basic actions are thought to be primitive actions in some strong generational sense: they are actions that an agent does not do *by* or through doing something else. On the other hand basic actions are usually supposed to generate all the other actions and thus to play the role in action theory 'basic cognitions' or 'protocol knowledge', etc., plays or was supposed to play in (at least empiricist) epistemology. Thus, for instance, Davidson summarizes his view by saying that "our primitive actions, the ones we do not do by doing something else, mere movements of the body – these are all the actions there are. We never do more than move our bodies: the rest is up to nature." (Davidson, 1971, p. 23.)

Our quotation brings out another aspect in addition to the generational one that basic actions are often thought to have: they are *bodily actions*, whatever that exactly means. A third aspect is that basic actions are something an agent has the power to do in standard circumstances. They belong to his *action repertoire*.

In spite of the heavy criticisms directed against most of the attempts to provide for a sharp criterion for classifying actions into basic and non-basic ones we shall below make another attempt which takes into account the above three features and which is not open to the criticisms made in the literature (cf. Baier, 1971).

Our aim in this paper is to show that in a suitable sense *all* actions are conceptually built out of simple or compound bodily actions through some kind of generation and that everything an agent *can do* is reducible to his simple and compound basic actions. Causal notions are employed essentially in carrying out this programme. I know of only one relatively detailed attempt along these lines. It is that by Goldman (1970). However, his attempt is inadequate for a number of reasons as I have argued in Tuomela (1974a).

We shall begin by presenting a definition of a basic action type primarily in terms of the notion of purposive causation and the notion of an action token (see Section IV for a more exact clarification of the latter notion). Much of the rest of this section is devoted to a discussion which purports to show that our definition is indeed adequate and that it captures the above three features of basic actions.

We define for a non-compound property X:

DEFINITION III.1.   Property X is a *basic action type* for A at **T** if and only if for any singular event u exemplifying X at **T** it is true that:

(1)   If u had the internal structure $\langle t, b \rangle$ and if A were in standard conditions with respect to X at **T** and if he knew that, then the causal conditionals '$D_1(t) \rhd\!\!\rightarrow D_2(b)$' and '$\sim D_1(t) \rhd\!\!\rightarrow \sim D_2(b)$' would be true in that situation. Here D_1 describes t as A's trying (by his bodily behavior) to exemplify X and D_2 expresses that b is a maximal simple overt bodily component-event of u and that b is the result event of u (as an X).

(2)   There is no statement D_2' which satisfies clause (1) as a substitute of D_2 and which is such that the conditional '$D_2' \odot_{\overrightarrow{2}} D_2(u)$' is true in this situation in virtue of semantic principles internalized by A. There are several clarificatory remarks to be made concerning this definition which characterizes a basic action type X in terms of some of its exemplifications. We start with its clause (1) and leave (2) and the definition of $\odot_{\overrightarrow{2}}$ until later. We assume that u has the structure $\langle t, ..., b \rangle$ or $\langle t, b \rangle$, for short. (In our later terminology u will be an action

token$_1$ in which the result event coincides with b.) The event t is describable as A's trying by his bodily behavior to exemplify X. We notice that even in this simple case of a basic action the event t activating A's conduct plan is not only an effective intending but it may also comprise (i.e. make true a statement about) the belief that trying to exemplify or bring about X causes a bringing about of X.

What are the standard conditions in clause (1)? They involve that (a) there is an (external) opportunity for action (e.g. one can try to raise one's arm only if it is not raised); (b) the agent is not in any way prevented by external forces (e.g. one's arm should not be tied, etc.). On the other hand, in the present case standard conditions are not assumed to include internal states of the agent's body. Thus one may try to raise one's arm even if it is paralyzed, and even if one knows that.

Our notions of intending and trying do involve conceptual elements (see Tuomela, 1974a, p. 55f.). As a matter of fact, on our strong interpretation of these notions the standard conditions in the above sense must hold before one can affectively intend (try) to bring about X. Furthermore, the agent's knowledge (in a rather weak sense of awareness) of the obtaining of standard conditions is also entailed by our notion of trying. Thus for our present purpose an explicit standard conditions-clause can be regarded as superfluous. But we use the standard conditions clause a) to leave room for other – weaker – interpretations of intending and trying and b) to allow for the conceptual possibility of more complex (with respect to standard conditions) basic action types than arm raisings and the like.

Our clause (1) not only clarifies the 'standard' action repertoire-aspect but it also in effect requires that a basic action be a bodily action. For it requires b to be described (or at least indirectly expressed) as a maximal simple overt bodily component event of u, and as the result event of u (as an X). First we notice that b is an *overt* bodily event – it cannot thus be a mental event. (Analogously we required above t to be A's trying to exemplify X *by his bodily behavior.*) But b must also be a maximal bodily event, while not being 'more' than a bodily event.

We require b to be a maximal bodily event in order to exclude from basic actions such examples as exerting oneself to raise one's arm and flexing one's arm muscles, which both cause the bodily event of the arm going up and which have been regarded as fully intentional actions (cf. McCann, 1972). If we accept them as basic actions then either we were wrong in assuming arm raisings to be basic actions or then we were mistaken in accepting that basic actions could not be causally generated by other actions. The assumption that b is a (spatiotemporally and compositionally) *maximal* event excludes these cases. For I take maximality to entail that for no other event $e \neq t$ in u is it the case that $C(e, b)$, and that suffices for our present purposes.

We also required b to be a simple (as opposed to a composite) event. Thus the bodily event involved in my shrugging my shoulder *and* bending my left little finger would be a composite (non-simple) event in this sense, whereas both the shrugging of the shoulder and the twisting of the left little finger would separately qualify as simple events and as b's. I admit that the present simple-composite distinction may be problematic, but for the purposes of this paper it is assumed to be clear enough.

Another term to be assumed sufficiently clear is 'bodily'. Thus we presuppose a distinction between the body of an agent and his (physical) environment, and that is hardly a seriously problematic distinction in our present discussion.

Basic action tokens u do not have any external result component r. Thus my opening the door involves the result event of the door becoming open, though opening the door (let's say by pushing it) might well belong to my standard action repertoire (cf. our later definition of this). Thus opening the door cannot be my basic action type according to the above definition. Still there is a closely related candidate for a basic action: my bodily action involved in (or required for) my opening the door. In other words, we may always pick as our b the bodily movements involved in the opening of the door, and surely that is possible. Thus, by means of the above kind of *functional* (indirect) characterization we can get hold of many bodily movements in our standard repertoire for which we do not have better names or descriptions.

One thing still to be considered is in what sense 'the bodily action involved in my opening the door' really is a description of an *action*. First we notice that the bodily movement in question may seem (in some sense) to differ generically from case to case. But even if that were the case and there were no intrinsic (non-functional) descriptions of this bodily action, in each singular case the bodily action 'is there' – you may, for instance, point at it. So it presents no great problem for us.

The bodily action $u^* = \langle t, b \rangle$, involved in the singular action $u = \langle t, b, r \rangle$, where $r =$ the door opening and which represents my opening the door, is a proper part of the latter action (recall our characterization of t). They differ only with respect to the presence of the result r. But r is in a sense the contribution of nature rather than by me. Thus, though u^* is obtained by means of a conceptual abstraction from u there is this much of 'real' difference. On the other hand, u and u^* are the same as far as my own direct contribution is concerned. Whatever self-knowledge I have of t and b in the case of u I also have in the case of u^*. Similarly, if t purposively causes b in the case of u it does that in the case of u^* as well (as is also seen from my definition in Section II). That b is the result of u^* should be clear from the very description of u^*. Thus, we conclude that if u satisfies clause (1) of our definition of a basic action type, u^* does it (cf. our later def. of action token$_1$).

We have not yet commented on the causal conditional statements occurring in clause (1) of our definition. What they say is that the trying event t is both a sufficient and a necessary purposive causal condition of b. In general, causes are not necessary in this sense (cf. Section II), but here we need that: otherwise an action token or satisfying clause (1) might not be intentional (which we exclude). In other words, when the behavior process is carved up as in our analysis these conditions are what the essence of our causal analysis of basic action comes to. (Note: instead of using the phrase 'in that situation' we could have explicitly spoken of a Tarskian structure $\mathcal{M}$ of an appropriate type as in our definition of purposive causation.) In the case u actually occurs (i.e. $(Ex)(x = u)$) we can just require the truth of '$C^*(t, b)$' instead of the causal conditionals. Thus we see that an essential element of (purposive) causation is presupposed by the notion of action. As we recall from Section II, our notions of purposive causation $C^*(-, -)$ and $\triangleright \rightarrow$ can be defined independently of action type concepts; thus no direct circularity lurks here.

In this connection a central aspect of our analysis has to be noted.

We have characterized basic actions primarily in terms of antecedent factors such as A's having a conduct plan and activating it in his trying. But one can also characterize basic actions in terms of their consequences. A most illustrative example is provided by the bodily action involved in opening the door. This bodily action (almost by definition) *causes* the opening of the door. Put more generally, an agent intervenes with the world by means of his bodily behavior thereby producing certain states of affairs and preventing other states from obtaining.

Till now we have been discussing only clause (1) of our definition of basic action type. In order to get to clause (2) we briefly define our notions of generation. We start by giving a backing law analysis of the notion of *factual* generation '$R_1(e_1, e_2)$', which reads 'singular event e_1 factually generates singular event e_2' (cf. Section II):

DEFINITION III.2. '$R_1(e_1, e_2)$' is true (in a given situation) if and only if (in that situation) there are descriptions D_1 of e_1 and D_2 of e_2 and a suitable factual theory T such that $\varepsilon(D_2, D_1 \& T)$.

What a *suitable* theory is we cannot here discuss – a *causal* theory or conjunction of causal theories is a paradigm example (cf. Tuomela, 1974a). It is required that T (model theoretically) entails that the relation R_1 is irreflexive and asymmetric. In the case the backing theory T is true and $(Ex)(x = e_1)$ and $(Ey)(y = e_2)$ we speak of *actual* R_1-generation, otherwise only of possible R_1-generation.

Completely analogously with our definition of indirect causation (IC) in Section II we can define a notion of *indirect generation* (IG) by using 'R_1' instead of 'C' in the definiens. A notion of *indirect purposive generation* (IG^*) is obtained from our definition of indirect purposive causation by using the relation 'IG' instead of 'IC' in its clause (a).

Let us now go to conceptual generation. Our concept of conceptual generation R_2 is defined to hold between an event u and its two different but conceptually (semantically) related descriptions D_1 and D_2: '$R_2(u, D_1, D_2)$' reads: 'u under D_1' conceptually generates 'u under D_2'. (If D_1 describes u as an X and D_2 as an Y we might also speak, in terms of intensional structured events, of u *as an* X'ing conceptually generating u *as a* Y'ing.) The bare bones of the relation R_2 then are as follows:

DEFINITION III.3. '$R_2(u, D_1, D_2)$' is true (in a given situation) if and only if (in that situation) there is a suitable meaning postulate T such that the formal conditions for $\varepsilon(D_2, D_1 \& T)$ are satisfied.

Here the 'meaning postulates' are supposed to cover a wide variety of semantic postulates and rules, norms, etc. (see Tuomela (1974a) and especially the good classification in Goldman (1970)). I cannot here discuss either what *suitable* meaning postulates are. $R_2(u, D_1, D_2)$ is assumed to be irreflexive and asymmetric with respect to the second and third arguments (due to T).

In addition, clause (2) speaks of the semantic principles *internalized* by A. This means, roughly: the principles have to be implicitly contained as a competence-component (cf. Chomsky) in A's operative conduct plan.

By and large, our generative relations R_1 and R_2 are tailored to capture the 'by'-relation of ordinary language. Thus e.g. A brings about the (intensional) result of an action e_2 by bringing about the result of an action e_1.

We can define a conditional $\bigcirc \underset{2}{\rightarrow}$ as follows: '$D_1 \bigcirc \underset{2}{\rightarrow} D_2$' is true if and only if '$R_2(u, D_1, D_2)$' is true. ($\bigcirc \underset{1}{\rightarrow}$ is of course analogously definable.) The conditional $\odot \underset{2}{\rightarrow}$ corresponding to purposive conceptual generation is defined on the basis of $\bigcirc \underset{2}{\rightarrow}$ completely analogously with our definition of $\triangleright \rightarrow$ on the basis of $\triangleright \rightarrow$ (cf. Section II).

We can now see that, for instance, signalling for a left turn (with one's left arm) does not become a basic action type because of clause (2) of our definition: each of A's signallings for left turn is (in this kind of circumstances) conceptually generated by A's left arm raising (according to A's conduct plan and some semantic rule such as: left arm raising in situation C means signalling for left turn).

What now will be the basic action repertoire of any given agent A according to our definition is a problem which cannot be solved a priori. Our definition involves references to the agent's operative conduct plan and to some causal processes, and all the relevant facts can ultimately be found out only by means of scientific research.

However, we can make 'educated guesses' about what a normal agent's basic action types are. I think the reader will agree with me that the following action types will usually be included: raising one's arm, moving one's finger, blinking one's eye, the bodily action involved in tying one's shoe laces, the bodily action involved in signing one's initials. On the other hand, among actions excluded from this list would probably be: the mental act of trying to raise one's arm, flexing one's muscles for raising one's arm, flipping a switch, tying one's shoe laces, signalling for a left turn, signing one's initials. (Note: I have omitted the discussion of actions of neglect, e.g., omissions, in the present context.)

Our definition of basic action type is relative to agent and to time. That this should be the case is clear from what we have said: the movements required for tying one's shoe laces do not belong to the standard repertoire of a two year old child, even if they do belong to mine presently and to that child's repertoire when he is five.

Till now we have been discussing only simple (or non-compound) properties (or types). But there are compound action properties like raising one's left arm *and* lowering one's right arm, or, to mention more interesting examples, performing a dance, or completing a Ph.D. degree in philosophy. The last example can be analyzed as involving passing such and such examinations, writing a dissertation, etc.

Generally speaking, we accept as action properties only simple ones (viz. X, not-X) and conjunctions of simple ones. Thus $X = X_1 \& X_2 \& \cdots \& X_m$ stands for an action property in general. Here each X_i is a positive or negated non-compound property. There will be no disjunctive, conditional or any other truth-functional combinations qualifying as action properties. (Note: in our *language* we may of course use disjunctive, conditional, etc. action-describing statements.)

On our way to a characterization of an arbitrary action property we start with compound basic action types. They are essentially conjunctions of simple basic action types such that the agent has the power to perform them jointly.

We again present the definition first and comment on it afterwards (cf. a somewhat related analysis in Goldman, 1970, p. 202):

DEFINITION III.4.   Property X is a *compound basic action type* for A at **T** if and only if

(1) X is the conjunction of some properties $X_1, X_2, ..., X_m$; and

(2)   it is physically possible for A to be in standard conditions at a single time with respect to each of $X_1, X_2, ..., X_m$; and

(3) for any exemplification u of X it is true that if u had the internal structure $\langle t, ..., b \rangle$ and if A were in standard conditions with respect to X_1, X_2, ..., X_m at $\mathbf{T}$ and if he knew that, then the following causal conditionals would be true in this situation for $i = 1, 2, ..., m$:

(a) $D_i(t_i) \triangleright\!\!\rightarrow D_i'(b_i)$

(b) $\sim D_i(t_i) \triangleright\!\!\rightarrow \sim D_i'(b_i)$

(c) $D''(t) \triangleright\!\!\rightarrow D_i(t_i)$.

Here t is described by D'' as A's trying (by his bodily behavior) to do X (and possibly as the relevant believings); each t_i, $i = 1, 2, ..., m$, is described by D_i as A's trying (by his bodily behavior) to do X; and D_i' expresses that b_i is a maximal simple overt bodily component-event of the respective event $u_i = \langle t_i, ..., b_i \rangle$ tokening X_i in this situation; b is the event composed of the events b_1, b_2, ..., b_m. In addition, b is assumed to be the result event of u (as an X).

Conditions (1) and (2) in this definition do not require special comments except that we allow $\mathbf{T}$ to be a set of time points and not only a single instance of time; hence 'at a single time' in (2) will have to be understood accordingly. If $\mathbf{T}$ is a set of points of time then the basic actions X_i in X can be performed e.g. in *sequence* rather than parallel in time.

Clause (3) is formulated to some extent in analogy with clause (1) of our definition of a simple basic action type. Thus (a) and (b) simply follow from the fact that each basic action type X_i, $i = 1, 2, ..., m$, is exemplified in the situation concerned.

Condition (c) of clause (3) in effect contains the requirement that t is a *purposive cause* of each t_i. It can be assumed here that each D_i is obtained from D'' by means of a suitable (presumably) causal intention transferral process. Thus we may assume that '$C^*(t, t_i)$' is true for each $i = 1, 2, ..., m$, and as a consequence of this (c) becomes true.

We do not presently know much about the ontological relationship between t and the t_i's. Could it be that $t = t_1 \circ t_2 \circ \cdots \circ t_m$ (summation '$\circ$' taken in the sense of the calculus of individuals)? In view of how these events are described in A's conduct plan K I do not think we are entitled to assume that a priori. But I would think that it is correct to require that t contains each t_i as its proper part.[4]

How about the bodily components b_i then? Here I think we can safely assume that $b = b_1 \circ b_2 \circ \cdots \circ b_m$: the total bodily behavior of A involved in his bringing about X is just the composition or sum of the bodily behaviors b_i, $i = 1, 2, ..., m$, the agent performs in the situation in question. Furthermore b will be the result of $u = \langle t, b \rangle$. Thus, even if results of compound actions are not in general summative, I think in the case of compound basic actions they are.

We notice that our definitions of a compound basic action type as well as of a simple basic action type are *epistemic* roughly in the sense that A is required to have adequate knowledge of what he is doing. The difference between the epistemic notions and the corresponding non-epistemic ones would be primarily the use of purposive versus mere causation in the definitions (cf. Goldman, 1970, for a discussion of a non-epistemic notion).

IV. ACTION TOKENS AND BODILY ACTIONS

Although we have discussed basic actions at length, the most interesting types of action are, however, (compound) actions with external *results* – at least if we think of man as a changer and moulder of the (external) world. The main aim of this paper is to outline and sketch in some detail an account of the structure of everything we do (doing understood in a task-performance sense). On our way to that we have to discuss in some more detail the notion of action token, on which our characterizations of various action types rely.

As we have emphasized we think of actions as achievements with suitable causal mental antecedents. They are a kind of 'responses' to *task* specifications. Actions also essentially involve a *public* element – a result, by which the correctness of the achievement (action performed) can be judged. We have construed actions as sequences of singular events of the general type $\langle v, ..., b, ..., r \rangle$. On the basis of what we have just said it should be obvious that the events b and r are intrinsically involved in an action. But according to the causal theory of action I have developed elsewhere (Tuomela, 1975) it is also essential to logically or intrinsically include a suitable mental cause in the action. For instance, the problem of which events count as results of actions does not seem to me to be solvable without recourse to a suitable causal mental antecedent (cf. Tuomela, 1975). Furthermore, if mental entities are semantically introduced according to conceptual functionalism then there will be conceptual connections between mental concepts and action concepts. This will ensure the tightness between mental entities and behavior we seem to need while logically permitting the existence of mental states, events, and episodes as ontologically genuine entities (which is needed for them to be proper causes of behavior).

As will be seen from our characterization of complex actions below, it is essential to any action type that it is exemplifiable at will, i.e. it must have exemplifications which are (a) action tokens and which are (b) intentional. In saying this I of course presuppose that action tokens can be distinguished from mere exemplifications. For instance, the property eye-blinking can be exemplified both as a reflex and as an action. Action tokens thus form a strongly restricted subclass of all the exemplifications of a given property. A pivotal role is assigned to action tokens which are intentional exemplifications of the action property. Our characterization of intentionality always intrinsically involves the activation of the agent's conduct plan by means of a trying event, which ultimately, through conscious or unconscious intention transferral, is about the action type the bodily event b cum the result r of the action token exemplify.

Let us now consider an example concerning an action token of a simple complex action. Assume that an agent A wants to get a cup of creamed coffee out of a slot machine. In order to get his coffee he must insert a coin in the slot, push a certain button for coffee brewing, and push another button for having the coffee creamed. Assume that A performed this complex action intentionally and got his coffee. We may now analyze this action token as consisting of the following complex event $u = \langle t, ..., b_1, b_2, b_3, ..., r_1, r_2, r_3, r \rangle$, where

$t = A$'s trying to do by his bodily behavior what is required for getting a cup of creamed coffee.

$b_1 = A$'s bodily movement involved in his inserting the coin in the slot.

$b_2 = A$'s bodily movement involved in his pushing the button for coffee brewing.

$b_3 = A$'s bodily movement involved in his pushing the button for cream.

$r_1 =$ the result-event of the coin becoming inserted in the slot.

$r_2 =$ the result-event of the button for coffee brewing becoming pushed.

$r_3 =$ the result-event of the button for cream becoming pushed.

$r =$ a cup of creamed coffee coming out of the machine.

Here the trying event may be assumed to contain as its proper parts the relevant subintentions t_1 (describable as A's trying to insert the coin), t_2 (describable as A's trying to push the button for coffee brewing), and t_3 (describable as A's trying to push the button for cream). The full action token u can in fact be said to contain the subactions $u_i = \langle t_i, ..., b_i, ..., r_i \rangle$, $i = 1, 2, 3$, although we do not explicitly write out the t_i's in u. Similarly corresponding to each u_i there is a bodily action $u_i^* = \langle t_i, ..., b_i \rangle$ which lacks the agent-external result r_i of u_i.

What is interesting and important about our example is the occurrence of r in it. This full result-event of u is a kind of causal 'interaction' effect of the r_i's. We can say here that the composite events $r_1 \circ r_2 \circ r_3$ caused r. Accordingly, A's actions of inserting the coin and his pushing the two buttons causally generated his action of bringing it about that a cup of creamed coffee came out of the machine.

Notice that, in general, in an action token of a complex action there may be bodily events which belong to basic actions and thus are results at the same time.

We now go on to propose the following definition for an arbitrary action token u. (We consider u as a series of events but (normally) short enough to be itself called an event.)

DEFINITION IV.1. Singular event

$$u = \langle v, ..., b_1, ..., b_m, ..., r_1, ..., r_k, r \rangle$$

is an *action token*$_1$ of A if and only if

(1) v is a singular mental event (or state or episode) activating a propositional attitude of A; v is a direct or indirect cause (i.e. causally contributing event) of the events $b_1, ..., b_m$ and an indirect cause or an indirect factual generator (via the events $b_1, ..., b_k, ..., b_m$) of the events $r_1, ..., r_k, r$;

(2)   each b_i, $i = 1, 2, ..., m$, is a bodily event which does not cause any other bodily events occurring in u;

(3)   each r_j, $j = 1, 2, ..., k$, is a public result-event and, especially, r is the result-event of the total event u.

This definition should almost be understandable on the basis of what we have said earlier in this section, but perhaps some additional remarks are needed.

First, we shall later define action tokens in another way, whence the subindex 1. Consider now clause (1). It is central that the propositional attitude (e.g. want, belief) which v activates be a realistically construed disposition (cf. Section I). Only then can the event v be describable as a trying, wanting, believing, etc. as the case may be. We note that as propositional attitudes in action contexts usually are about actions, there is in our notion of action token some conceptual dependence on action types. However, we do not rely on any *systematic* notion of action type but only on an intuitive idea of an action type.

Notice furthermore that if propositional attitudes and other mental entities are introduced according to conceptual functionalism there will be a still stronger dependence on the (presystematic) notion of action type. In fact, for many interpretations of v the claim that v causes a b_i will be circular.

Causation has to be understood as excluding cases of overdetermination (e.g. by states of shock or the like). In those cases when u consists of subactions $u_i = \langle v_i, ..., b_i, ..., r_i \rangle$, v's causing the b_i's must go somehow via the v_i's (which are causes of their b_i's). In these cases at least '$IC(v, v_i)$' must be true (usually).

Clause (2) simply requires the bodily events b_i to be (causally) maximal within u. All the overt bodily behavior involved in u can simply be regarded as the composite event $b = b_1 \circ \cdots \circ b_m$.

Concerning our notion of result in the above definition, it should also be taken in a presystematic sense (which e.g. does not clearly say whether r_j is extensional or intensional). Except for cases of bodily actions, the results occur 'outside' A. Whereas the r_j's, $j = 1, 2, ..., k$, are simple events (as are the b_i's. $i = 1, 2, ..., m$), r may be a composite event.

If we do not have to display the full inner structure of u we will below sometimes write $\langle v, ..., b, ..., r \rangle$ for u; this shows that u is of the same general type as action tokens of elementary actions.

In principle our definition of an action token$_1$ covers such negative actions as omissions and acts of negligence in which the overt events are missing. But here we shall not discuss them in any detail.

Let us still emphasize that we are in our definition trying to give some precision to a vague notion of an action as a singular behavioral achievement event. We need the present explication in order not to make our definitions of action types directly circular. Having this much of a definition of an action token is of course much better than an ostensive definition or an enumerative list, etc. We shall later characterize action tokens somewhat more precisely in terms of action types.

The real test for our definition of action token$_1$ is of course what it covers and what it does not. First we notice that it obviously covers intentional basic action tokens and actions factually and conceptually generated by them (see our earlier relevant definitions). The claim holds true independently of whether the basic actions concerned are simple or compound.

Secondly, examples like that about the slot-machine fit our characterization. In fact I claim that any typical token of a complex action (yet to be defined exactly) fits our definition. It is of course impossible here to try and go through even representative examples. Let me, however, mention two cases here. As even our slot-machine example shows, *parts* of action tokens can be action tokens$_1$: each u_i, $i = 1, 2, 3$ is an action token$_1$ in that example. Next, *sequences* of action tokens$_1$ can be action tokens$_1$ (cf. the hammering example to be discussed later, in Section V).

As a third large group, our notion of an action token$_1$ will cover lots of cases of non-intentional actions. Thus standard examples of mistakes (such as misspelling one's name), which are intentional under some other descriptions, are included. But according to our characterization, contra Davidson, not every action token is intentional under some description. For instance, a person may accidentally break a vase when doing something else in his room. His breaking the vase may satisfy our definition, for we recall that in clause (1) v (be it a wanting to flip the switch or whatever) need not of course purposively cause the breaking of the vase. Another type of case of an action which is not intentional under any description would be my action of scratching my right cheek while writing this sentence. It was indirectly and non-purposively caused by my intention (of writing) in this situation, we may assume. Still another example would be a deprived alcoholic's emptying the glass of whisky brought in front of him. He perhaps did nothing intentionally at the moment. His largely uncontrolled desire caused his drinking the whisky, which thus is seen to be an action token$_1$.

A somewhat special group of behaviors, most (or many) of whose exemplifications are action tokens$_1$, are thinkings-out-loud and (typical) pattern governed behaviors in the sense characterized by Sellars (1973).

We may also consider what our definition excludes. First, e.g. stumbling or falling down due to external factors would be excluded – these do not satisfy clause (1). Similarly clear cases of reflexes like the patellar reflex fail to satisfy clause (1) and, I think, (3), too.

There may be borderline cases which our broad definition is not sharp enough to handle in its present form, but we shall not here try to qualify our definition further.

Given our definition of an action token$_1$ we can characterize the notion of a *bodily action* more exactly than before, given, however, an 'understanding' of the bodily events b_i. We recall that a basic action type of A is something which is in (1) the *repertoire* of A; and its tokens are (in standard circumstances) (2) *bodily actions*, (3) *intentional*, and (4) *non-generated*. Now we want to get hold of a concept of bodily action which does not at least always satisfy (1), (3), and (4) though it trivially must satisfy (2). In fact, we want to respect (4) in that only causally maximal bodily events b_i should be included. But no requirements concerning conceptual generation are needed. This latter remark is true because we are now looking for a non-epistemic notion; hence we can speak of mere causation instead of purposive causation (or generation).

But now obviously our notion of action token$_1$ will be helpful in characterizing a bodily action type: we just require each result r_j to coincide with one of the b_i's, in the general case. To start from simple actions we first get for a simple X (using our earlier symbols):

DEFINITION IV.2. X is a simple *bodily action type* for A at **T** if and only if every exemplification u of X which is an action token$_1$ of A is of the form $u = \langle v, ..., b \rangle$ such that b is the result of u.

Note that a bodily action type will include many tokens outside A's standard repertoire (e.g. various maximal or accidentally successful performances such as some complicated gymnastic movements). Thus (simple) basic action tokens form a proper subclass of the class of bodily action tokens as defined. There is one important qualification to our definition which has to be read into it: it must be conceptually possible that X be exemplifiable at will. Thus, even if A may never de facto intentionally exemplify X we understand that, at least for some agent, v can occasionally be an effective intending.

Somewhat analogously to our notion of a compound basic action type we now have:

DEFINITION IV.3. Property X is a *compound bodily action type* for A at **T** if and only if

(1) X is the conjunction of some simple bodily action types X_1, $X_2, ..., X_m$ ($m \geqslant 1$); and

(2) It is conceptually possible that each of X_1, $X_2, ..., X_m$ is tokened$_1$ at a single time.

We note that (2) is weaker than the corresponding clause for compound basic action types. The present clause does not require it to be satisfiable for A, but it only requires exemplifiability in principle (in some 'possible world').

An action token for a compound bodily action type will look like $u = \langle v, ..., b_1, b_2, ..., b_m \rangle$. If there were a $b \neq b_1 \circ b_2 \circ \cdots \circ b_m$ (corresponding to the 'full' result r) it would have to be caused by some of the b_i's, it seems. But that would contradict the causal maximality requirement for the b_i's. Therefore such a b will not occur in u.

V. COMPLEX ACTIONS

Now we are ready to proceed to a characterization of complex actions. We are going to define the notion of a compound action type. By means of this notion and the notion of (simple or compound) bodily action type we seem to be able to exhaust all actions. That is, the result will be an exhaustive characterization of the structure of everything we do (or what is in principle capable of being done) in our task-action sense of doing.

Keeping in mind the slot machine example and our earlier discussion of complex actions we can directly proceed to the characterization of an arbitrary compound action type, which contains a non-compound or simple action type as its special case. We consider a conjunctive property X and we assume, to simplify our exposition that all the conjuncts X_i, $i = 1, ..., m$, are simple, i.e. X has been analyzed as far as it goes.

DEFINITION V.1.   Property X is a *compound action type* for A at **T** if and only if

(1) X is the conjunction of some simple properties $X_1, ..., X_m$ ($m \geqslant 1$) at least one of which is not a bodily action type for A at **T**;

(2)   it is conceptually possible that each X_i, $i = 1, ..., m$, has exemplifications which are action tokens$_1$.

(3)   (a) Each action token$_1$ $u = \langle v, ..., b, ..., r \rangle$ of X is (directly or indirectly) generated by some tokens of the action types X_i, $i = 1, ..., m$, occurring as conjuncts in X; and

(b)   for each conjunct X_i such that X_i is not itself a simple bodily action type nor does X_i occur as a conjunct in any compound bodily action type consisting of conjuncts of X, it is true that every action token$_1$ of X_i is generated by simple or compound bodily action tokens$_1$ of A.

In (3)(a) v is assumed to be describable as an initiating event as required in clause (1) of the definition of action token$_1$; $b = b_1 \circ b_2 \circ \cdots \circ b_m$; and r is the (possibly composite) full result-event of X.

Clause (1) of this definition guarantees that X is not a compound bodily action type. It is allowed that $m = 1$, in which case X is a non-basic and 'non-bodily' simple action type.

Clause (2) is intended to guarantee that each conjunct X_i in X is an action type: it must be in principle exemplifiable by an action token$_1$. (According to our earlier reading in Section IV, X is then going to have some intentional action tokens$_1$.)

The central clause in our definition is clause (3), which tries to clarify in exactly which sense everything we do is, somehow or other, generated by our bodily actions, if not by basic actions. We shall below go through (3) in terms of an example, which should be sufficiently complex to be immediately generalizable.

But let us first note some general aspects of the situation. Clause (3) claims that for each token $u = \langle v, ..., b, ..., r \rangle$ two generational claims are true. To see their content better let us write out the full internal structure of this token u as

$$\langle v, v_1, ..., v_m, ..., b_1, ..., b_m, ..., r_1, ..., r_k, r \rangle,$$

where now even the v_i's (presumably parts of v) are included. Let me emphasize strongly that at present we know at best that it is conceptually possible that such structures exist as ontologically real. The factual existence of the b_i's and r_i's is hardly in doubt here, though. The question is rather about the v_i's. To be sure, we can in principle conceptually introduce as many v_i's as our linguistic resources allow. But the interesting question here is whether v in some sense (e.g. the part-whole sense of the calculus of individuals) ontologically contains the v_i's (as we have in our example cases claimed and assumed). On a priori grounds we cannot say much, it seems. The problem is ultimately a scientific and a factual one. (It also depends on how future neurophysiology is going to carve up its subject matter and build its concepts.) Similarly, the question whether u can always be said to contain the subtokens $u_1 = \langle v_1, ..., b_1, ..., r_1 \rangle$ (cf. X_1), $u_2 = \langle v_2, ..., b_2, ..., r_2 \rangle$ (cf. X_2), etc. in an ontologically genuine sense over and above mere conceptual abstracta is primarily a scientific and an a posteriori problem.

We have to clarify the following two generational claims in the case of clause (3):

(1)　　　the u_i's ($i = 1, ..., m$) jointly generate u; and

(2)　　　each u_i ($i = 1, ..., m$) is generated by some of A's (simple or compound) bodily action tokens.

We shall illustrate these generational problems by means of an example (also cf. our slot-machine example). It is presented by the following diagram representing an action token u for a complex action property $X = X_1 \& X_2 \& \cdots \& X_5$:

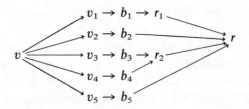

In this diagram the arrows stand for (direct or indirect) factual generation. Here we can thus use as an explicate our *IG*, which also covers direct generation. The event v activates one or more propositional attitudes of A. For example, v might be describable as a wanting cum believing. In any case, v will somehow be about or directed towards the action X with the result r. The event v is assumed to generate each v_i, and each v_i in turn causally produces b_i. The behavioral events b_i in turn generate the results r_1, r_2, and r as shown in the diagram.

The conceptually possible subtokens of $X_1, X_2, ..., X_5$ here are:

$$u_1 = \langle v_1, ..., b_1, ..., r_1 \rangle, \qquad u_2 = \langle v_2, ..., b_2 \rangle,$$
$$u_3 = \langle v_3, ..., b_3, ..., r_2 \rangle, \qquad u_4 = \langle v_4, ..., b_4, ..., r_2 \rangle,$$
$$u_5 = \langle v_5, ..., b_5 \rangle.$$

(Here especially the genuine existence of u_3 and u_4 is problematic: perhaps we only should have $u_{3,4} = \langle v_{3,4}, ..., b_3, b_4, ..., r_2 \rangle$.) In order to solve our first generation problem we note three things. First, we can take $b = b_1 \circ b_2 \circ \cdots \circ b_5$. Secondly, our definitions for causation and generation in principle apply to any events, hence for any *composite* events, too. Thirdly, we are here concerned with non-epistemic notions of generation only. Thus we do not here need our notion of purposive generation but only that of mere generation (*IG*). Thus, when discussing the generation of results of behavior all the v's can be ignored for our present purposes. (They are relevant only in discussing purposive generation, intentional and voluntary actions, etc.)

Accordingly we can accept the following principle of action token generation for two arbitrary action tokens$_1$ $u_1 = \langle v_1, ..., b_1, ..., r_1 \rangle$ and $u_2 = \langle v_2, ..., b_2, ..., r_2 \rangle$:

(P)   If $b_1 = b_2$, then $IG(u_1, u_2)$ if and only if $IG(r_1, r_2)$.

Principle (P) is an obvious meaning postulate for token generation in the cases of simple (elementary) or non-composite tokens. But here we want to apply it to compositions of action tokens as well. Let us use the symbol '$\circ$' for the composition (or sum) of action tokens, too, as they are events, after all. Then, for the cases we have been discussing (where we have an action token u of a compound action such that u_1, u_2, ..., u_m are its subtokens) we accept:

(P*)   If $b = b_1 \circ b_2 \circ \cdots \circ b_m$ then $IG(u_1 \circ u_2 \circ \cdots \circ u_m, u)$ if and only if $IG(r_1 \circ r_2 \circ \cdots \circ r_m, r)$.

(Here we have m results r_j, some of which may be b_i's.)

Armed with (P*) we can now take a look at our example. As we see from the diagram, r_1, b_2, r_2, and b_5 generate r. As $b = b_1 \circ b_2 \circ \cdots \circ b_5$ we conclude on the basis of (P*) that '$IG(u_1 \circ u_2 \circ \cdots \circ u_5, u)$' is true. This conclusion presents the answer to our first generation problem.

Let us now discuss clause (3)(b) and the second generation problem. This clause is to be regarded as a constraint on X: only those complex actions are to be called compound actions which satisfy the following 'reducibility' condition. Each token $u_j = \langle v_j, ..., b_j, ..., r_j \rangle$ of a conjunct X_i of X such that X_i is not a bodily action type or part of a compound bodily action type must be generated by (possibly composite) bodily action tokens$_1$ of A.

In our present example we have to reduce u_1, u_3, and u_4 in this way to bodily actions. We can safely assume that u_2 and u_5 are tokens of bodily actions. X_2 and X_5 are hence taken care of.

In the case of u_1 the generation is trivial: b_1 generates r_1, and we have $IG(b_1, r_1)$. Now the token $u_1^* = \langle t_1, ..., b_1 \rangle$ clearly qualifies as a bodily action as it is a proper part of the action token $u_1 = \langle v_1, ..., b_1, ..., r_1 \rangle$. The bodily action type X_1^* consisting of the bodily action tokens$_1$ involved in X_1:ing will be the bodily action type u_1^* tokens here. Although we are not requested to specify that bodily action type, we always seem to be able to do it. We can now apply our principle (P) trivially to get the result that u_1^* generates u_1.

In the case of u_3 and u_4 the generational situation is a little different. For, to make the situation interesting, we assume that neither b_3 nor b_4 alone generates r_2, although jointly they do generate it. The properties X_3 and X_4 clearly must be treated together as they have this kind of 'interaction' effects. Thus, instead of considering u_3 and u_4 separately we consider $u_{3,4} = \langle v_{3,4}, ..., b_{3,4}, ..., r_2 \rangle$. Now look at the bodily actions $u_3^* = \langle v_3, ..., b_3 \rangle$ and $u_4^* = \langle v_4, ..., b_4 \rangle$ involved in u_3 and u_4. As $b_{3,4} = b_3 \circ b_4$ then principle (P*) gives us $IG(u_3^* \circ u_4^*, u_3)$, as required. (We could as well have said here that '$IG(u_{3,4}^*, u_{3,4})$' is true, where $u_{3,4}^* = \langle v_{3,4}, ..., b_{3,4} \rangle$.) We have gone through our example.

Does our definition of compound actions really cover all complex actions and is our partial definition of action token generation sufficient? 'Completeness theorems' are very hard to 'prove' in cases like this. To see some of the problems involved we consider an example where the agent levels a nail by striking it with a hammer. He holds the hammer in his right hand and keeps the nail straight with his left hand. Each time the nail is struck it goes down a few millimeters. After n strikes the nail has been levelled. We can analyze the situation in the following terms.

Let X_1 be the non-bodily action property of striking the nail (with a hammer), and let X_2 be the non-bodily action property of holding the nail straight. Each action token of X_1 involves a bodily component b_1 (striking-movement) plus a result component r_1 of the nail becoming struck. Each token of X_2 involves a bodily component plus a result r_2 of the nail staying straight. Now r_1 and r_2 cause r_3 – the nail's going down a few millimeters. So we call $X = X_1 \,\&\, X_2$ the action type of hammering. Tokens of X have the general form $u = \langle v, ..., b_1, b_2, ..., r_1, r_2, r \rangle$.

The action Y of levelling the nail by hammering consists of n successive tokens of X. Let us write this out with obvious symbolism as $Y = X^1 \,\&\, X^2 \,\&\, \cdots \,\&\, X^n$. (Notice that in other examples the conjuncts of Y could of course be non-identical properties.)

We can easily see that the tokens of Y are of almost the same general type as those of X. Their general structure is in full

$$u' = \langle t, \, ..., \, b_1^1, \, b_2^1, \, ..., \, b_1^n, \, b_2^n, \, ..., \, r_1^1, \, r_2^1, \, r_3^1, \, ..., \, r_1^n, \, r_2^n, \, r_3^n, \, r \rangle.$$

(The notation should be self-explanatory; r is the result of u under Y.) Nothing in our definition of action token$_1$ precludes such causal relationships between results as $C(r_1^i \circ r_2^i, r_3^i)$, of course.

But we seem to be able to say even more. If we substitute for $X_1 \& X_2$ in the definition of Y we can presumably claim that the following identities hold:

$$\begin{aligned}
Y &= X^1 \, \& \, \cdots \, \& \, X^n \\
&= (X_1 \, \& \, X_2)^1 \, \& \, \cdots \, \& \, (X_1 \, \& \, X_2)^n \\
&= X_1^1 \, \& \, X_2^1 \, \& \, \cdots \, \& \, X_1^n \, \& \, X_2^n.
\end{aligned}$$

But according to the last identity Y is a compound action type. It obviously satisfies clauses (1) and (2) of our definition. It clearly satisfies (3)(a), for r will be indirectly (via the r_3^i's) generated by the r_1^i's and r_2^i's (or $r_1^i \circ r_2^i$'s) as is easily seen; and this essentially suffices. It is as easily seen, on the basis of our description of the example, that the r_1^i's and r_2^i's are generated by the bodily actions as required by (3)(b) (cf. our treatment of the previous example).

One interesting feature about our present example is that the performance of each X^{i+1} ($i = 1, \, ..., \, n - 1$) has as its *causal precondition* the exemplification of X^i. This kind of relative dependence feature is very prominent in many sequential complex actions such as performing a dance, writing a book or in various social ceremonies and rituals. (Note that the time factor **T** will in these cases usually be a set of time points rather than a single instance.)

Our example concerning a complex sequential action proved to be analyzable as a compound action, and my general conjecture, supported by a variety of examples, is that no more complex types of actions are needed (such as 'conjunctive compound action types'). We can thus go on and define:

DEFINITION V.2. X is a *complex action type* (an arbitrary *action type*) for A if and only if X is either 1) a simple or compound bodily action type or 2) a compound action type.

On the basis of what has been argued we can now say that *all actions are complex actions* independently of their other special characteristics (such as their being 'parallel', 'hierarchical', 'sequential', etc.). My thesis (or conjecture) and its earlier clarifications also tells us how it indeed can be so that "we never do more than move our bodies: the rest is up to nature" (Davidson).

There is one counterclaim to my conjecture which has to be answered immediately. I have defined my notions of causation and generation only for *deterministic* cases. If indeterminism is true, we must allow for *probabilistic* causation and generation as well. This small addition can indeed be made, but I shall not here discuss it.

We can now, if our above conjecture is true, characterize action tokens somewhat more informatively. We simply say that:

DEFINITION V.3. u is an *action token$_2$* of A if and only if (1) u is an action token$_1$ and (2) u exemplifies either a (simple or compound) bodily action type or a compound action type.

Notice that there are action tokens$_1$ which are not action tokens$_2$. For instance, perceptual takings (notably takings-out-loud) and inferences (notably inferrings-out-loud) are not action types as they cannot be exemplified at will. Still they have action tokens$_1$ among their exemplifications.

What is conceptually possible for A to do may of course consist of much more than what he has the power to do (intentionally) or what belongs to his 'action repertoire'. This latter notion can be clarified by defining a concept of a compound action in an *epistemic* sense (cf. Section II). This also gives us a definition of A's power to do a complex action X.

The new epistemic concept has the following features to be noticed. First, as we want to get hold of what A can do we must reduce his complex doings to basic actions rather than to mere bodily actions. Secondly, we must preserve intentionality in generation for those situations where A intentionally exercises his causal powers (i.e. brings about intentional basic action tokens).[5] We thus arrive at the following characterization for an arbitrarily complex X:

DEFINITION V.4. Property X belongs to A's *action repertoire* at **T** if and only if

(1) X is the conjunction of some simple action properties $X_1, ..., X_m$ ($m > 1$);

(2)   it is physically possible for A to be in standard conditions at a single time with respect to each of X_1, X_2, ..., X_m; and

(3)(a)   each action token$_1$ $u = \langle v, ..., b, ..., r \rangle$ of X is (directly or indirectly) generated by some tokens of the action types X_i, $i = 1, ..., m$, occurring as conjuncts in X, such that when u is intentional, purposive generation is required; and

(b)   for each conjunct X_i such that X_i is not itself a simple basic action type nor does X_i occur as a conjunct in any compound basic action type consisting of conjuncts of X, it is true that each action token$_1$ of X_i is generated by standard (simple or composite) basic action tokens$_1$ of A. In the case of intentional action tokens$_1$ of X_i purposive generation is required.

In clause (3)(a) r is assumed to be describable as an initiating event as required in clause (1) of the definition of action token$_1$; $b = b_1 \circ b_2 \circ \cdots \circ b_m$; and r is the (possibly composite) result-event of X.

In clause (1) we require the X_i's to be simple action properties in the sense of our definition of a compound action type (for $m = 1$). In clause (3) we speak of intentional action tokens – see note 5 for a remark concerning it.

We can now plausibly think that an agent *can* do an arbitrary X just in case X belongs to A's action repertoire. If we, following 'causalist' thinking, analyze 'can' in terms of A's causal powers we arrive at the following characterization for an arbitrary (simple or purely complex) action property X:

DEFINITION V.5.   *A can do X intentionally at $\mathbf{T}$ if and only if A has the power* to do X intentionally at $\mathbf{T}$ if and only if X belongs to A's action repertoire at $\mathbf{T}$.

Space does not permit discussing powers and cans at any length here. Let us however notice one connection to the literature. Davidson (1973) analyzes cans in terms of causal powers as we do, though in less technical detail. His basic formula is this: A can do x intentionally (under a given description D) means that if A has desires and beliefs that rationalize x (under D), then A does x. (Here x is a variable for action tokens.)

Our above analysis is clearly compatible with Davidson's analysis and can be regarded as a further elaboration of this type of analysis.

Notice especially that in my analysis a rationalizing desire-cum-belief-complex must be an event t (viewable as an effective intending cum believing) described according to the agent's operative conduct plan K.

This concludes the presentation of the basic concepts of our causal theory of complex actions. This theory seems to have interesting applications at least to the analysis of collective and other social actions and to the study of the foundations of decision and game theory. A detailed investigation of these prospects must, however, be left for another occasion.

University of Helsinki

NOTES

[1] Psychologists have done much work in studying the rise and change of cognitive systems. Still they have not paid sufficient attention to the important differences between various types of behavior. They have hardly even tried to distinguish clearly actions from non-actions, not to speak of making a difference between intentional and non-intentional, deliberate and non-deliberate actions, etc. (For instance, the recent elegant theory of motivation by Atkinson and Birch, 1970, is very unsatisfactory in this respect.)

[2] Recent neurophysiological findings can be interpreted to support my view on the nature and the ontological aspects of trying and action. Thus, in an experiment subjects were asked to (intentionally) push a rod in a tube at irregular intervals (see Becker *et al.*, 1973). Simultaneously EEG recordings were made from several positions of the head, including the mid-vertex position C_z, which is central for limb motion. The results show that there exists a very clear readiness potential about 0.5–0.8 seconds before the pushing movement (depending on the rapidity of the movement). This indicates that the action of pushing the rod starts a little before the pushing movement. If the pushing movement were blocked we would be left with this specific readiness potential together with its dramatic reduction right at the cerebral 'beginning' of the movement. It is not yet quite clear whether we have here a strict and specific necessary and sufficient cause of arm movement. In any case the readiness potential and its reduction seems *necessary* for intentional arm movement.

It is now tempting to suggest that we are here dealing with an *intending* to push the rod (readiness potential at C_z) which becomes a trying to push the rod. (Also see Pribram, 1969, where 'intention waves' are discussed.) But still one should not overinterpret these experimental results. They clearly indicate the presence of certain activity related rather specifically to rod pushing, but we are of course very far from having anything like a specific neurophysiological account of the nature of intending and trying.

[3] For an exact characterization of the notion of a conduct plan see Tuomela (1975). Intuitively speaking, a conduct plan is the agent's 'plan' (widely understood) for action in a behavior situation. A conduct plan description is *our* description of the agent's plan involving his relevant aims and cognitions together with both their conceptual and dynamic relationships.

[4] Notice that although we claim that the 'subtryings' t_i corresponding to t do exist in the case of compound *basic* action tokens we do not make the corresponding general claim.

If an agent intends to do something, then, normally through conscious or unconscious practical inference, he will intend and, ultimately, try to do by his bodily behavior whatever is required for the satisfaction of that intention. Still there does not have to be an ontologically separate trying event corresponding to each of the required behaviors in the full action: one may effectively intend to play a melody (and play it) without intending or trying (in an ontologically genuine sense) to play each note.

[5] Space does not permit a discussion of intentionality here. In Tuomela (1974a, 1975) I have characterized intentionality roughly as follows. Suppose an agent A performed an action token u. Then, u is intentional if and only if there is a (complete) conduct plan K such that A brought about u because of K ('because of' suitably causally understood). This characterization can be made precise for an arbitrarily complex action in terms of the relation $IG^*(-, -)$.

BIBLIOGRAPHY

Atkinson, J. and Birch, D.: 1970, *The Dynamics of Action*, John Wiley & Sons, New York.

Baier, A.: 1971, 'The Search for Basic Actions', *American Philosophical Quarterly* **8**, 161–170.

Becker, W., Iwase, K., Jürgens, R. and Kornhuber, H.: 1973, 'Brain Potentials Preceding Slow and Rapid Hand Movements', paper presented at the 3rd International Congress on Event Related Slow Potentials of the Brain, held in Bristol, August 13–18, 1973.

Davidson, D.: 1966, 'The Logical Form of Action Sentences', in N. Rescher (ed.), *The Logic of Decision and Action*, Univ. of Pittsburgh Press, Pittsburgh, pp. 81–95.

Davidson, D.: 1971, 'Agency', in Binkley *et. al.* (eds.), *Agent, Action and Reason*, Toronto Univ. Press, Toronto, pp. 1–25.

Davidson, D.: 1973, 'Freedom to Act', in T. Honderich (ed.), *Essays on Freedom of Action*, Routledge and Kegan Paul, London, pp. 139–156.

Goldman, A.: 1970, *A Theory of Human Action*, Prentice-Hall, Englewood Cliffs.

Kim, J.: 1976, 'Intention and Practical Inference', in J. Manninen and R. Tuomela (eds.), *Essays on Explanation and Understanding*, Synthese Library, Reidel, Dordrecht, pp. 249–269.

McCann, H.: 1972, 'Is Raising One's Arm a Basic Action', *Journal of Philosophy* **69**, 235–249.

Pribram, K.: 1969, 'The Neurophysiology of Remembering', *Scientific American*, January Issue, 73–86.

Sellars, W.: 1956, 'Empiricism and the Philosophy of Mind', in H. Feigl and M. Scriven (eds.), *Minnesota Studies in the Philosophy of Science*, Vol. I, Univ. of Minnesota Press, Minneapolis, pp. 253–329.

Sellars, W.: 1973, 'Actions and Events', *Nôus* **VII,** 179–202.

Tuomela, R.: 1972, 'Deductive Explanation of Scientific Laws', *Journal of Philosophical Logic* **1**, 369–392.

Tuomela, R.: 1973, *Theoretical Concepts*, Library of Exact Philosophy, Springer-Verlag, Vienna.

Tuomela, R.: 1974a, *Human Action and Its Explanation*, Reports from the Institute of Philosophy, University of Helsinki, No. 2. (To appear, strongly revised and expanded, as *Human Actions and Their Explanation*, Synthese Library, D. Reidel, Dordrecht and Boston, 1977.)

Tuomela, R.: 1974b, 'Causes and Deductive Explanation', forthcoming in R. Cohen, C. Hooker, A. Michalos, and J. Van Evra (eds.), *PSA 1974. Selected Proceedings of the 1974 Biennial Meeting of the Philosophy of Science Association*, D. Reidel, Dordrecht and Boston, pp. 301–336.

Tuomela, R.: 1975, 'Purposive Causation of Action', forthcoming in R. Cohen and M. Wartofsky (eds.), *Boston Studies in the Philosophy of Science*, Vol. XXXI, D. Reidel, Dordrecht and Boston.

von Wright, G. H.: 1971, *Explanation and Understanding*, Cornell University Press, Ithaca.

V. P. ZINCHENKO

METHODOLOGICAL ASPECTS OF ANALYSIS OF ACTIVITY

As follows from the history of psychology, the psychological science has always been connected in this or that way with the study of activity (Tätigkeit). This is quite natural since psychics turns into 'a thing for us' to the extent it manifests itself through the activity of an individual.

However, the activity represents a reality too complex to be reflected directly in knowledge. Therefore it was only a certain projection of activity considered as a whole, as the concept representing in some sense the total human world, that was taken as a basis by the majority of psychological schools.

Without going into detail one can identify four main trends in psychology, in which the explanation of psychological reality was founded on a certain interpretation of the concept of activity. First of all it is the behaviouristic tradition that has realized with sufficient clarity that psychics, the inner world of man may be a subject to study only to the extent it manifests itself externally, i.e. in the activity. Basing their conception on the unquestionable value of the so called objective scientific method, the behaviourists assumed that psychology could deal only with those manifestations of the psychics which allow application of the available scientific methods. It was exactly for this reason that a certain projection of activity, namely, the behaviour, rather than the activity itself became the subject of psychology for behaviourism. This resulted in a number of methodological limitations that had been treated in detail in the world literature on psychology.

The second tradition has originated from some ideas of French psychology. This trend made an accent on the social determination of the activity, and, therefore, of the psychics. True, the representatives of this tradition interpreted the social determination in the socio-psychological way rather than from the point of view of the broad social-historic perspective. Nevertheless such an approach allowed to

Butts and Hintikka (eds.), Foundational Problems in the Special Sciences, 205–213.
Copyright © 1977 by D. Reidel Publishing Company, Dordrecht-Holland. All Rights Reserved.

identify very important typological aspects of psychic activity; it also outlined the general trends of analysis of its functional structure.

The third tradition represents psychological ideas of Freudism and, to some extent, those of neo-Freudism. Rather than discussing the essence of these concepts we consider their methodological aspect which is of interest for us in this case. From the methodological point of view Freud's attempts at explaining the activity itself as something whole resulted in his formulation of the thesis of the three-level structure of the psychics. One consequence of this was that the activity and the psychics cannot be depicted in a linear way, i.e. in one plane.

Finally, the fourth tradition has been developed within Soviet psychology, particularly in the works of L. S. Vygotsky and a number of his followers. Apparently, in their formulation the development of the psychological concept of activity has been the most broad and consistent. This was due, on one hand, to the application of the principles of dialectical and historical materialism in psychological research (L. S. Vygotsky, A. N. Leontjev, S. L. Rubinstein), and on the other—to the creative approach to the previous experience of the development of psychology. From the point of view of methodology it appears important that in the frame of Soviet psychological tradition practically all important aspects of the three preceding traditions were expressed in a constructive way. The formulation of the psychological theory of activity that up to now plays the role of the methodological foundation of Soviet psychological research has confirmed this fact. At the same time many general psychological problems remain unsolved inside this sufficiently developed psychological trend, a trend applying the concepts of activity, action, operation. The main of these problems is that of the structure of already developed (interiorized) forms of psychic activity. Because of the difficulties associated with this problem, there is a tendency for reduction of the psychics to physiological mechanisms, logical structures, etc.

Recently there appeared the real possibilities for the study of the structure of the interiorized forms of activity as purely psychological formations (V. M. Gordon, V. P. Zinchenko). This was due to the expansion of the techniques of the theory of activity by means of applying the methods of microstructural, microgenetic, psycho-physiological and phenomenological analysis. The present paper

presents the results of the functional-structural analysis of the perceptual, mnemonic, intellectual and performance actions that occur both
in the outer, external form and in the reduced one, including the
internal, interiorized one.

In order to study in detail the structure and functions of the activity
of an individual, to determine the functional relations between its
elements and those properties that they acquire in the structure of the
whole of activity, it would be useful to consider the categories and
methods of system-structural approach. The fundamental issue of the
system-structural approach is the study of the specific attributes of
complex objects, including, certainly, the human activity. We have
used the category of functional structure, which is defined as a principle of association between the functional components of the object
under study. In system-structural approach the term 'components'
means the activities, localized in space and time. In our case these are
the goal-oriented actions, operations and other units of the activity
analysis that may be interpreted not only as elements, but also as the
stages of activity. Let us recall that even before the appearance of the
system-structural method N. A. Bernstein considered the motion of a
living organ exactly from this standpoint. According to him, the
analogy of motion with the anatomical organs or tissues is based on its
two fundamental properties: first, the living motion reacts; second, its
evolution and involution has a character of regularity. It was also N. A.
Bernstein who came up with a descriptive characteristic of the living
motion as a 'biodynamic tissue'. The components of cognitive activity
(perceptual, mnemonic, intellectual) have the same properties as the
living motion; i.e. they react and evolve regularly. Besides, in various
forms of cognitive activity the biodynamic tissue of the living motion
plays an important role. For instance, as far as the perceptual activity is
concerned, it is also possible to use the terms 'biodynamic' and
'sensory tissue'.

Thus, we arrived at a point of interpreting the psychic activity as a
morphological object characterized by a developed functional structure, a certain object content and a meaning structure. Because of all
this it becomes necessary to expand considerably the psychological
conceptions concerning activity; in particular, the taxonomy of the
units of psychological research needs to be completed and defined

more precisely. Undoubtedly, the problem of such units merits a particular discussion since its methodological role in the development of the modern psychology is extremely important. But in this case we shall present only a few general considerations on this topic. As is well known, the problem of the units of psychological research appeared very urgent already for the followers of behaviourism. In particular, A. Tolmen treated it as one of the central ones. Among the French psychologists, a special attention to this problem is paid by J. Piaget whose concept of the unit represents an operation considered within the frame of the system of operations. The Freudian tradition adopted an essentially different standpoint: according to the concept of the complex level structure of the psychics the representatives of this trend actually reject the concept of the universal unit of research; they propose to construct a definite taxonomy of such units so that each level would correspond to its own kind of units.

The Soviet psychology is also characterized by the 'taxonomy' approach to the problem of units. As it is known, A. N. Leontjev, in particular, has identified three levels of the organization of activity, and respectively, three types of psychological units–the activity on the whole, the action and the operation. As far as the methodological standpoint is concerned, such an approach appears as the most perspective for the modern psychology. However, while evaluating it one should take into account that the present day practice of psychological research requires further refinement and development of the initial scheme of units. Our own research, in particular, dictated the necessity to supplement the A. N. Leontjev's scheme with at least one kind of units. Finally, our modified version of this scheme appears as follows:

> Motive   – Activity
> Goal   – Action
> Functional property – Operation
> Object property   – Functional block.

The above types of the units of analysis as well as their determinants constitute together four levels of analysis.

The activity is guided by its object–the motive, which is always backed by the subject's need. It is possible to determine the structure of activity in terms of actions that constitute it; each of these actions is directed at solution of one or another goal. The latter is related in a regular way (or coincides) with the motive of activity.

Action represents a process guided by the notion of result that has to be achieved, i.e. a process, subjected to a conscious goal (A. N. Leontjev). There is also the operational aspect of action, in addition to the intentional one (and the ideal). The operational aspect is determined, rather than by the goal itself, by the functionally meaningful properties of reality. The selection and realization of operations which constitute action are determined by the tasks related to the identification of these properties. Thus, one may interpret action in terms of the functional structure of the constituting operations directed at identification of the functionally meaningful properties of the object of activity, necessary for the achievement of the goal. There is a certain contribution from each of the operations into the reaching the result of action. Depending on the possibility of identification in a real situation of the functionally meaningful properties, the same action may be characterized by different kinds of the functional structure, i.e. it can be realized through various operation compositions or through various versions of their coordination. This definition of the operation differs in some respect from A. N. Leontjev's interpretation, namely, the operations are produced and determined only by the functionally meaningful properties present in the conditions of action, rather than by the conditions of the action realization as such. But the conditions are specified by the object properties of reality, and the number of those is considerably greater than that of the functionally meaningful properties.

As a result of the differentiation of conditions into functional and object properties, the handling of operation as a unit of the activity analysis is changed; this is accompanied by introduction of one more level of analysis, determined by the situation object properties. The psychological unit of analysis relevant to the object properties of reality represents a functional block. Accordingly, one may interpret operation through the functional structure of blocks designed for transforming the input information. The transformation is determined by the goal, the task and the object content of the activity. In analogy with the frequently observed mutual transitions between action and

operation, there may be coincidence or mutual transitions of operations and functional blocks during the overlapping or coincidence of
the functional and object properties of reality.

The above four levels of the analysis of activity form its functional
structure. As an essential addition to the macro-analysis of activity, we
consider the microstructural analytic level. One may thus identify the
real psychological meaning of the microstructural research that was
initially developed within the frame of information-cybernetic approach. The study of the basic levels of cognitive activity by means of
the microstructural analysis is nothing but the study of the regularities
of psychic reflection in dependance from the real, object properties of
the situation. In the absence of such a study, it would be impossible to
understand the processes of identification of the functionally meaningful properties in a situation, the processes of attaching the real object
content to activity, i.e. the processes of formation of the goals and the
motives of activity.

In principle, one may identify in the frame of psychological research
still more fractional units and levels of analysis, in particular, an
analytic level by means of which it would be possible to describe the
structure of functional blocks. This problem is solved in microgenetic
research. Apparently, its solution would require further modification of
the scheme of activity analysis, discussed above. Therefore, this
scheme is open from the taxonomic point of view; this is also true with
respect to the whole problem of taxonomy in the psychological analysis
of activity. However, it's of more importance now to consider the
problem of relations between various levels of the analysis of activity.

As follows from the description of the levels, each successive level
represents a means for explaining the preceding one. The analysis
acquires a certain quality of reality, a certain texture, and the activity—
a definite object content through interpretation of activity in terms of
actions, operations, blocks, and possibly subblocks. This process of
filling, figuratively speaking, occurs from below, from the more
elementary activity units that are directly connected with reality. Each
of the levels is characterized by a certain product, or result, which does
not remain stable. During the transition from one level to another it
evolves and expands. Undoubtedly, the levels identified, as well as the
units inside each level interact with each other through certain causal

relations. Those relations are not incidental; they may be both functional and genetic. The establishment, tracking and classification of these relations appear to be a task as important as that of further fractionation of activity into units that are smaller but not more elementary.

Although most important, the determination of the higher levels by the lower ones represents only one aspect of the issue related to activity acquiring the object character. It is not just its object quality (Gegenständlichkeit) but the spiritual ones as well that are reflected in the activity products. From this standpoint, a statement to the effect that the higher levels serve as a means of determination and explanation of the lower ones, is also justified. It is the higher levels that impart to the lower ones the unique subjective shade: motives, goals and meanings. For this reason any minimal unit of activity whether it is an operation or a functional block should be analyzed as a psychological unit but not as a physiological one. This is how we obtain the two most important categories of traditional as well as modern psychology—those of the object quality and meaningfulness. The research is transferred from the psychological plane into a different one (logical, physiological, etc.) if any of these categories is neglected. At the same time, it must be understood that these both properties of activity originate from different sources. It is as if these properties move towards each other. Because of their meeting the activity is generated, and as well as the subject matter of psychological research. Apparently, a natural assumption is that at each of the identified levels a different relation between the object quality and the meaningfulness is possible. Besides, these properties themselves are manifested in different forms. But the activity cannot be realized or reduced to the level of reactions (which is not infrequent in psychological laboratories' practice) if these properties are not present at least in the form of echo.

The object quality and the meaningfulness of activity are not independent categories. The transformations of the object functional structures of activity into meaningful ones in the course of activity can and do occur in reality. They have been recorded by A. N. Leontjev. According to him, the goal is a motive that has acquired the object property. It seems that meanings evince themselves in objects and objects generate new meanings. It is possible to find such transformations at any of the above levels of the activity analysis, and it appears to us a very important task to study them. As follows from the analysis of the microstructure of short-term processes, this problem may be solved. These studies resulted in the identification of such blocks as motor instructions' programs formation, the manipulator and the semantic processing block, which to a considerable extent carry out the functions of the meaning formation.

As far as the level of the psychological analysis of activity is concerned, one cannot ignore the issue of the actual physiological mechanisms providing for activity. Naturally, each level is based on the corresponding physiological functional system. In the context of this statement the psychology again formulates quite difficult problems for the physiology. There should be a correspondence between the psychic activity levels and the levels of organization of physiological functional systems. The study of the latter independently from the object content of activity, from its meaning structure is doomed to failure. A sufficient evidence supporting this is provided by the experience of the world psychology.

The construction of the subject of every experimental investigation constitutes the necessary condition for conducting it. This can be done with different degrees of theoretical awareness. The task of constructing the subject of psychological analysis is unbelievably complex. There is a considerable danger to exaggerate one or another aspect of psychological reality. Even when one considers a certain action (perceptual, mnemonic, etc.) as the subject of investigation one frequently exaggerates either its operation-technological aspect or the object qualification, or the motivation-emotional tint or the neurophysiological mechanisms. Although all these aspects of the study of action are equally necessary, when one exaggerates any of them and neglects others, various forms of psychological reductionism, existing in modern psychology, arise. Most frequently it is the information-cybernetical one, the neurophysiological, the logico-pedagogical, and the sociocultural reductionism. In the particular, concrete research, when one cannot practically take into account the overall diversity of the object of study, any of the above forms of reduction is possible. In the construction of the theory of a phenomenon, the theory which accounts for the experience of numerous and variable scientific experiments, the reductionism should be eliminated. As far as the theory is concerned, the reductionism manifests itself, first of all, in that the theory turn out to be limited in the application in various spheres of practice. It is exactly for this reason that the collision of the fixed research sets with the resisting reality results frequently in illusions and disappointments.

The problem of the human activity design, its control and improved organization are in the focus of attention of modern psychology. These problems are concerned not only with the kinds of activity characterized by explicit outer forms of expression, but also with those usually called the inner psychic activities. Similar problems have been solved up to now (and not unsuccessfully) in the intuitive form. It seems that there are already means in contemporary psychology that allow to solve these problems, if not better than before, then at least on a greater scale.

Moscow State University

BIBLIOGRAPHY

Leontjev, A. N.: 1974–1975, 'The Problem of Activity in Psychology', *Soviet Psychology*, **13,** No. 2.
Zinchenko, V. P.: 1974–1975, 'Productive Perception', *Soviet Psychology*, **13,** No. 2.
Zinchenko, V. P. and Gordon, V. M.: 1975, 'Methodological Problems of Psychological Analysis of Activity', in *Systems Research Yearbook*–1975, Moscow, Nauka (in Russian).

V

THE STATUS
OF LEARNING THEORIES

PATRICK SUPPES

A SURVEY OF CONTEMPORARY
LEARNING THEORIES

In the two and one-half decades from 1940 to 1965, the concepts, methods, and viewpoint of learning theory dominated the scene in psychology. There were, of course, significant and important developments in perception, the cognitive and social development of children, etc. All the same, it is a fair summary to think of these decades as the decades of the intellectual ascendancy of learning theory within psychology. This has not been true in the last decade, in fact in the period that began with the appearance of Chomsky's review (1959) of Skinner's *Verbal Behavior* (1957). The dominance of behavioristic theories of learning is certainly now over and has been for some years. The cognitive psychology of Anderson and Bower (1973), Atkinson and Wescourt (1975), Newell and Simon (1972), Rumelhart *et al.* (1972), Schank (1972), Winograd (1972), and many others represents the high points of innovation in psychology in the past decade. The influence of Piaget in the theory of child development is consonant with this rising ascendancy of cognitive psychology. The new theories of perception that have been much influenced by work in artificial intelligence are also closely connected with the developments in cognitive psychology characteristic of the decade. The picture grammars of Rosenfeld (1971) and Shaw (1969) represent one kind of development. Another more characteristic of artificial intelligence itself is represented by the group of articles in Winston (1975).

But learning theory is not dead. In this paper I survey different contemporary approaches, each of which in its own way is active and contributing something conceptually new. There are several good reasons for believing that learning theory is ready for a new period of vigorous development. The cognitive psychology that followed on the ascendancy of behavioristic learning theory was a natural development. It has represented a natural thrust for a deeper consideration of structure and a deeper consideration of more complex and more

Butts and Hintikka (eds.), Foundational Problems in the Special Sciences, 217–239.
Copyright © 1977 by D. Reidel Publishing Company, Dordrecht-Holland. All Rights Reserved.

cognitively developed aspects of behavior. That exploration has reached the point that a theory of how complex behavior is acquired is once again in order and will, I predict, be of increasing importance and significance in the psychology of the coming decade. To some extent the predicted return to learning will be a reflection of the fact that the attempt to characterize structures has for the moment exhausted itself.

I have identified seven different contemporary approaches to learning. In many cases the lines between them are fuzzy, and in all likelihood I have omitted some variant approaches that others would want to classify as separate and distinct. All the same, the very identification of seven approaches to learning, as broadly varying as they are, is in itself a reflection of the vigor of the search for appropriate theories. I emphasize deliberately *theories* rather than *theory* because the approaches that I describe are in many cases addressing very different clusters of problems. It is not surprising that different approaches have developed for different problems, when the variation is from biological adaptation to restricted methods of pattern recognition or statistical inference. In the survey of various approaches, the literature references are restricted primarily to sources that provide a more detailed survey or that constitute important systematic statements. I have not attempted to give a detailed historical account of developments, although where possible I have tried to indicate the earliest significant conceptual work.

1. PERCEPTRONS AND PATTERN RECOGNITION

The concept of the perceptron was originally introduced by Rosenblatt (1959, 1962), and its appeal led to its conceptualization as a simplified model of elementary properties of the peripheral nervous system. Linear devices similar to the mathematical nature of perceptrons have also been widely used in one form or another in pattern recognition and have today a range of practical applications. One of the best known perceptron-type linear devices is the highly successful checker player programmed by Samuel (1959, 1967) in the 1950s. For a variety of reasons, an extension of his approach to chess has been notably unsuccessful.

The deepest and most interesting work on perceptrons is that of Minsky and Papert (1969). I summarize here in an intuitive fashion the concept of a perceptron learning model and state the perceptron convergence theorem essentially in the form given by Minsky and Papert. The theorem does not originate with them but has a rather tangled history, which they describe in some detail and which I will not attempt to repeat here.

The conceptual apparatus is of the following sort. There is a set D of perceptual displays and a subset G that we want the device to learn to select correctly, that is, to identify correctly membership or non-membership in G. More particularly, on each trial the device responds *yes* to the presented display d if it classifies d as a member of G; otherwise it responds *no*. If the answer is correct, the device receives positive reinforcement (e_1), and, if the answer is incorrect, it receives negative reinforcement (e_2). Notice that the reinforcement is symmetric with respect to yes-no responses and depends only on the correctness or incorrectness of the response. This type of reinforcement is what I shall term standard non-determinate reinforcement. Later on I shall discuss more complex and more informative reinforcement structures. The heart of the idea of the perceptron is that there is a set Φ of elementary predicates. Each of these predicates is assigned a weight, and the perceptron combines in linear fashion the weighted 'answers' of the predicates to answer more complex questions. We can state the basic definition and the basic theorem without specifying more exactly the character of the elementary predicates because the results are relative to the set Φ of predicates. For notational purposes, let A be a k-dimensional vector of real numbers that gives the weights assigned to each of the k-elementary predicates with a_i being the weight assigned to elementary predicate φ_i. The set Φ can also most easily be represented as a k-dimensional vector of the k-elementary predicates. For each perceptual display d in D, $\varphi_i(d)$ has the value 1 if d has the property expressed by φ_i and 0 otherwise. We can thus use standard inner product notation for vectors so that in place of $\sum a_i \varphi_i(d)$ we can write $A \cdot \Phi(d)$. It is understood that the response of the perceptron learning model is *yes* if this inner product is greater than 0 and *no* otherwise. To refer to a particular trial, the vector A_n of coefficients can also be referred to as the state of conditioning of the perceptron learning model at the beginning of trial n, and d_n is the object presented on trial n.

In the present context, finally the sample space X consists of all possible experiments, with each experiment, of course, being a sequence of trials. A trial in the present case is simply a triple (A, d, e), where A is the state of conditioning as described already, d is a perceptual display that is a member of D, and e is a reinforcement.

We thus have the following definition.

DEFINITION 1. *A structure* $\Lambda = (D, \varphi, E, X)$ *is a perceptron learning model if and only if the following axioms are satisfied.*
  (i) *If* e_1 *occurs on trial n, then* $A_{n+1} = A_n$.
  (ii) *If* e_2 *occurs on trial n and* $A_n \cdot \Phi(d_n) \leqslant 0$, *then* $A_{n+1} = A_n + \Phi(d_n)$.
  (iii) *If* e_2 *occurs on trial n and* $A_n \cdot \Phi(d_n) > 0$, *then* $A_{n+1} = A_n - \Phi(d_n)$.

Note that the main feature of the learning of perceptrons is that learning occurs only when an error is made, that is, when an e_2 reinforcement occurs. The vector expressing the state of conditioning changes in one of two directions, depending upon the type of error.

In terms of these concepts we can then state the standard theorem.

THEOREM 1 (Perceptron Convergence Theorem). *For any set D and any subset G of D, if there is a vector A such that* $A \cdot \Phi(d) > 0$ *iff* $d \in G$, *then in any perceptron learning model there will only be a finite number of trials on which the conditioning vector changes.*

What is particularly important to note about this theorem is the ergodic property of convergence independent of the particular choice of weights A_1 at the beginning of the experiment. The finite number of changes also implies that the perceptron learning model will make only a finite number of mistakes. The hypothesis about G expressed in the theorem is equivalent to saying that G and its complement with respect to D are linearly separable.

The perceptron convergence theorem is asymptotic in character. The more important question of rate of learning is not settled by the theorem, although there exist some results in the literature that are of some interest. It is also important to note that many of the significant results of Minsky and Papert consist of describing in explicit detail the kinds of problems that perceptrons cannot satisfactorily solve or learn the solutions to. It is in fact because of the numerous negative results obtained by Minsky and Papert that there has been a tendency to prematurely dismiss perceptrons as theoretically interesting devices. Minsky and Papert themselves take a more moderate view of the matter, and an even more promising view of the as yet undeveloped potential of perceptrons is to be found in the review by Block (1970) of Minsky and Papert's book.

The work on perceptrons and other linear devices has been vigorously pursued to the level of practical application in a great deal of research that is broadly within mathematically oriented engineering. Much of this work within the disciplinary framework of engineering is mathematically sophisticated and uses a wide variety of mathematical tools in approaching problems of pattern recognition and machine learning. An excellent example of the methods used and the kinds of results obtained is to be found in the monograph by Fu (1968).

In somewhat different directions, often pointing toward biological work, a number of Russian authors have produced a good deal of interesting literature; a good example of the Russian work is the book on learning systems by Tsypkin (1973)—the reference is to the English translation. This work also contains extensive references to the Russian periodical literature on learning systems.

2. LEARNABLE FUNCTIONS

In one sense, the very interesting work of Ehrenfeucht and Mycielski on learnable functions can be regarded as an important generalization of the approach characteristic of the literature on pattern recognition. An account of this work and references to earlier publications is contained in the article by Ehrenfeucht and Mycielski contributed to this symposium, and consequently I shall not attempt to summarize their results here.

I do want to remark that at this preliminary stage the approach of Ehrenfeucht and Mycielski seems to have a good deal of promise. In their examination of the organization of memory, for instance, they are directly facing and attempting to solve some of the more difficult problems of any algorithm for answering new questions on the basis of past experience. Basically what they are concerned to do is to devise procedures that work when the dimension of the space in which information is stored is quite large, for example, greater than 100. As is well known, classical statistical methods and classical mathematical methods of interpolation and approximation work very poorly on such large-dimensional spaces. What they attempt to do is to exploit the natural assumption that few of the parameters are important for solving any given question.

3. BIOLOGICALLY ADAPTIVE SYSTEMS

Within biology, it is in many respects more natural to talk about adapting populations than adapting organisms. The kind of conceptual thinking that goes into the theory of evolution and the concept of adaptation of a population has been applied to develop mathematical models of learning in the past decade or so. Perhaps the best known work is that of John H. Holland and his associates, summarized in Holland (1975). Holland has attempted to generalize the concept of a schema in biology as applied to an interacting adaptive set of genes, and also to generalize the familiar genetic operators of crossing-over, inversion, and mutation. From a mathematical standpoint, the theory has been especially aimed at providing methods of optimization that can deal with situations not easily handled by classical mathematical methods, for example, situations in which interactions are non-additive and functions are highly non-linear in character.

In many respects I see the two most important general concepts of learning to be that of biological adaptation and the corresponding modification of gene pools, on the one hand, and that of the psychological concept of conditioning, on the other. Broadly speaking, I would classify approaches to learning as having basic affinity with one of these two main approaches, although advocates of particular methods would undoubtedly find this bifurcation too crude and simplistic. I am also somewhat suspicious of it myself, even though it may have some heuristic merit, because all of the approaches I survey in this article seem to come closer to the psychological than to the

biological model if a forced choice is made. The approaches to problem solving, for example, that of Newell and Simon (1972), that use
the concept of a problem space do have close affinities to Holland's
approach, but I have not surveyed that problem-solving literature here
because it is primarily not oriented toward learning. All the same,
some of the earlier work should be mentioned as related, for example,
Newell *et al.* (1958). Probably the most extensive effort to relate the
Holland type of adaptive scheme to mathematical learning models as
developed in psychology is to be found in Martin (1973).

I describe informally one of the principal results of Martin in order
to give a sense of the kind of theorems that come out of the biologi-
cally oriented adaptive approach to learning. To begin with, a large
class of algorithms for adaptation in solving a given class of problems is
defined. The class of algorithms can be viewed as treating adaptation
as a problem in statistical decision theory. There is a set of possible
strategies and a probability distribution over the strategies. There is
also an evaluation function, which in the decision-theory framework
could be a utility function, that measures the 'worth' of a strategy at a
particular time. An adaptive procedure sequentially chooses a strategy
according to the probability distribution. The procedure does not have
direct access to the evaluation function—that is, the utility function is
not known—but the procedure receives the resulting value of the
evaluation function applied to the strategy chosen. This value is used
to update the probability distribution for the next choice. The aim is to
have the probability of the 'best' strategy approach one as the number
of trials increases. If a procedure is such that the probability of
choosing any particular strategy goes to one with a number of trials,
the procedure is said to converge to that strategy. The class of
algorithms is also called a class of reproductive plans.

The problem of showing convergence is non-trivial in the problem
domains being investigated by Holland and his associates. Martin
proved the following theorem.

THEOREM 2 (Martin, 1973).   *In a suitably restricted problem domain*
there are non-trivial subclasses of reproductive plans that converge, but
convergence is not fast enough to achieve finite loss, where loss is defined
as in statistical decision theory.

The presence of infinite loss is one way of formulating the fact that the procedure of convergence is often far too slow for practical application. Generally speaking, problems about the rate of convergence are important and critical to practical applications of the kind of genetic models that originate with Holland and his group. On the other hand, this problem about the rate of convergence is a problem throughout the domain of learning, and is discussed again later.

4. ARTIFICIAL INTELLIGENCE APPROACHES

In many respects, the early work of Newell, Shaw, and Simon (1958) marks one of the first efforts to introduce some form of learning into research on artificial intelligence. The kind of effort represented in the general problem-solver program introduced in that 1958 study is described in detail in Ernst and Newell (1969). A more recent typical example of the artificial intelligence approach to learning is to be found in the dissertation of Patrick Henry Winston, published in 1975. The book in which Winston's dissertation appears is also edited by Winston and contains several significant pieces of work on constructing computer programs that can recognize perceptual scenes. There are a variety of intriguing and perplexing problems that have now received a great deal of attention. These problems range from recognizing when one object obtrudes another to recognizing shadows and using texture for depth perception. The literature on these matters is now quite large, and a great deal of interesting work has been done. Surprisingly, there has been very little emphasis on learning, and Winston's work constitutes one of the exceptions. However, even in the case of Winston's work, the emphasis is much more upon clever analysis by fixed programs of perceptual features or of subtle conceptual distinctions rather than of powerful learning procedures. In Winston's description of his own work, the few pages on learning as such receive the least detailed and least technical development. Although details are lacking, Winston makes a number of good general points about learning. First, he emphasizes that what appear to be superficially different kinds of learning will probably come to have a much more common invariant description. Thus he believes that the differences between learning by example, learning by being told, learning by imitation, and learning by reinforcement are not nearly so distinct as they seem to be on the surface. Secondly, and even more importantly in my judgment, he emphasizes the importance for either machine or human learning of extremely good and carefully worked out training sequences of examples.

In contrast to the other approaches being examined here, it is extremely difficult to evaluate the exact accomplishments of Winston's work. On the one hand, there are no mathematical theorems that express the power of the system. Certainly it should not be a necessary condition of an approach to learning that general mathematical theorems be proved from the assumptions of the theory. If this were so, most of the work in psychology that has traditionally been called learning theory would not qualify. On the other hand, there is a rigorous empirical tradition in psychology of offering experimental evidence of the kinds of things that can be learned and under what conditions, and of the kinds that cannot. What is disappointing is that Winston does not present any systematic data about what the program can or cannot do, and consequently it is very hard to judge in any detail how powerful the results are.

It seems to me that there is much that is promising about Winston's approach, but the promise is exhibited in the routines for recognizing perceptual features of a scene and it is much less apparent that the learning part of the program is itself nearly as clever and imaginative.

5. LANGUAGE ACQUISITION DEVICES

The recent flourishing of linguistics has led to a focus on language learning in a new and more intense way. Psycholinguistics as such is one of the more active branches of psychology, and a wide variety of empirical studies about the acquisition of various features of language, especially by children in the early years of their language learning, has been undertaken. Most of these studies are remarkably empirical in character, I mean by this that they are not testing any very specific, at least any very explicit, theoretical ideas, especially ideas of a quantitative or mathematical sort. Consequently, a survey of the range of these studies consists primarily of the task of summarizing a large number of empirical studies. Concurrently, however, and to a large extent separately, there has been an active pursuit of purely theoretical results regarding language acquisition.

As in the case of stimulus-response theory as applied to language learning, as yet the most interesting results are asymptotic results. Perhaps the first definite results in this area were obtained by Gold (1967), who made an interesting and important analysis of what it means to have asymptotic identifiability of a language. Gold restricted himself to the learnability of the grammatical features of the language without consideration of semantics, on the one hand, or phonology, on the other. He supposed discrete trials are available to the learner (in contrast to the more realistic assumption of continuous time). On each trial the learner receives some specific information about the language he is to be learning, and on each trial the learner guesses the identity of the language he is learning. Clearly, this is not a task that is identical with that of a child learning, but the results of Gold, as will be seen, are of interest all the same. A class of languages is said to be asymptotically identifiable if there is an algorithm such that, given any language in the class, there is a finite time after which the guesses will all be correct. Notice that the formulation of this definition is similar to the condition regarding finite number of errors expressed in the perceptron learning theorem.

One important method of presenting information that Gold studied was that of *text presentation*. This method is characterized by the presentation on each trial of positive instances of sentences in the language. Gold proved the following theorem.

THEOREM 3 (Gold, 1967).   *If the class of languages is the class of context-sensitive, the class of context-free, or the class of finite-state languages, then a language from any one of these classes is not asymptotically identifiable under the method of text presentation.*

In *informant presentation*, as contrasted to text presentation, the negative instances of expressions of the terminal vocabulary that are not sentences of the language are identified as such negative instances. This is called informant presentation because it corresponds to asking an informant whether a given sequence of words of the language is or is not a well-formed sentence. On the basis of the much more powerful method of informant presentation, Gold was able to prove the following theorem.

THEOREM 4 (Gold, 1967).   *Under informant presentation each of the classes of languages mentioned in Theorem 3 and also the class of primitive recursive languages are asymptotically identifiable.*

It has been commented several times in the literature that, unfortunately, text presentation corresponds more closely to what the child is confronted with than does informant presentation, because the number of positive instances a child hears of a language seems to far outweigh any sampling of negative instances, especially identified negative instances. The negative results of Gold's theorem about text presentation have also been a basis for suggesting that the context-free base of a transformational grammar must be a universal context-free base and that this context-free base or deep structure is known a priori, that is, in the present context, is genetically determined. What the child must learn are the transformations that take him to the surface of his particular language.

It should be noted that consideration of a class of languages is not as unrealistic as it seems. This clearly can be treated as an extensional version of an essentially intensional approach to the problem of learning in which the child is learning the distinguishing properties of the language that he hears around him. I have not seen the Gold-type theorems formulated in such intensional fashion, but it seems unlikely that there would be any major difficulties in giving such a formulation.

Extensions of Gold's approach to identifiability in the limit are to be found in Feldman (1969) and Feldman *et al.* (1969).

The same general kinds of questions also arise about the learning of transformational languages. Especially as a complement to Gold's results, the learning of transformations with a given base has been pursued. Negative results on this question are to be found in Wexler and Hamburger (1973). Related results are to be found in Peters and Ritchie (1973).

Some positive asymptotic results on learning transformational grammars or languages have recently been given by Hamburger and Wexler (1975). Hamburger and Wexler take as basic data pairs (b, s) where b

is a phrase marker or derivation tree of the base grammar and s is the surface structure resulting from applying the appropriate transformations to b. The problem is to learn the transformational grammar that will correctly carry any phrase marker b to its corresponding surface structures. The information content is conveyed entirely by the couple (b, s) on a given trial, but this is of course a very powerful piece of information to give the learner. The learning procedure consists of making a guess at each point in time of a set of transformations. This set is used together with the current datum on a given trial to produce two other sets of transformations: a set of hypothetical candidates and a set of candidates for rejection. A member of the union of the two sets of candidates is selected and is added to the set of guesses if it is hypothesized or removed if it is from the other set of candidates. Again, from a realistic learning standpoint it would seem desirable to think of these sets in intensional rather than extensional terms, as remarked already above about Gold's approach. The technical details of Hamburger and Wexler's setup will not be outlined here. The intuitive idea should be clear from what has been said. On this basis they are able to prove the following theorem.

THEOREM 5 (Hamburger and Wexler, 1975). *Given the basic data and learning procedure outlined above, transformational grammars, subject to relatively mild restrictions, operating on a context-free base can be asymptotically learned with probability one.*

The results of Hamburger and Wexler basically assume that the meaning of an utterance is given to the listener by a phrase structure tree of a context-free grammar. The assumption that meaning is given in this form is an extraordinarily strong one and almost certainly not empirically sound. All the same, it is important to see exactly what additionally is required in order to get to the surface structure of language where the surface is characterized by transformations of base structures.

Certainly the results achieved on language acquisition are not intuitively fully satisfactory as yet, but it is important to recognize that significant progress is being made and that our conceptual understanding of what is and what is not a feasible way of thinking about these matters has certainly been clarified by the approaches described above and by the relevant general theorems that have been proved. Further developments in the directions the literature has already taken should be extremely interesting, even if results that are of an empirically testable sort are not obtained for some time.

6.   RATIONAL CHANGES OF BELIEF

Closely related to concepts of learning theory is a rather different literature concerned with rational changes of beliefs. Broadly speaking, any conceptual contribution to the theory of statistical inference might be regarded as a contribution to the theory of rational changes of belief and thus indirectly to learning theory. On the other hand, it does not seem to me desirable to define learning theory so broadly as to include Bayesian approaches or the recent literature concerned with rational changes in strongly held beliefs as exemplified in Harper (1976). It is characteristic of learning theory not to be concerned with deliberate rational changes in belief. The subject of how one should make rational changes is, of course, an important one and may conceptually be regarded as a limiting case of learning theory. De Finetti (1964, p. 146) expresses very well this point by emphasizing that the rational man never corrects his initial evaluation of probabilities but only changes them by appropriate conditionalization as new information is made available to him. For a number of reasons I am skeptical of this pure Bayesian view as being adequate to even the theory of *rational* behavior in complex circumstances. My principal reasons for skepticism have been stated previously (Suppes, 1966), but in the present context it will be useful to rephrase them. Moreover, the difficulties suggest that a satisfactory theory of rationality must move closer to the kinds of problems that are characteristic of learning theory, and leave the relatively simple framework of conditional probability suggested by de Finetti.

The central problems are those of attention to perceptual input and internal data and selection of evidence for conditionalization. At any given instant there is a welter of potential information impinging on the sensory organs of a person, and, at the same time, there is an extraordinarily complex monitoring of internal data as generated by thinking and musings of all sorts. In inductive logic and statistics, these matters do not directly arise, because problems for consideration are already formalized and appropriate fixed decisions about attention and selection of data are made prior to formalization. On the other hand, it is exactly such mechanisms that are the focus of a great deal of learning theory.

A second different, but related, difficulty arises over the use of a fixed conceptual apparatus. Insofar as the basic framework is formulated within a fixed sample space, the introduction of new concepts must be restricted to that which can be expressed in the assumed algebra of events. I find it difficult to think of any genuinely new concepts in the history of scientific or ordinary experience that can be accounted for in such terms. The introduction of new concepts—or what is termed, within the framework of learning theory, concept formation—cannot be accounted for simply in terms of conditional probabilities but requires much more fundamental changes of conceptual apparatus. There is, it seems to me, a strikingly unrealistic line of talk on these matters by Bayesians like de Finetti and Savage, and the absence of an explicit formulation of the problems that need to be solved in developing a more adequate theory of rationality. The impression is often left by their writings that there are no fundamental conceptual difficulties in the Bayesian approach to experience and how it is to be absorbed to generate a new belief structure. Although it is not appropriate to push these matters in more detail here, it is my conjecture that the kinds of problems that have been of central concern in learning theory for a long time provide both the source of problems that must be solved and in some cases at least partial methods of solution in order to have a more adequate theory of rationality.

7.   STIMULUS-RESPONSE THEORY

I have, in that adage of American cooking, saved the best for last, and so I turn to stimulus-response theory of learning. It is controversial to claim it is the best but certainly of the theories I have surveyed it has historically had the greatest influence as a theory of learning within psychology. The history of what I would call the modern stimulus-response theory, which means the development of a mathematical and quantitative theory, goes back to the early 1950s. Perhaps the two most important historical references to list are the early influential paper of Estes (1950) and the book of Bush and Mosteller (1955). Mathematical and foundational studies of the theory occur only a few years later and are represented by Estes and Suppes (1959a, 1959b), Karlin (1953), and Lamperti and Suppes (1959). The mathematical properties of the theory, especially the asymptotic theory, have been thoroughly studied in great detail in more recent years, especially in long monographs by Iosifescu and Theodorescu (1969) and Norman (1972). The stochastic processes that arise as particular models of stimulus-response theory present difficult and, in some cases intractable, mathematical problems. The two monographs just referred to provide a very thorough introduction to the mathematical or, more particularly, the probabilistic methods that are currently useful in attacking mathematical problems that are, in many cases, of direct experimental interest. In this period running from 1950 to 1972, I have listed only a few references, and all of these references are either of conceptual or mathematical interest. It is important to emphasize that a large number of additional articles of a conceptual nature were written during this period and an even larger number of experimental studies testing a variety of applications of the theory were published.

The possibility or impossibility of reducing one branch of science to another has been of perennial philosophical interest. For many years I have advocated the view that for advanced branches of science such reduction will almost necessarily take the form of representation theorems, or proofs of the impossibility of such theorems. Roughly speaking, the idea of such representations is this. Given a theory T of one part of science, we regard it as *reduced* to another theory T' if given any model of T we can construct from a model of T' an isomorphic representation of the model of T. A familiar and well-known example of such reduction is Stone's representation theorem for Boolean algebras: Any Boolean algebra is isomorphic to an algebra of sets. Within stimulus-response theory itself a natural question of such representation arose in the 1950s, and although the question answered seems now to be of mainly historical interest, because of my own involvement in proving the appropriate representation theorem I sketch here the results as an example of such reduction. The initial and early work of Estes mentioned above rested upon a theory of stimulus sampling and conditioning by means of reinforcement. Building upon earlier qualitative and vaguer theoretical developments, he developed a formulation of stimulus sampling and conditioning by means of reinforcement that could be given a firm and exact mathematical formulation. This was done, for example, in great detail in Estes and Suppes (1959b). In contrast, the theory that was developed early by Bush and Mosteller depended upon a simpler set of concepts, namely, just that of response and reinforcement. Linear models for the probability changes in responses as a result of reinforcement were developed and tested extensively. The question naturally arose whether the stimulus-sampling models that originated with Estes included as a special case the linear models of Bush and Mosteller. More particularly, given any linear model, could one obtain a representation of the linear operators in terms of the richer psychological models of stimulus sampling and conditioning? The basic theorem restricted to one-parameter linear models is the following.

THEOREM 6 (Estes and Suppes, 1959b). *Given any one-parameter linear model, there exists a sequence of models of stimulus-sampling theory such that as the number of stimuli goes to infinity there is convergence to the linear model.*

The proof of this theorem as originally given is long and complicated. Using more powerful methods, Norman (1972) was able to give a much simpler proof for the restricted case when the reinforcement on a given trial is contingent only upon events, for example responses, happening on that trial and not on any previous trial. By and large, the advanced methods used in Norman's monograph work only when the reinforcement is so restricted. In general it remains difficult and complicated to get results when the reinforcement schedule extends contingently into the past.

A conceptually more interesting and also more controversial question of representation centers around the issues of language learning. Without attempting a complete theory of language learning it is still close to the core of the theoretical issues to ask to what extent the kinds of automata that are powerful enough to generate or recognize grammars of a given sort can be represented by stimulus-response models. I focus here on a theorem I proved about the representation of finite automata within stimulus-response theory in 1969 and the criticisms subsequently made of that theorem. Because of the interest of these matters for many of the broad issues of contemporary learning theories, I give some of the details but, in the spirit of this article, without introducing an apparatus of technical notation necessary for exact statement of the theorems and their proofs. The ideas introduced here also correspond closely to the conceptual formulation found in the earlier versions of stimulus-sampling theory already referred to. Roughly speaking, the organism begins a trial in a certain state of conditioning. The state of conditioning consists of a connection between stimuli and responses. This connection is not further characterized but is represented by an arbitrary set of ordered pairs. A set of stimuli is presented to the organism and from this set is then sampled a subset. On the basis of the sampled subset a response is made following the response rule that the probability of response is proportional to the number of sampled stimuli conditioned to a given response. After a response a reinforcement occurs and, on the basis of the reinforcement, conditioning relations of sampled stimuli may be changed, a new state of conditioning is entered, and the organism is ready for the beginning of a new trial.

The axioms that embody these ideas taken from the formulation in Suppes (1969) may be given the following verbal formulation.

Sampling Axioms:

S1. *On every trial a set of stimuli of positive measure is sampled with probability one.*

S2. *If the same presentation set occurs on two different trials, then the probability of a given sample is independent of the trial number.*

S3. *Samples of equal measure that are subsets of the presentation set have an equal probability of being sampled on a given trial.*

S4. *The probability of drawing a particular sample on a trial, given the presentation set of stimuli, is independent of any preceding pattern of events.*

Conditioning Axioms:

C1. *On every trial with probability one each stimulus element is conditioned to at most one response.*

C2. *The probability is c of any sampled stimulus element's becoming conditioned to the reinforced response if it is not already so conditioned, and this probability is independent of the particular response, trial number, or any preceding pattern of events.*

C3. *With probability one, the conditioning of all stimulus elements remains the same if no response is reinforced.*

C4. *With probability one, the conditioning of unsampled stimuli does not change.*

Response Axioms:

R1. *If at least one sampled stimulus is conditioned to some response, then the probability of any response is the ratio of the measure of sampled stimuli conditioned to this response to the measure of all the sampled conditioned stimuli, and this probability is independent of any preceding pattern of events.*

R2. *If no sampled stimulus is conditioned to any response, then the probability of any response is a constant guessing probability that is independent of the trial number and any preceding pattern of events.*

Note that there are no reinforcement axioms, because the schedule of reinforcement itself is determined externally by the experimenter or, in the more general non-experimental applications of the theory, by the environment and the response history of the individual.

On the basis of these axioms, the following representation theorem for finite automata may be proved. (Some related results are to be found in Feichtinger, 1970.)

THEOREM 7 (Suppes, 1969). *Given any connected finite automaton, there is a model of stimulus-response theory that asymptotically becomes isomorphic to it. Moreover, the stimulus-response model may have all responses initially unconditioned.*

A direct criticism of the significance of this theorem is that it is restricted to finite automata, which are capable of generating only regular languages. A richer device is required to obtain context-free or context-sensitive languages, not to speak of transformations. This issue I discuss later.

A second criticism, especially one that has been made recently in the literature (Nelson, 1975; Kieras, 1976), is that the theorem proved is actually more restricted than the formulation given above. Nelson and I are rather close in our viewpoints, and in Suppes (1975) I have expanded upon my own ideas in response to his article. However, the issues raised about finite acceptors and finite transducers and what Kieras calls S-machines are too technical to enter into here. I am writing a comment on Kieras's article for publication elsewhere and will not enter into the details. For our present discussion, the central point is that it is important to consider methods of reinforcement that are more restricted in character than the ones discussed in my 1969 article. The reinforcement used there is what is called in the literature of learning theory determinate reinforcement, that is, when an incorrect response is made, an exact and explicit correction indicating a correct response is given. At the opposite extreme is non-determinate reinforcement, in which the organism is only given the information that the response is correct or incorrect but is not given any indication about the correct response. Weaker results, but conceptually interesting ones, can be obtained with non-determinate reinforcement. The results are weaker in the sense that the same isomorphism of structure internally cannot be obtained, but behavioral equivalence is possible. This weaker theory is developed in Suppes and Rottmayer (1974) and discussed again in Suppes (1975). The basic theorem is the following.

THEOREM 8.   *If a set of objects, for example, perceptual displays, has a subset that can be recognized by a finite automaton, then there is a stimulus-response model that can also learn to recognize the subset, with performance at asymptote being behaviorally equivalent to that of the automaton.*

An extension of the theorem just stated to partial recursive functions of one variable and to an internal structure that permits the construction of internal programs is given in Suppes (in press).

From a conceptual standpoint, the theorem just stated is appealing because of its use of a weak form of reinforcement, a form of reinforcement that seems comparable to the strength encountered in many non-experimental but significant environmental situations. On the other hand, a weakness of the setup is that the rate of learning is slow, and, without the introduction of further structure, is far too slow to represent a viable theory psychologically. A detailed analysis of some of the improvements that can result from introducing hierarchies both internally in terms of subproblems or subroutines and externally in terms of the sequence of problems is in Suppes (in press). The virtue of the first theorem with determinate reinforcement and resulting isomorphism of structure is that the rate of learning is feasible and is in principle as fast as could be possibly expected on the assumption that no major aspects of the learning task had been previously learned.

8. CONCLUDING REMARKS

Thirty years ago, a survey of learning theories would have been mainly restricted to stimulus-response theory and rather closely related variants developed within experimental psychology. The variety of approaches characteristic of the present survey is testimony to the fact that interest in learning theory is in no sense mainly restricted today to psychology but is in fact receiving much of its impetus from theoretical and conceptual concerns far removed from the range of problems characteristic of experimental psychology. At the same time that this breaking away of learning theory from the framework of psychology has taken place, the influence of learning theory within psychology has decreased and it is unlikely that it will return to the ascendant position it occupied two or three decades ago.

In the next decade or so, I would in fact predict greater progress in the development of learning theories outside of psychology than within it. As yet, the use of learning-theoretic ideas in artificial intelligence has been minimal, and there is also great skepticism in theoretical-linguistics circles about the possibility of having a reasonable theory of language learning in the near future, But the learning problems in artificial intelligence or linguistics have in many instances the kind of formal definiteness and concreteness that make them amenable to detailed conceptual and mathematical analysis. For this and other reasons I will close by predicting that the next decade will witness a renaissance of theoretical interest in learning.

Stanford University

BIBLIOGRAPHY

Anderson, J. R. and Bower, G. H., *Human Associative Memory*, Winston, Washington, D.C., 1973.
Atkinson, R. C. and Wescourt, K. T., 'Some Remarks on a Theory of Memory', in P. M. A. Rabbitt and S. Dornic (eds.), *Attention and Performance V*, Academic Press, London, 1975.
Block, H. D., 'A Review of Perceptrons: An Introduction to Computational Geometry', *Information and Control* **17**, (1970) 501–522.
Bush, R. R. and Mosteller, F., *Stochastic Models for Learning*, Wiley, New York, 1955.
Chomsky, N., 'Review of B. F. Skinner, *Verbal Behavior*', *Language* **35**, (1959) 26–58.
De Finetti, B., 'La prévision: Ses lois logiques, ses sources subjectives', *Annales de l'Institut Henri Poincaré* **7**, (1937) 1–68. English translation in H. E. Kyburg and H. E. Smokler (eds.), *Studies in Subjective Probability*, Wiley, New York, 1964.
Ernst, G. W. and Newell, A., *GPS: A Case Study in Generality of Problem Solving*, Academic Press, New York, 1969.
Estes, W. K., 'Toward a Statistical Theory of Learning', *Psychological Review* **57**, (1950) 94–107.
Estes, W. K. and Suppes, P., 'Foundations of Linear Models', in R. R. Bush and W. K. Estes (eds.), *Studies in Mathematical Learning Theory*', Stanford University Press, Stanford, Calif., 1959a.
Estes, W. K. and Suppes, P., *Foundations of Stimulus-sampling Theory, II: The stimulus sampling model* (Tech. Rep. 26, Educ. and Psych. Ser.), Stanford University, Institute for Mathematical Studies in the Social Sciences, Stanford, Calif., October 1959b.
Feichtinger, G., 'Lernprozesse in stochastischen Automaten', *Lecture Notes in Operations Research and Mathematical Systems* **24**, (1970) 1–66.
Feldman, J., *Some Decidability Results on Grammatical Inference and Complexity* (Memo AI–93), Stanford University, Artificial Intelligence Project, Stanford, Calif., 1969.
Feldman, J., Gips, J., Horning, J., and Reder, S., *Grammatical Complexity and Inference* (Tech. Rep. CS 125), Stanford University, Department of Computer Science, Stanford, Calif., 1969.
Fu, K. S., *Sequential Methods in Pattern Recognition and Machine Learning*', Academic Press, New York, 1968.
Gold, E. M., 'Language Identification in the Limit', *Information and Control* **10**, (1967) 447–474.
Hamburger, H. and Wexler, K. N., 'A Mathematical Theory of Learning Transformational Grammar', *Journal of Mathematical Psychology* **12**, (1975) 137–177.

238   P. SUPPES

Harper, W. L., 'Rational Belief Change, Popper Functions and Counter-factuals', in W. L. Harper and C. A. Hooker (eds.), *Foundations of Probability Theory, Statistical Inference and Statistical Theories of Science* (Vol. 1: *Foundations and Philosophy of Epistemic Applications of Probability Theory*), D. Reidel, Dordrecht, 1976.

Holland, J. H., *Adaptation in Natural and Artificial Systems*', University of Michigan Press, Ann Arbor, 1975.

Iosifescu, M. and Theodorescu, R., *Random Processes in Learning*, Springer-Verlag, New York, 1969.

Karlin, S., 'Some Random Walks Arising in Learning Models, I', *Pacific Journal of Mathematics* **3**, (1953) 725–756.

Kieras, D. E., 'Finite Automata and S-R Models', *Journal of Mathematical Psychology* **13**, (1976) 127–147.

Lamperti, J. and Suppes, P., 'Chains of Infinite Order and Their Application to Learning Theory', *Pacific Journal of Mathematics* **9**, (1959) 739–754.

Martin, N., *Convergence Properties of a Class of Probabilistic Adaptive Schemes called Reproductive Plans* (Tech. Rep. 210, Educ. and Psych. Ser.), Stanford University. Institute for Mathematical Studies in the Social Sciences, Stanford, Calif., July 1973.

Minsky, M. and Papert, S., *Perceptrons*, MIT Press, Cambridge, Mass., 1969.

Nelson, R. J., 'Behaviorism, Finite Automata and Stimulus-Response Theory', *Journal of Theory and Decision* **6**, (1975) 249–268.

Newell, A. and Simon, H. A., *Human Problem Solving*, Prentice-Hall, Englewood Cliffs, N. J., 1972.

Newell, A., Shaw, J. C., and Simon, H. A., 'Elements of a Theory of Human Problem Solving', *Psychological Review* **65**, (1958) 151–166.

Norman , F., *Markov Processes and Learning Models*, Academic Press, New York, 1972.

Peters, P. S., Jr. and Ritchie, R. W., 'Nonfiltering and Local-filtering Transformational Grammars', in K. J. J. Hintikka, J. M. E. Moravcsik, and P. Suppes (eds.), *Approaches to Natural Language*, D. Reidel, Dordrecht, 1973.

Rosenblatt, F., *Two Theorems of Statistical Separability in the Perceptron* (Proceedings of a symposium on the mechanization of thought processes), HM Stationery Office, London, 1959.

Rosenfeld, F., *Principles of Neurodynamics*, Spartan Books, New York, 1962.

Rosenfeld, A., 'Isotonic Grammars, Parallel Grammars and Picture Grammars,' in B. Meltzer and D. Michie (eds.), *Machine Intelligence* **6**, American Elsevier, New York, 1971.

Rumelhart, D. E., Lindsay, P. H., and Norman, D. A., 'A Process Model for Long-term Memory,' in E. Tulving and W. Donaldson (eds.), *Organization and Memory*, Academic Press, New York, 1972.

Samuel, A. L., 'Some Studies in Machine Learning Using the Game of Checkers', *IBM Journal of Research and Development* **3**, (1959) 210–229.

Samuel, A. L., 'Some Studies in Machine Learning Using the Game of Checkers, Part II, *IBM Journal of Research and Development* **11**, (1967) 601–618.

Schank, R. C., 'Conceptual Dependency: A Theory of Natural-Language Understanding', *Cognitive Psychology* **3**, (1972) 552–631.

Shaw, A. C., 'A Formal Picture Description Scheme as a Basis for Picture Processing Systems', *Information and Control* **14**, (1969) 9–52.

Skinner, B. F., *Verbal Behavior*, Appleton, New York, 1957.

Suppes, P., 'Probabilistic Inference and the Concept of Total Evidence', in J. Hintikka and P. Suppes (eds.), *Aspects of Inductive Logic*, North-Holland, Amsterdam, 1966.

Suppes, P., 'Stimulus-Response Theory of Finite Automata', *Journal of Mathematical Psychology* **6**, (1969) 327–355.

Suppes, P., 'From Behaviorism to Neobehaviorism', *Theory and Decision* **6**, (1975) 269–285.

Suppes, P., 'Learning Theory for Probabilistic Automata and Register Machines, with Applications to Educational Research', in H. Spada and W. Kempf (eds.), *Structural Models of Thinking and Learning*, in press.

Suppes, P. and Rottmayer, W., 'Automata', in E. C. Carterette and M. P. Friedman (eds.), *Handbook of Perception* (Vol. 1: *Historical and Philosophical Roots of Perception*), Academic Press, New York, 1974.

Tsypkin, Y. A. Z., *Foundations of the Theory of Learning Systems*, Academic Press, New York, 1973.

Wexler, K. N. and Hamburger, H., 'On the Insufficiency of Surface Data for the Learning of Transformational Languages', in K. J. J. Hintikka, J. M. E. Moravcsik, and P. Suppes (eds.), *Approaches to Natural Language*, D. Reidel, Dordrecht, 1973.

Winograd, T., 'Understanding Natural Language', *Cognitive Psychology* **3**, (1972) 1–191.

Winston, P. H., 'Learning Structural Descriptions from Examples', in P. H. Winston (ed.), *The Psychology of Computer Vision*, McGraw-Hill, New York, 1975.

N. J. MACKINTOSH

CONDITIONING AS THE PERCEPTION
OF CAUSAL RELATIONS

There was a time, not very long ago, when learning theory arrogantly claimed for itself a central place in psychology, and learning theorists were confident they could explain any aspect of human behaviour from principles supposedly derived from experiments on conditioning in animals. Although Skinner and his follows may still believe this, there is a rather general consensus that learning theory has been dethroned from this exalted position, and the majority of learning theorists, as befits men under siege, have retrenched, either into developing miniature theories for miniature arrays of data, or (the final ignominy) into wondering whether there is any such general process as learning for them to study.

Without being too imperialistic, I believe that learning theory can do better than this – but only if we profit from past mistakes. At least part of the trouble was due to one rather simple reason: the particular theory which was extrapolated with such confidence to such a wide range of new data did not even provide a satisfactory explanation of the limited range of data on which it was supposedly based. It owed its general acceptance as an account of conditioning experiments, not to any spectacular success in predicting the outcome of those experiments, but to its close conformity to the generally received conception of a scientific account of animal behaviour. The set of assumptions that go to make up what is usually understood by the term 'learning theory' in this context has gradually turned into a dogmatic and restrictive theory which bears little relation to any realistic experimental or theoretical analysis of conditioning.

At one time this learning theory must have seemed an adequate account of the data available from animal experiments. The problem was that most of these early data were fortuitously restricted to those cases that could be handled by the theory. Thorndike (1911) placed cats in a puzzle box, and observed that they learned an arbitrary

Butts and Hintikka (eds.), *Foundational Problems in the Special Sciences*, 241–250.
Copyright © 1977 by D. Reidel Publishing Company, Dordrecht-Holland. All Rights Reserved.

response, such as pressing a panel or catch, which gained their release from the box, and access to food. Pavlov (1927) restrained dogs on a conditioning stand, and observed that they learned to salivate to a bell which regularly preceded the delivery of food. In each case the experimenter recorded an increase in the probability of a particular response occurring in the presence of a particular set of stimuli. The theory which was applied to these data was, of course, stimulus-response (S-R) theory, and it is easy to see why it should have been so applied. If the experimenter records an increase in the probability of a response to a stimulus, then it seems reasonable to assume that what has been learned is precisely the establishment of some set of connections or associations between that response and that stimulus.

This is a plausible account of what happens in a particular experimental situation. But it is obviously not the only possible account, and its rapid acceptance must surely be attributed to other factors. Of most importance was that it conformed to a prevailing view of what constituted a correct scientific account of behaviour. That view was classical reflex theory – the view that animals are reflex machines, whose behaviour consists of responses to stimuli and is therefore to be understood by tracing pathways (the reflex arc) between stimulus and response. In early versions of this theory, animal behaviour is understood by analogy with the workings of simple mechanical toys – a particular stimulus automatically produces a particular response. The possibility that an animal without a soul might be capable of learning implies a certain departure from this scheme, for no one will dispute that learning may cause a change in the response made to a particular stimulus. The simplest departure, however, appeared to be one which allowed that learning simply consisted of the establishment of *new* S-R pathways.

Although Thorndike himself advanced S-R theory cautiously, and Pavlov never at all, the theory rode the wave of Watson's behaviourism, and soon became established dogma – to the point where learning has often been *defined* as the formation of S-R connections. It is not, however, very difficult to find instances of learning which, at least on the face of it, do not fit this S-R paradigm, and no less easy to see that even those paradigm cases of S-R learning such as Thorndike's and Pavlov's experiments can be described in other terms. Why should we not say that Thorndike's cat learned that food and escape from the puzzle box depended on pressing the catch, and that Pavlov's dog learned that the delivery of food depended on the ringing of the bell? Whole classes of animal experiments on learning, indeed, create immense problems for an S-R view. For if learning is defined as the establishment of an S-R connection, what are we to say when, as an apparent consequence of learning, we observe not an increase in the probability of a particular response, but its disappearance? Pavlov's dog, that showed an increase in the probability of salivating to the bell when the bell was followed by food, soon stopped salivating when the bell was presented without food. Pavlov termed this extinction, and there seems no reason why one should not say that just as the dog starts salivating because it learns that the bell signals the delivery of food, so it stops salivating because it learns that the bell no longer does so. But the opening chapter of a standard textbook of conditioning, discussing the problem of defining learning, dismisses certain solutions as inadequate because 'none of them distinguishes adequately between learning and the loss of learning, or extinction' (Kimble, 1961, p. 5). Because it does not fit the paradigm, extinction is defined as not being learning.

Similarly, Thorndike's cats, or their modern equivalents, Skinner's rats, who learn to press a lever to obtain food, will learn to stop doing so, if, in addition to food, pressing the lever also produces a mild electric shock. The result is hardly surprising. The symmetrical effects of rewards and punishments seem to require a symmetrical theory of learning: we could say that the rat can learn that a particular response has certain consequences; when those consequences are beneficial the probability of that response will increase; when they are aversive the probability of that response will decrease. For many years, however, S-R theorists either ignored or denied the existence of punishment, so difficult was it for them to come to terms with a situation in which learning resulted in the disappearance of a response. If they did accept the experimental evidence, they resorted to a bizarre and tortuous redescription of the data, whereby the reason why the rat stopped pressing the lever when punished for doing so, was because it learned, for reasons that remain obscure, to perform some other response, and this new response interfered with lever pressing.

It is time to accept a different account of conditioning. It is generally accepted that conditioning experiments can be operationally defined as arrangements of correlations or contingencies between events. Thorndike (and Skinner) arranged a contingency between a particular response (pressing a lever) and an event of significance or reinforcer (e.g. food or shock). Pavlov arranged a contingency between a particular stimulus (the bell) and a reinforcer. The most natural interpretation of their experiments, therefore, is that animals detect these contingencies: exposed to these particular relationships they learn to associate these correlated events.

Not only does this view have the advantage over S-R theory of providing a unified account of conditioning which includes those cases which create problems for S-R theory, it has the further virtue that it brings the theory of conditioning into the general fold of traditional association theory. No one has doubted that conditioning is a form of associative learning, but few conditioning theorists have made much of this connection. If it is accepted, however, that conditioning is a matter of associating those events which are correlated together in a conditioning experiment, i.e. stimuli or responses with reinforcers, then it is easy to show that these associations obey generally accepted associative laws. Both temporal and spatial contiguity between stimulus or response and reinforcer in conditioning experiments have profound effects on conditioning. A rat will come to press a lever if rewarded, or desist from pressing if punished, but such learning, which will proceed rapidly if food or shock is immediately contingent on lever pressing, will at best be slow and erratic if a delay of 10 seconds intervenes between response and reinforcer (Skinner, 1938). Similarly, the association between stimulus and reinforcer in a Pavlovian experiment depends not only on the temporal contiguity between the two, but on their spatial contiguity (Testa, 1974). Finally, constant conjunction is an important ingredient of successful conditioning: any degradation of the contingency between stimulus and reinforcer in a Pavlovian experiment, resulting either from decreasing the probability of the reinforcer in the presence of the stimulus, or from increasing the probability of the reinforcer in the absence of the stimulus, is likely to interfere with successful conditioning (Rescorla, 1968).

As one might well expect, however, neither perfect contiguity nor

constant conjunction are necessary for conditioning. Under appropriate circumstances, rats can associate events separated by several
minutes, and quite a low correlation between response and reinforcer
is sufficient to produce reliable instrumental learning. There is, however, an important constraint which limits the occurrence of conditioning under these circumstances, and which, as far as I am aware, has
escaped the notice of traditional associative theories. An animal may
indeed associate a stimulus or response with a particular reinforcer in
spite of a long interval or poor correlation between the two – but only
provided that there is no other event more closely related to that
reinforcer. The presence of such an event – in effect of a better
predictor of reinforcement – will prevent the less valid predictor from
being associated with that reinforcer.

This conclusion has been documented in a very large number of
experiments (reviewed in Mackintosh, 1974). Two brief illustrations
may be worth giving here. We can show that a rat has associated a light
with a brief electric shock, by arranging a contingency between the
two, and then turning on the light while the thirsty rat is licking at a
water tube. If it has associated light and shock, it will stop licking when
the light is turned on. Using this procedure, we can show that a single
trial on which the light signals the shock is sufficient to establish
significant conditioning, and will do so even though there is a 30-
second interval between the termination of the light and the delivery of
shock. But if that interval is filled with another stimulus, a tone, then
the temporal contiguity between tone and shock is greater than that
between light and shock; the tone is a better predictor of shock than is
the light, and the presence of the tone prevents conditioning to the
light.

Similarly, we can show that a hungry rat originally trained to press a
lever for occasional food, will associate pressing the lever with the
delivery of shock; we arrange a suitable contingency between responding and shock and observe a decline in lever pressing. Conditioning
occurs even though lever pressing is punished only occasionally, say
once every 5 minutes. But we can so arrange matters that each shock is
signalled by a brief tone, which occurs only on each punished trial,
coincident with the occurrence of the response that is just about to be
punished; now the tone is a better predictor of shock than is the

response of lever pressing, for lever pressing occurs throughout the experimental session without being punished more than once every 5 minutes. The presence of the tone significantly interferes with the association between responding and shock, and we can observe a rat who fails to learn that its lever pressing produces shock, and who therefore continues to press the lever in spite of being shocked for doing so.

Conditioning, as these results suggest, occurs selectively, in favour of better predictors of reinforcement at the expense of worse predictors. That this should be so is not entirely surprising: it is indeed the only rational way for the rat to learn. For consider: the rat is capable of associating two events separated by 30 seconds or more, and of learning this association in a single trial. Over a series of trials, significant conditioning occurs even when only one response in 50 or more is actually followed by reinforcement. Even in well-controlled conditioning experiments, therefore, let alone the ill-controlled natural environment, there will be numerous events whose relationship to a reinforcer would be sufficient to produce significant conditioning, but which are only accidentally, not causally, related to that reinforcer. The world, even the world of a laboratory rat, is full of chance conjunctions of events. If our laboratory animals attributed the occurrence of a reinforcer to an event that only occasionally preceded it, and that by a long interval of time, when some other event regularly and immediately preceded that reinforcer, they would have failed to grasp the causal structure of the world. Of course, if there is no other potential cause, then they must latch onto something; but, as a matter of fact, causes do more often than not immediately and regularly precede their effects, and an efficient associative system will be built to take advantage of this fact. By conditioning selectively in the way we have seen, laboratory rats succeed in attributing the occurrence of reinforcers to their most probable causes.

The subjects of a conditioning experiment, by detecting the set of contingencies in effect between various events in their environment, learn what causes those events of significance to them which psychologists call reinforcers. We are dealing with a primitive form of causal analysis – and one very much easier to study in animal subjects than in humans, whose concept of causality is inextricably interwoven with the notion of *understanding* the relationship between events. A rat treats event A rather than event B as the cause of C if A is better correlated with C than B is. A human adult may not do so unless he has a theory which explains the relationship between A and C. If his theory of the world implies that B and C are related while A and C are not, this may be sufficient to override A's superior correlation with C. Most people have formed a rather elaborate picture of the world by the time they enter the psychologist's laboratory. Because these preconceptions can influence their detection of new causal relationships, there are substantial advantages in the use of animal subjects in the study of basic associative learning.

The view that conditioning is a primitive form of causal analysis may seem far removed from the careful, precise analysis of experimental data that we expect of the practising scientist. I do not, of course, offer it as an *explanation* of any of the facts of conditioning; I do believe, however, that it provides an equally valid and rather more interesting way of thinking about conditioning than that offered by traditional S-R theory. It also provides a more plausible starting point for any attempt to extrapolate from conditioning to supposedly more complex processes than does a view which has encouraged the spread of the popular cliché of Pavlov's dog helplessly salivating at the sound of the bell, or the laboratory rat learning the way through the maze by blind trial and error.

If we want to extrapolate conditioning theory to cases of apparently more complex behaviour, it is surely foolish to start with a theory that is only uncomfortably applied to the behaviour of dogs, rats and pigeons in conditioning experiments. Whether it is worse than foolish is open to question. However implausible a theory may be, its survival can, with sufficient ingenuity and determination, usually be ensured for a depressingly long time. With a suitable redescription of the situation and the response that is learned, and with suitably elastic definitions of such key terms as stimulus and response, S-R theory can, in principle, be applied not only to most of the data of animal conditioning experiments, but also to a wide range of human behaviour. Most attempts to prove the impossibility of such an endeavour, as Professor Suppes has noted (Suppes, 1969), amount to little more than negative dogma rather than negative proof. They are arguments about plausibility, theoretical parsimony and elegance rather than formal proofs – however often disguised as the latter. This is only to say that they are the sorts of argument most commonly found in science. In psychology, at any rate, one would have to search long and hard to find an example of a theory disproved by a logically compelling argument. More typically, psychologists have been, more or less slowly and grudgingly, persuaded by an accumulation of evidence favouring one theory and creating problems for another. Logically tenable theories are discarded because they are cumbersome or inelegant, and fail to capture the salient features of the facts they are supposed to explain.

So has it been in the long debate on the adequacy of S-R theory. It may be impossible to disprove S-R theory, but we may be willing to agree with a distinguished former proponent of the theory that 'it seems time to face up squarely to the question of whether greater overall simplicity might not be achieved within a theory which conceives learning to be basically a matter of learning relations between stimulus events rather than the strengthening and weakening of the connections of responses to stimuli' (Estes, 1969, p. 165). Even Estes' formulation seems unnecessarily restrictive: when an experimenter arranges a contingency between two stimuli (a bell and food) in a Pavlovian experiment, it is reasonable to suppose that the subject associates these two events; but when the experimenter arranges a contingency between a particular response of the subject (pressing a lever) and a stimulus (food), it seems equally reasonable to suppose that the subject may associate its response with the reinforcer. Conditioning may not simply depend on learning the relationship between stimulus events. It is quite certain, indeed, that the delivery of a reinforcer may be associated not only with currently occurring stimuli and responses but also with events that occurred on earlier trials in a conditioning session: we must assume that animals remember these earlier events and associate their memories of those events with the outcome of the present trial.

As an associative theory of conditioning, S-R theory creates unnecessary problems by insisting that associations are established only between stimuli and responses. It is more realistic to suppose that animals are capable of associating any correlated events in their environment. It remains, however, to be seen whether *any* purely associative theory is sufficient to encompass the data of conditioning experiments. The major stumbling block for such a theory is the fact that conditioning occurs selectively. Animals do not associate all available events with the occurrence of a reinforcer, but condition selectively to a better predictor of reinforcement at the expense of a worse predictor. It is no accident that this principle should have been unrecognised by traditional associative theory. Traditionally, the formation of an association between a particular pair of events depends solely on the relationship between them – their similarity and contiguity, and the degree of conjunction or correlation between them. In fact, however, it is also affected by the relationship between these events and others. A given correlation between a light and food may or may not result in significant conditioning to the light, depending on whether there is a better predictor of food available.

At first sight, therefore, the fact of selective association suggests that an adequate theory of conditioning may have to incorporate some complex, non-associate principles. Here again, however, it would be extraordinarily rash to suppose that associative theories had been formally proved to be inadequate. Indeed, Rescorla and Wagner (1972) have recently proposed a model of conditioning which is able to explain with great elegance and parsimony the great majority of data on selective association with only one rather minor departure from a simple associative model. It is too early to say whether their solution to these problems is satisfactory (I have argued elsewhere – Mackintosh, 1975 – that it is not), but we may at least be sure that the decision will not be taken by logical fiat.

The debate on the current status of learning theory, it may be suggested in conclusion, has all too often been conducted at the wrong level. We are led to believe that conditioning is a process of great simplicity, and that there exists a well-defined, generally accepted, theory which accounts for the data of conditioning experiments in terms of simple S-R associations. Nothing could be further from the truth. Thus the question whether that well-defined simple theory applies to apparently more complicated situations should not arise. Questions about the status of learning theory or conditioning theory are, in the first instance, questions about the adequacy of our current conceptions of what happens in conditioning experiments. I suggest that those conceptions are not particularly adequate, for conditioning is itself a more complex process than has usually been supposed. We may then go on to consider further questions. The biologist or comparative psychologist might ask what sort of continuities or discontinuities exist between the behaviour of different animal groups, or between animals and men. A theoretical psychologist might want to ask whether the principles suggested by a careful analysis of conditioning experiments are relevant to the analysis of any aspects of human behaviour. I am confident that the answer to that question is yes: but I am equally confident that the success of any such extrapolation depends on the adequacy of the original theory.

University of Sussex

BIBLIOGRAPHY

Estes, W. K.: 1969, 'New Perspectives on Some Old Issues in Association Theory', in N. J. Mackintosh and W. K. Honig (eds.), *Fundamental Issues in Associative Learning*, Dalhousie University Press, Halifax, pp. 162–189.
Kimble, G. A.: 1961, *Hilgard and Marquis' Conditioning and Learning*, Appleton-Century-Crofts, New York.
Mackintosh, N. J.: 1974, *The Psychology of Animal Learning*, Academic Press, London and New York.
Mackintosh, N. J.: 1975, 'A Theory of Attention: Variations in the Associability of Stimuli with Reinforcement', *Psychological Review* **82**, 276–298.
Pavlov, I. P.: 1927, *Conditioned Reflexes*, Oxford University Press, London.
Rescorla, R. A.: 1968, 'Probability of Shock in the Presence and Absence of CS in Fear Conditioning', *Journal of Comparative and Physiological Psychology* **66**, 1–5.
Rescorla, R. A. and Wagner, A. R.: 1972, 'A Theory of Pavlovian Conditioning: Variations in the Effectiveness of Reinforcement and Nonreinforcement', in A. H. Black and W. F. Prokasy (eds.), *Classical Conditioning II: Current Research and Theory*, Appleton-Century-Crofts, New York, pp. 64–99.
Skinner, B. F.: 1938, *The Behavior of Organisms*, Appleton-Century-Crofts, New York.
Suppes, P.: 1969, 'Stimulus-Response Theory of Automata and Tote Hierarchies: a Reply to Arbib', *Psychological Review* **76**, 511–514.
Testa, T. J.: 1974, 'Causal Relationships and the Acquisition of Avoidance Responses', *Psychological Review* **81**, 491–505.
Thorndike, E. L.: 1911, *Animal Intelligence: Experimental Studies*, Macmillan, New York.

ANDRZEJ EHRENFEUCHT AND JAN MYCIELSKI

LEARNABLE FUNCTIONS

The general nature of the problem is that an organism must learn to make the 'right', or appropriate, response to its inputs. Typically the inputs are large amounts of data, so that the machine must learn to recognize the similarities between different inputs which call for the same response, contrasted with the distinctions that call for different responses. (Minsky and Selfridge, 1961.)

To describe a Boolean function $f: \{0, 1\}^n \to \{0, 1\}$ we need in general 2^n bits of information. Therefore for fairly large n most f's are not learnable in any sense, not even practically describable. Thus good learning algorithms, i.e. methods for interpolation or extrapolation, may exist only for very special classes of f's. One such class consists of linear threshold operations. For this class we have the perceptron learning algorithm (Duda and Hart, 1973; Minsky and Papert, 1972). A more traditional approach is to identify every argument $(x_1, ..., x_n) \in \{0, 1\}^n$ with a real number whose dyadic expansion is $0 \cdot x_1 x_2 ... x_n$ and to consider the class of continuous real valued functions with a sufficiently small modulus of continuity. Then the classical approximation theory is available (Lorentz, 1966; Natanson, 1964).

Now we will talk about a different approach which was initiated in Ehrenfeucht and Mycielski (1973a, b). Theorem 1 which we prove here was announced at the end of Ehrenfeucht and Mycielski (1973b).

Let I be the unit interval and $n = \{0, 1, ..., n-1\}$. We assume that the domain of f is the n-dimensional cube I^n. The points $x = (x(0), x(1), ..., x(n-1)) \in I^n$ represent the states of the environment of an organism and $f(x)$ is the correct response when the environment is in state x. We assume that n is fairly large (say $n = 200$) but that $f(x)$ can be computed from not more than k coordinates of x and k is quite small (say $k = 5$). The k-tuple of coordinates of x required to compute $f(x)$ may depend on x. Thus we make the following definition. First for any set S let $|S|$ denote the number of elements of S.

Butts and Hintikka (eds.), Foundational Problems in the Special Sciences, 251–256.

DEFINITION 1. f is k-continuous, in symbols $f \in C_k$, iff for every $x \in I^n$ there exists a set $S(x) \subseteq n$ such that $|S(x)| \leq k$ and for every $y \in I^n$ if $y \restriction S(x) = x \restriction S(x)$ then $f(y) = f(x)$.

For several facts concerning C_k see Ehrenfeucht and Mycielski (1973a). Given $f \in C_k$ and $x \in I^n$ it may be quite hard to find the appropriate $S(x)$ due to the enormous values of $\binom{n}{k}$. Thus we are led to the concept of a function which does this search. To define it first recall the conventions $I^{\varnothing} = \{\varnothing\}$, $n + 1 = \{0, ..., n\}$ and $S \subset T \Leftrightarrow (S \subset t$ and $S \neq T)$.

DEFINITION 2. A function

$$a: \bigcup_{K \subseteq n} I^K \to n + 1$$

is called an addressing function for f if it has the following four properties

   (i) $a(x \restriction K) \notin K$ for all $x \in I^n$ and $K \subseteq n$.
   (ii) If $a(x \restriction K) = n$ then $a(x \restriction L) = n$ for all $K \subseteq L \subseteq n$ and $x \in I^n$.
To state the remaining two properties we put

$$K_a(x) = \{a_0, ..., a_{m-1}\}, \quad \text{for all} \quad x \in I^n,$$

where $a_0, ..., a_{m-1}$ are defined by the following conditions $a_i \in n$, $a(x \restriction \{a_0, ..., a_{m-1}\}) = n$, $a_0 = a(\varnothing)$ and $a_{i+1} = a(x \restriction \{a_0, ..., a_i\})$. The third and fourth properties are the following:

   (iii) If $L \subset K_a(x)$ then $a(x \restriction L) \in K_a(x)$.
   (iv)   There exists a function g such that

$$f(x) = g(x \restriction K_a(x)) \quad \text{for all} \quad x \in I^n.$$

The intuitive meaning of this definition is the following. If we want to evaluate $f(x)$ we have to scan x, and a does this scanning. First if f is a constant, no scan is needed and then $a(\varnothing) = n$ will do. Otherwise $a(\varnothing)$ is some index such that we have to look at $x(a(\varnothing))$, then we look at $x(a(x(a(\varnothing))))$, $x(a(x(a(\varnothing))$, $x(a(x(a(\varnothing))))))$, etc. The procedure stops as soon as a takes on the value n (which is outside of the domain of x).

DEFINITION 3. If a is an addressing function for f then we define the height of a to be

$$h(a) = \max\{|K_a(x)|: x \in I^n\}.$$

We also define the height of f to be

$$h(f) = \min\{h(a): a \text{ is an addressing function for } f\}.$$

PROPOSITION. f is $h(f)$-continuous, i.e. $f \in C_{h(f)}$.

The proof is obvious.

THEOREM 1. If $f \in C_k$ then $h(f) \leq k^2$ and there are $f \in C_k$ such that $h(f) = k^2$.

Proof. (1) $f \in C_k \Rightarrow h(f) \leq k^2$. This is obvious for $k = 0$. Suppose it is valid for $k - 1$. Let x_1, $x_2 \in I^n$ and $f(x_1) \neq f(x_2)$. We put $S_0 = S(x_1) \cup S(x_2)$, where S is given by Definition 1. Thus $|S(x_1)| \leq k$ and $|S(x_2)| \leq k$ and $S(x_1) \cap S(x_2) \neq \varnothing$. (Otherwise $x_1 \upharpoonright S(x_1) \cup x_2 \upharpoonright S(x_2)$ would be a function which could be extended to an element y of I^n. Then by Definition 1 we would have $f(x_1) = f(y) = f(x_2)$ which violates our assumption $f(x_1) \neq f(x_2)$.) Therefore

$$|S_0| \leq 2k - 1.$$

Now let $x \in I^n$ and consider the cylinder $C(x \upharpoonright S_0)$ in I^n with basis $\{x \upharpoonright S_0\}$. $C(x \upharpoonright S_0)$ is a cube. f restricted to $C(x \upharpoonright S_0)$ is $(k-1)$-continuous since, for every $y \in I^n$, $S(y)$ intersects S_0 (this is readily verified by an argument similar to the above one which established $S(x_1) \cap S(x_2) \neq \varnothing$). Hence, by the inductive assumption, $f \upharpoonright C(x \upharpoonright S_0)$ has an addressing function $a_{x \upharpoonright S_0}$ with $h(a_{x \upharpoonright S_0}) \leq (k-1)^2$. Now let a be an addressing function for f which first scans completely $x \upharpoonright S_0$ and then continues scanning x following the instructions provided by $a_{x \upharpoonright S_0}$. Hence we have

$$h(a) \leq 2k - 1 + (k-1)^2 = k^2.$$

Therefore $h(f) \leqslant k^2$.

(2) $\exists f \in C_k$ with $h(f) = k^2$. f is defined as follows. Its domain is the space of matrices of dimension $k \times k$ with entries from I (which can be regarded as a cube I^{k^2}) and

$$f(x_{ij}) = \begin{cases} 0 & \text{if each column of } (x_{ij}) \text{ contains 0;} \\ 1 & \text{otherwise.} \end{cases}$$

Then $f \in C_k$ since to compute $f(x_{ij})$ it is enough to know a column of non-zeros in (x_{ij}) (if one exists) or a selector of zeros from each column (if one exists), that is, in each case, only k entries of (x_{ij}). Now let a be an addressing function for f. Then let us define a special matrix $(\varepsilon_{ij}) \in I^{k^2}$ such that whenever a scans an entry ε_{ij} it turns out that $\varepsilon_{ij} = 1$ unless ε_{ij} is the last unscanned entry of the jth column in which case $\varepsilon_{ij} = 0$. Thus a defines the matrix (ε_{ij}). It is clear that $h(a) = k^2$ since the value $f(\varepsilon_{ij})$ is unknown as long as not all ε_{ij} have been scanned. Hence $h(f) = k^2$. ■

Remark. The above result cannot be extended to all functions f whose domain is a subset of I^n, although the concept of k-continuity and heights can be extended to this situation, see Ehrenfeucht and Mycielski (1973a). The claim that $f(x_i) \neq f(x_2) \Rightarrow [S(x_1) \cap S(x_2) \neq \varnothing$ and $\forall x \in \text{dom}\,(f)\ [S(x) \cap (S(x_1) \cup S(x_2)) \neq \varnothing]]$, which was used in our proof, depends on the assumption that dom $(f) = I^n$. However, a weaker assumption suffices: For every pair of disjoint sets S_1, $S_2 \subseteq n$ with $|S_1|$, $|S_2| \leqslant k$ and every $x_1, x_2 \in \text{dom}\,(f)$ there exists an $x \in \text{dom}\,(f)$ such that

$$x \upharpoonright S_1 = x_1 \upharpoonright S_1 \quad \text{and} \quad x \upharpoonright S_2 = x_2 \upharpoonright S_2.$$

Let now the domain of f be any set $D \subseteq I^n$. We extend the Definitions 2 and 3 by substituting all occurrences of I^n by D. Let a be an addressing function for f. Given any sequence of lessons of the form

$$(\mathscr{L}) \qquad x_1, K_a(x_1), x_2, K_a(x_2), \ldots, x_t, K_a(x_t),$$

in which $x_i \in D$ and x_{i+1} differ from x_i by one coordinate only, we shall define an algorithm for computing certain values of a. Some rules (ℓ), $(\ell\ell)$ and $(\ell\ell\ell)$ will constitute this algorithm.

Given $(\mathcal{L})$ we define a function $\tilde{a}$ according to the following rules.

(ℓ) If $K_a(x_i) \subseteq L \subseteq n$ then

$$\tilde{a}(x_i \upharpoonright L) = n.$$

$(\ell\ell)$ If $j \in K_a(x_i)$ then

$$\tilde{a}(x_i \upharpoonright (K_a(x_i) - \{j\})) = j.$$

$$\tilde{a}(x_i \upharpoonright (K_a(x_i) - \{j\})) = j.$$

$(\ell\ell\ell)$ If $K_a(x_{i+1}) \neq K_a(x_i)$ and j the index of this (unique) coordinate which makes $x_{i+1} \neq x_i$ then

$$\tilde{a}(x_i \upharpoonright [(K_a(x_i) \cap K_a(x_{i+1})) - \{j\}]) = j.$$

Remark. Notice that under the assumptions of $(\ell\ell\ell)$ we have $j \in K_a(x_i) \cap K_a(x_{i+1})$ since otherwise we would have $K_a(x_i) = K_a(x_{i+1})$.

THEOREM 2.   *$\tilde{a}$ is a function and $\tilde{a} \subseteq a$.*

Proof. The assumptions of each of the rules (ℓ), $(\ell\ell)$, $(\ell\ell\ell)$ and the properties of the addressing function a (see Definition 2) force the consistency of $\tilde{a}$ with a. For each rule the proof is immediate. ∎

Remark. If D is finite then $\tilde{a} = a$ may happen.

What is the rôle of addressing functions in real life recognition and decision processes? Man and animals have mechanisms searching and scanning the important features of every situation and the addressing functions of this paper constitute an idealised mathematical model of such mechanisms.

Are there real life situations in which organisms learn their addressing functions from data in the form $(\mathcal{L})$ using the rules (ℓ), $(\ell\ell)$ and $(\ell\ell\ell)$? We can say only that this schema is reminiscent of some forms of real learning but literal interpretation is probably impossible. We have been attracted by this learning schema mostly because of its efficiency (although it produces an interesting $\tilde{a}$ only if the sets $K_a(x_i)$ in $(\mathcal{L})$ have a kind of independence), its deterministic character and the fact that the function which is learned does not exceed in height the original one. A related statistical learning algorithm is described in Ehrenfeucht and Mycielski (1973b).

University of Colorado, Boulder, Colorado

BIBLIOGRAPHY

Duda, R. O. and Hart, P. E.: 1973, *Pattern Classification and Scene Analysis*, Wiley and Sons.
Ehrenfeucht, A. and Mycielski, J.: 1973a, 'Interpolation of Functions over a Measure Space and Conjectures about Memory', *Journal of Approximation Theory* **9**, 218–236.
Ehrenfeucht, A. and Mycielski, J.: 1973b, 'Organisation of Memory', *Proc. Nat. Acad. Sci. U.S.A.* **70**, 1478–1480.
Lorentz, G. G.: 1966, *Approximation of Functions*, Holt Rinehart and Winston.
Minsky, M. and Papert, S.: 1972, *Perceptrons, an Introduction to Computational Geometry*, MIT Press (second printing).
Minsky, M. and Selfridge, O. G.: 1961, 'Learning in Random Nets, Information Theory', *Fourth London Symposium*, Butterworths, pp. 335–347.
Natanson, I. P.: 1964, *Constructive Function Theory*, F. Ungar Publ. Co.

VI

FOUNDATIONS
OF THE SOCIAL SCIENCES

V. Zh. KELLE AND E. S. MARKARIAN

THE METHODOLOGY OF SOCIAL KNOWLEDGE AND THE PROBLEM OF THE INTEGRATION OF THE SCIENCES

In this paper we decided to select the problem of the methodological analysis of the integration processes which are today taking place at the junction of the social and natural sciences. The main goal of the paper is to draw attention to certain new problems which confront methodological knowledge in connection with the change and transformation of the general strategy of the development of modern social and natural sciences and in connection with the construction of integration types and levels of scientific knowledge common to them. But before embarking on this topic we would like to clarify the nature of methodological knowledge because the term 'methodology' has various meanings.

We can single out two main meanings of this term as applied to scientific knowledge. First is the meaning which expresses the totality of basic theoretical principles for the cognition of the object. This type of methodological knowledge can be tentatively named theoretical methodology. Traditionally theoretical methodology was as a rule associated with philosophical knowledge. But in reality with the same effect it may express the special scientific principles of investigation as well.

Here we deal with a theory, already considered not as a result of cognition, but as a means of getting new knowledge. And that is why we named this type of methodological knowledge 'theoretical methodology'. Theory in this case appears as performing scientific technological functions. But as distinct from theory, 'theoretical methodology', while carrying out these functions, should not be rigidly attached to a given fragment of reality, since such an attachment would lead to methodology's loss of the necessary degree of generality.

Historical materialism is the marxist methodology of social cognition. But historical materialism is also a theory – a theory of social development, a philosophical-sociological theory of marxism. The reality reflected by this theory is the historic process as a whole, its most

Butts and Hintikka (eds.), Foundational Problems in the Special Sciences, 259–268.

general laws and moving forces. In the case of a society considered at this level historical materialism itself appears as a meaningful theory carrying out the explanatory function. The materialistic interpretation of history being based on cognition of the laws of social development provides an explanation of the essence of the historic process and the logic of its progressive development; it provides an explanation of the general interrelations existing between the various parts and components of the social whole; and it analyzes and interprets the dependence of all aspects of social life on the method of production of material goods, etc. But historical materialism does not and cannot explain the concrete processes, phenomena, relations of social reality. With respect to these phenomena its role is methodological.

The purpose and designation of scientific methodology is to provide an objective approach to reality. Its use allows a correct orientation of human activity based on the data of social cognition, an analysis of the historic process and of the modern social reality; and its use makes it possible to detect and generalize the essential changes taking place in life itself as well as in its cognition.

Not only the development but also the application of scientific methodology is a creative process since the social reality is flowing and infinitely variable. To penetrate into it does not mean to reduce the facts to a ready made scheme. It is known that Marx was absolutely against any transformation of his methodology of cognition of historic reality into a universal scheme that is forced on history.

The role of methodology is not to replace the process of cognition but to make it more efficient and fruitful.

Methodology is necessary not as a source of ready answers and solutions but as a powerful means for their search.

The condition for the preservation of the scientific character of methodology is its inseparable unity with the very process of cognition. Methodology varies in the process of cognition. It even exists as methodology only as long as it is applied and promotes the analysis of reality and generation of new knowledge.

In its second meaning the term 'methodology' is to express a type of knowledge of the totality of the cognitive means of science, i.e. the means of the study of the objects, construction and expression of scientific knowledge. In other words science itself, being considered as a totality of cognitive means (scientific theories, various methods, modes of the construction of knowledge, the language of the science etc.) becomes here an object of a special scientific investigation (Markarian, 1973). We can say that this is the proper meaning of the term because 'methodology' in its literal sense means a teaching of method.

It is this type of methodological knowledge that is meant when we speak about the methodology of science (or sciences). The methodology of science is usually correlated and considered in unity with the logic of science, because they investigate a common object, following, however, different goals.

As distinct from the logical approach to the systems of scientific knowledge, which abstracts from their contents, the methodological analysis of sciences does not follow the way of such an abstraction. It's sphere of interest comprises the really functioning systems of scientific knowledge considered from the point of view of their cognitive potential.

Such are the two main meanings of the term 'methodology' as applied to scientific knowledge. Both of them became established in the literature and, depending on the context, are equally appropriate.

The general aim of the fundamental empirical sciences up to the most recent decades of the 20th century was mainly to study specific laws of various levels of the organization of reality.

This task determined the differentiated and in many aspects separate development of the main groups of sciences: physics, biology, social sciences. The integration tendencies between them were expressed very slightly. These tendencies were mainly of a local nature, i.e. they took place within the limits of different groups of sciences. This general strategy of the scientific development in the past determined in particular the gap between the social and the natural sciences.

Today the general strategy of scientific development radically changes. If in the past the isolation of the investigating efforts of the main groups of sciences corresponded somewhat to the needs of social practice and technical progress; now, on the contrary, it becomes necessary to bring together as close as possible the main groups of sciences, and the cooperation of research efforts in the various spheres of knowledge, including the natural and social sciences.

These changes in the general strategy of scientific development in our times were caused by the enormous complication of the objects included in the social practice. From the point of view of exercising effective influence over the complex and supercomplex systems with which modern society deals, the knowledge of only the specific laws of the individual fundamental levels of organization became quite insufficient. Therefore the task of solving the problems facing society now requires broader and more profound theoretical generalizations, and the cognition of the invariants of various spheres of reality, capable of becoming a base for the integration of the fundamental ideas and conceptual schemes worked out in the major scientific disciplines.

The problems facing modern society nowadays require a considerable expansion in the spheres of usage of social knowledge. This expansion is first of all the result of the development of new applied functions of social knowledge, directed to the scientific control and regulation of social processes, including the processes of interaction between society and nature.

It is necessary to stress in this connection that the applied functions performed by the systems of knowledge may be quite different depending on the mode of obtaining and using information. This problem helps us to reveal a very important specific feature of the social knowledge which traditionally in addition to the purely scientific functions has humanitarian and education functions as well. These functions are directed towards the formation of socially important ideals and values.

The practice of regulating the social processes brings into the foreground the task of using the data of science for decision making. It was exactly these processes in the sphere of social knowledge which led to the distinct separation between the purely scientific components of social knowledge (social sciences) and the humanities. These two types of social knowledge are equally important for the functioning and the development of modern society, but their roles are different (Kelle, 1972).

The interrelations between these parts of social knowledge are very complex and complicated and we are not going to analyse them in detail. Our task consists only in abstracting from the sphere of modern social knowledge those components which are today directly related to the process of integration with the different branches of natural sciences.

Many authors, when speaking on the integration of social and natural sciences, mean the penetration of certain moral and other values into the practice of natural sciences. This problem is, of course, extremely important today. But, in this connection it is necessary to take into account the following. Firstly, it is not the social knowledge that develops those fundamental values which has to influence modern science. These values are produced by more general systems of society and culture and only find their rationalized, systematic and meaningful expression in corresponding spheres of social knowledge (usually intermediate between theoretical and practical consciousness).

Secondly, the problem of values in the development of science, as a special problem, which expresses complicated relations between the strategies for carrying out the scientific research and fundamental goals of the society, arises not only in the process of development of the natural sciences, but of the social sciences as well. The respective systems of values are equally important for these two branches of scientific knowledge, because in both of them the proper scientific and value criteria may not coincide.

Though the problem of the integration of the social and natural sciences and the problem of the application of the systems of values to the development of science relate to each other, nevertheless they are not identical, equivalent problems.

If we want to formulate the problem of the integration of the social and natural sciences in a correct way it is necessary to correlate equivalent components of these branches of knowledge. From this point of view it would be right to say that the process of integration with the natural sciences is first of all specific for those components of social knowledge which are in some way related to the problem of the scientific control and regulation of social processes. This is the fundamental problem that stimulates and determines the close interaction between the social and the natural sciences.

The basic goal of the social sciences on the whole during the last few centuries consisted in the theoretical self-determination in the system of scientific knowledge. This goal could be achieved only through solving the problem of the specificity of human society as an object of scientific investigation, by discovering its characteristic laws.

The working out of the theory of historical materialism played a decisive role in the processes of the theoretical self-determination of the social sciences. This theory, giving the explanation of the fundamental laws of the genesis of human society and its further development, has formulated the main methodological principles of understanding the history of mankind as a natural historical process. The most important property of these principles is that they make it possible to express the natural dynamics of society in the categories of consistent scientific thinking.

In contrast to the widespread practice of uncritical use of the experience of the natural sciences and the direct transference of their principles and concepts into the sphere of social sciences for the purpose of explaining sociocultural events, historical materialism explained the latter by revealing the inner specific features of the human society as a phenomenon of reality. And just due to this contribution, the necessary preconditions and premises for the theoretical self-determination of the social sciences were created.

F. Engels in his time compared the contribution made by K. Marx to the understanding of human history with the role C. Darwin played in the understanding of the biological evolution (Marx and Engels, 1961, p. 383). Many scholars of different theoretical orientations are induced to recognize the correctness of such a comparison. For example the well known contemporary American anthropologist H. Marvin writes in this connection:

In my opinion Engels was correct when he attributed to Marx the 'discovery of the law of evolution in human history'... Marx formulated a scientific principle at least as powerful as Darwin's natural selection – a general principle that showed how a science of human history might be constructed (Marvin, 1968, p. 19).

The theoretical self-determination of the social sciences was not only the most important general result of their previous developments, but also the necessary precondition for their fruitful interaction and integration with the natural sciences. Without this precondition, which created quite new methodological potentials for the social sciences, their fruitful interaction with the natural sciences would be impossible. Instead of interaction we would have fruitless, onesided uncritical perception of the ideas and principles of the natural sciences by the social sciences as so often took place in the past.

Analysing the problem of the interaction of the social and natural sciences we have in mind exactly these potentials of historical materialism. It is well known that social science has not up to now been unified in its theoretical principles like physics, but there are various schools and directions in this sphere of scientific knowledge which are far from being equivalent in their cognitive possibilities. We suppose that it is the methodological potential of historical materialist theory that determines the present rebirth of great interest in it by the representatives of non-Marxist social thought.

The integration of the social and the natural sciences should be understood not as a single overwhelming process, but as a dominating tendency with a wide perspective. This tendency is expressed in the processes of some aspects and components of the social and natural sciences growing closer to each other along with the emergence of various problems, goals and means of research common to them.

The demand for the integration of the social and natural sciences is ever growing. But because of their traditional independent development they are not yet sufficiently ready for close interaction. In this situation the logic and methodology of science have a very important role to play, because one of the conditions of the fruitful development of science now is its self-comprehension at the logical and methodological level. The problem of the interaction of the social and natural sciences should become one of the central problems for the modern logic and methodology of science. Meanwhile this problem and the problems of interaction between the different groups of sciences in general do not yet find sufficient space in logical and methodological research.

Even those logical and methodological trends whose representatives fully understand the necessity to investigate the processes of scientific development, pay almost no attention to the problem of the interaction of the sciences. They usually consider the problem in an abstract way, speaking of science in general; of they limit the interpretation of scientific development to the boundaries of individual groups of sciences. But today the problem of the interaction of the sciences becomes one of the most complicated and important problems in the understanding of scientific development. The practice of life and the practice of science both urge the necessity of a manysided and detailed study of the phenomenon of the interaction of different groups of sciences.

Using the term of T. Kuhn, it is possible to say that we are witnesses to the working out of a new scientific paradigm. But this paradigm is developed not within the limits or under the influence of this or that science but has a general character embracing practically all the main groups of sciences. It is a paradigm of the scientific and technical revolution taking place now. Due to its general nature this paradigm must be based on the new types and levels of the construction of scientific knowledge.

The first expressions of the paradigm of the general scientific and technical revolution are: cybernetics, general system theory, information theory, theory of organization, and some other new interdisciplinary branches of knowledge. They enable one to perceive the intellectual field formed within this paradigm. The main conclusion to be drawn concerning the nature of this field is that modern interdisciplinary integration syntheses are reached not by means of neglecting the specificity of the sciences but by constructing new levels of scientific abstraction.

The method of reductionism, although it can be effective in some cognitive situations, cannot in this case be used (Markarian, 1975). In this respect one cannot but stress the surprising agreement between the dialectical materialistic principles of approaching scientific knowledge and the fundamental requirements of the integration of different groups of sciences. These principles are based on the idea of the unity of scientific knowledge; but at the same time, unlike positivistic reductionism, they take fully into account the specificity of the different sciences.

The social sciences are very much interested in the logical and methodological studies of the problems of interaction between different groups of sciences. This is because one of the conditions of the further fruitful development of the social sciences is the accumulation and assimilation by them of many important scientific ideas of a general nature which up to now were considered mainly from the point of view of the natural sciences, although they are in fact also directly related to the social sciences. However, it is wrong to say that now only the natural sciences can positively influence the social sciences. There is every reason to believe that in the forthcoming decades the positive influence of the social sciences upon the natural sciences will become much greater. The influence of the social sciences upon the natural as well as the technical sciences is determined by the nature of the fundamental problems of modern scientfic knowledge, which urgently demand its humanization. Although in past centuries science always had to serve the interests of man, in many cases the disproportional and onesided development of science and material technology resulted in their objective relations to human demands being often either lost or superficially perceived.

This becomes obvious when one considers the example of the contemporary ecological situation. This situation is especially interesting to us because ecology belongs to those fundamental branches of science the nature of which urgently demands the integration of the ideas of the physical, biological and social sciences. Today at the junction of the problems of the optimal interaction of human society and nature an integrative scientific problem of grandiose character appears. It is apparent that the solution of this problem will become one of the most important theoretical bases of the integrative science of the future.

All this of course will influence the further development of the logic and methodology of science. Up to now they express mainly the traditional classification of sciences. Now, when this classification changes in many respects, it becomes necessary to form new dominant goals for logical and methodological studies. These goals have to reflect the processes of interaction of the main groups of sciences and their common problems and means of analysis.

Academy of the Sciences of the USSR
Academy of the Sciences of the Armenian SSR

BIBLIOGRAPHY

Kelle, V. Zh.: 1972, 'Znachenie i funkcii socialnogo znania pri socializme', *Voprosi Philosophii*, No. 5 (in Russian).
Markarian, E. S.: 1973, 'O fundamentalnikch aspectach podchoda k nauke' ('On the Fundamental Aspects of Approaching to the Science'), *Philosophskie Nauki, Dokladi Visshey Shkoli*, No. 4 (in Russian).

VII

JUSTICE AND SOCIAL CHANGE

AMARTYA SEN

WELFARE INEQUALITIES AND RAWLSIAN
AXIOMATICS*

ABSTRACT. This paper is concerned with ordinal comparisons of welfare inequality and its use in social welfare judgements, especially in the context of Rawls' 'difference principle'. In Section 1 the concept of ordinal inequality comparisons is developed and a theorem on ordinal comparisons of welfare inequality for distributional problems is noted. Section 2 is devoted to Harsanyi's (1955) argument that a concern for reducing welfare inequalities among persons must not enter social welfare judgements. In Section 3 an axiomatic derivation of Rawls' lexicographic maximin rule is presented; this relates closely to results established by Hammond (1975), d'Aspremont and Gevers (1975), and Strasnick (1975). In the last section the axioms used are examined and some alternative axioms are analysed with the aim of a discriminating evaluation of the Rawlsian approach to judgements on social welfare.

1. ORDINAL EQUALITY PREFERENCE

Usual measures of economic inequality concentrate on income, but frequently one's interest may lie in the inequality of welfare rather than of income as such.[1] The correspondence between income inequality and welfare inequality is weakened by two distinct problems: (i) welfare – even in so far as it relates to economic matters – depends not merely on income but also on other variables, and (ii) even if welfare depends on income alone, since it is not likely to be a linear function of it, the usual measures of inequality of income will differ from that of welfare. The first problem incorporates not merely the basic difficulties of interpersonal comparison of welfare, but also those arising from differences in non-income circumstances, e.g. age, the state of one's health, the pattern of love, friendship, concern, and hatred surrounding a person. The second reflects the fact that the usual measures, such as the coefficient of variation, or the standard deviation of logarithm, or the Gini coefficient, or the inter-decile ratio, will not be preserved under a strictly concave transformation of income as the welfare function is typically assumed to be, when it is taken to be cardinally

Butts and Hintikka (eds.), *Foundational Problems in the Special Sciences*, 271–292.
Copyright © 1977 by D. Reidel Publishing Company, Dordrecht-Holland. All Rights Reserved.

measurable. And when welfare is measurable only ordinally, then the usual measures of inequality are not even defined.

There is an obvious need for investigating inequality contrasts when welfare comparisons are purely ordinal. There is likely to be much greater agreement on the *ordering* of welfare levels of different persons than on a particular numerical interpersonal welfare function unique up to a positive linear transformation. However, the meaning of more or less inequality is not altogether clear when comparisons of welfare levels are purely ordinal.

There are, nevertheless, some unambiguous cases. Let (x, i) stand for the position of being person i in social state x. Taking two persons 1 and 2 and two states x and y, consider the following strict descending orders:

1	2	3	4
$(y, 2)$	$(y, 1)$	$(y, 2)$	$(y, 1)$
$(x, 2)$	$(x, 1)$	$(x, 1)$	$(x, 2)$
$(x, 1)$	$(x, 2)$	$(x, 2)$	$(x, 1)$
$(y, 1)$	$(y, 2)$	$(y, 1)$	$(y, 2)$

Note that irrespective of the relative values of the 'differences', in each case y displays more inequality than x in an obvious sense. This type of comparison will be referred to as 'ordinal inequality comparison'.[2]

To formalize this criterion, let $\tilde{R}$ stand for an agreed 'extended ordering'[3] over the Cartesian product of X (the set of social states) and H (the set of individuals), i.e. over pairs of the form (x, i). The meaning of $(x, i)\tilde{R}(y, j)$ is that i is at least as well off in x as is j in y. $\tilde{P}$ and $\tilde{I}$ stand for the corresponding concepts of 'strictly better' and 'indifference'. Let ρ stand for any one-to-one correspondence between the pair of persons to itself.

Two-person ordinal inequality criterion (TOIC): For any pair of social states x, y, for a two-person community, if there is a one-to-one correspondence ρ from the pair of persons (i, j) to the same pair such that: $(y, i)\tilde{P}(x, \rho(i))$, $(x, \rho(i))\tilde{R}(x, \rho(j))$, $(x, \rho(j))\tilde{P}(y, j)$, then x has less ordinal inequality than y, denoted $x\theta y$.[4]

The criterion can be extended to n-person communities also, by requiring the additional antecedent that all persons other than these two are equally well off under x and y. This implies an assumption of 'separability', which is however more debatable, and will be debated (see Section 4).

Strengthened two-person ordinal inequality criterion (STOIC): If for any n-person community with $n \geqslant 2$, for any two persons i and j, and any two social states x and y for some ρ: $(y, i)\tilde{P}(x, \rho(i))$, $(x, \rho(i))\tilde{R}(x, \rho(j))$, $(x, \rho(j))\tilde{P}(y, j)$, and for all $k \neq i, j$: $(x, k)\tilde{I}(y, k)$, then x has less ordinal inequality than y, denoted $x\theta^*y$. The transitive closure of θ^* is θ^{**}.

Of course, for a two person community $\theta = \theta^*$, STOIC implies TOIC in this sense.

Note that no condition of constancy of total welfare has been used in the definitions, and indeed no such concept is definable given utility comparisons that are purely ordinal. It is, however, possible to use these definitions in the particular context of ranking alternative distributions of a fixed total income. Indeed, for that particular problem of 'pure distribution', ordinal inequality comparisons can be linked with some well-known results in the normative approach to inequality measurement based on Lorenz curve comparisons (see Kolm, 1966; Atkinson, 1970; Dasgupta *et al.*, 1973; Rothschild and Stiglitz, 1973). Our motivation here differs, however, from those exercises in the sense that our current concern is to look at welfare inequalities as such without necessarily saying anything about social welfare by invoking some group welfare function, and this contrasts with comparing values of social welfare given by a group welfare function, or a class of such functions.

Let the ranking relation λ stand for strict 'Lorenz domination', i.e. $x\lambda y$ if and only if the Lorenz curve of x is nowhere below that of y and somewhere strictly above it.

(T.1) *In the 'pure distribution' problem with welfare rankings preserving the order of income rankings, if $x\lambda y$, then $x\theta^{**}y$.*

The proof follows immediately from a well-known result of Hardy *et al.* (1934), which in this context implies that $x\lambda y$ holds if and only if x can be obtained from some inter-personal permutation y° of y through a finite sequence of transfer operations with income being transferred from a richer person to a poorer one without reversing their income ranking.[5] Since in each of these operations taking us from x^s to x^{s+1}, the incomes of others except the two involved (say, $1(s)$ and $2(s)$ respectively) in that operation remain the same, and since more income implies higher welfare, clearly:

$$(x^s, 1(s))\tilde{P}(x^{s+1}, 1(s)), \quad (x^{s+1}, 1(s))\tilde{R}(x^{s+1}, 2(s)),$$
$$(x^{s+1}, 2(s))\tilde{P}(x^s, 2(s)).$$

Thus $x^{s+1}\theta^* x^s$. Since x and $y°$ are the two extreme members of this sequence, and by virtue of STOIC it makes no difference whether we start from $y°$ or from y, clearly $x\theta^{**}y$.

Notice that (T. 1) provides a welfare basis for comparisons of inequality which is not dependent on taking 'more social welfare' to be 'less unequal' and in this sense departs from the normative approach of Kolm (1966), Atkinson (1970), and others. Indeed, nothing is said about social welfare as such, and this welfare interpretation of inequality simply looks at the inequality of the welfare distribution (in ordinal terms).

Note also that no assumption of concavity (or quasi-concavity, or S-concavity) of welfare functions is required in establishing (T.1) in contrast with the earlier results on Lorenz comparisons referred to above, e.g. Kolm (1966), Atkinson (1970), Dasgupta et al. (1973), Rothschild and Stiglitz (1973).

It is, however, possible to introduce the additional assumption that less welfare-inequality is socially preferred, or at least regarded to be as good. Let R stand for the weak relation of social preference, with P and I its asymmetric and symmetric parts: 'strict preference' and 'indifference' respectively.

Two-person equality preference (TEP): In a 2-person community for any x, y, if $x\theta y$, then xRy.

Strengthened two-person equality preference (STEP): In any community, for any x, y, if $x\theta^* y$, then xRy.[6]

Again, STEP obviously implies TEP. Note, however, that STEP has been defined in terms of θ^* and not θ^{**}. Of course, if R is transitive, then the two are equivalent.

How appealing a condition is STEP? That would seem to depend on three types of considerations. The first applies to STEP only, while the last two apply to both STEP and TEP.

(1) STEP involves a 'separability' assumption, being based on θ^* (rather than θ in a 2-person community). A reduction of ordinal inequality between two persons is rather more definitive for a community of those two persons than for a community where there are others also, even though they are equally well off under x and y.[7] In Section 4 we shall examine the far-reaching consequences of this extension from a 2-person to an n-person comparison in the presence of other conditions, e.g. 'independence of irrelevant alternatives'.

(2) TEP and STEP both give overriding importance to the reduction of welfare inequality without bringing in any consideration of relative gains and losses of different persons.[8] How disturbing this criticism is will depend partly on the 'informational basis' of welfare comparisons, i.e. on the measurability and interpersonal comparability assumptions.[9] If individual welfare is not cardinal, or if interpersonal comparisons must be ordinal only (whether or not individual welfares are cardinal), then clearly the concept of 'gains' and 'losses' in welfare lose meaning. If, however, cardinal interpersonal comparisons can be made, then one can consider a choice in which x involves less ordinal inequality than y, but the loss of person 2 is so much and the gain of person 1 is so little, that a reasonable case can be made for the choice of y. With 'full comparability' this conflict of ethics (e.g. *vis-à-vis* utilitarianism) must be faced, but with 'level comparability' only, this objection to STEP or TEP cannot be sustained.[10]

(3) The approach of STEP or TEP shares with utilitarianism and other needs-based ethics, a disregard of non-welfare considerations, e.g., the concept of desert (see Sen 1973, Chapter 4) on the contrast between needs-based and deserts-based approaches). Arguments such as 'person i is better off than j both in x and in y and gains less than j loses, but the additional gain is his just desert', are not entertainable in this approach.[11]

2.   HARSANYI'S CRITICISM OF CONCERN FOR WELFARE INEQUALITY

In the context of social evaluation taking note of welfare inequalities, we should consider an objection of John Harsanyi (1975) to attempts at using social welfare functions that are non-linear on individual welfares. We know, of course, that with some assumptions of interpersonal comparability (e.g. 'unit comparability'), individual welfare levels cannot be interpersonally compared even though gains and losses can be compared (see Sen, 1970, Chapter 7). Harsanyi's attack is, however, not based on any subtlety of the comparability assumption.[12] It takes mainly the form of quoting his justly celebrated result that if individual preferences and social preference can both be given von Neumann-Morgenstern cardinal representation, and if Pareto indifference must imply social indifference, then social welfare must be a *linear combination* of individual welfares (Harsanyi, 1955), and then of defending the acceptability of the von Neumann–Morgenstern axioms. Harsanyi thus sees social welfare simply as an average welfare (unweighted if a further assumption of symmetry is made), and there is no question of reflecting a concern for welfare equality in the value of social welfare by choosing a non-linear form.

The first question to ask is whether the von Neumann–Morgenstern axioms are acceptable, especially for social choice. Diamond (1967) raised this question effectively, especially questioning the use of the strong independence axiom. Hansanyi (1975) has analysed the issue (pp. 315–8), but it seems to me to take little account of Diamond's main concern, viz, that our assessment of alternative policies from an *ethical* point of view, which is what Harsanyi means by 'social preference', may not depend only on the *outcomes* but also on the fairness in the *process* of interpersonal allocation. Harsanyi may well be right in claiming that 'when we act on behalf of other people, let alone when we act on behalf of society as a whole, we are under an obligation to follow, if anything *higher* standards of rationality than when we are dealing with our own private affairs' (Harsanyi, 1975, p. 316), but the bone of contention surely is *whether* the strong independence axiom represents a 'higher' standard of rationality in the social context than a rule that takes note of the allocational *process.*

The strong independence axiom is, of course, not the only axiom of the von Neumann–Morgenstern system that has been questioned. The continuity postulate raises difficulties that are well-known, and even the assumption of there being a complete social ordering over all lotteries is a fairly demanding requirement.

But suppose the von Neumann–Morgenstern axiom system is obeyed in social choice as well as in individual choices. In what sense does this rule out non-linear social welfare functions? Obviously, the von Neumann–Morgenstern values – let us call them the V-values – of social welfare will be a linear combination of the V-values of individual welfares. But when someone talks about social welfare being a non-linear function of individual welfares, the reference need not necessarily be to the V-values at all. The V-values are of obvious importance for predicting individual or social choice under uncertainty, but there is no obligation to talk about V-values only whenever one is talking about individual or social welfare.

What gives Harsanyi's (1955) concern with V-values a central role in his own model of social choice, is his concept of 'ethical' preference (the 'social preference of a person') being derived from the *as if* exercise (done by that person) of placing oneself in the position of everyone in the society with equal probability. (Note that Rawls' (1958) concern with 'ignorance' as opposed to 'equi-probability' is different and makes it impossible to define the von Neumann–Morganstern 'lotteries' for social choice except with some additional axiom, e.g. 'insufficient reason'.) These are lotteries that apply to a person's 'social' (or 'ethical') preference only, and need not figure in his actual preferences – what Harsanyi calls their 'subjective' preferences. It will, of course, still remain true that if the social preference follows the von Neumann–Morgenstern axioms, then welfare numbers W_i will be attributed to individuals in the V-value system for social preference such that social choice will be representable in terms of maximizing $W = (1/n)\sum_i W_i$. But this linear form asserts very little, since $W(x)$ is simply the value of the lottery of being each person i with $1/n$ probability in state x, and the set of W_i need not necessarily have any other significance.

Consider the following conversation:

1: "Let (x, i) be the position of being person i in state x. Tell me how you would rank $(x, 1)$, $(x, 2)$, $(y, 1)$, $(y, 2)$, please."

2: "The best is $(y, 2)$. Then $(x, 2)$. Then $(x, 1)$. Worst $(y, 1)$."

1: "And the welfare gaps between each pair of adjacent positions? Scale them with $(y, 2)$ being 10 and $(x, 1)$ marked zero."

2: "I can't on weekdays, when I feel ordinal."

1: "So on weekdays you are lost and don't know whether to recommend x or y as your ethical judgement for society?"

2: "No, I would recommend x. I even accept TEP on weekdays."

1: "On weekends you are not so ordinal?"

2: "On weekends, on your normalization, I would put 10 for $(y, 2)$, 5 for $(x, 2)$, 2 for $(x, 1)$ and 0 for $(y, 1)$, though I don't like making the 'origin' quite so arbitrary."

1: "Never mind the origin! Since the welfare sum is 10 with y and 7 with x, you clearly will recommend y on weekends?"

2: "No, no, I would recommend x."

1: ".So you don't follow von Neumann–Morgenstern axioms in these choices?"

2: "On Saturdays not. But on Sundays yes."

1: "But on Saturdays what do these cardinal welfare numbers stand for? What meaning can we attach to them since they are not von Neumann–Morgenstern numbers?"

2: "They reflect my views of the welfare levels and gaps. I can axiomatize them in many different ways.[13] The welfare numbers have quite nice properties."

1: "But I can't relate them to your observed behaviour."

2: "I should think not. Nor can I relate your von Neumann–Morgenstern numbers over *interpersonal* choices to your observed behaviour; there is not much to go by. No these numbers reflect my introspection on the subject as do yours, I presume."

1: "Okay, forget the Saturdays. But on Sundays you say your hypothetical interpersonal choices satisfy the von Neumann–Morgenstern axioms. Then you must choose y since social welfare must be the *sum* of these individual welfare numbers."

2: "No, not of these; social welfare is non-linear over these values. It is linear over the V-values, of course. The V-values, which take my distributional attitude into account (to the extent it is possible to do this within the von Neumann–Morgenstern system), are, with the normalization suggested by you: 10 for $(y, 2)$, 7 for $(x, 2)$, 4 for $(x, 1)$ and 0 for $(y, 1)$."

1: "I am relieved. I thought you were about to take social welfare to be a non-linear function of the V-values in von Neumann–Morgenstern representation."

2: "You must be joking."

1: "Anyway, I am so glad that on Sundays you are a utilitarian as far as V-values are concerned."

2: "I am also glad that your pleasures are inexpensive."

I end this section with two final comments. First, "Sen's utility-dispersion argument", to which Harsanyi (1975) makes extensive references (pp. 318 – 324), and which according to him "shows a close formal similarity ... to the view that the utility of a lottery ticket should depend, not only on its *expected* (*mean*) *utility*, but also on some measure of *risk*" (p. 320), and which "is an illegitimate transfer of a mathematical relationship for money amount, for which it does hold, to utility levels, for which it does not hold" (p. 321), is – in that form – a figment of Harsanyi's imagination. There is, alas, no 2-parameter "Sen's theory which would make social welfare depend, not only on the mean, but also on some measure of *inequality*, i.e. of *dispersion*" (Harsanyi, 1975, pp. 319 – 320). More importantly, there is no proposal, which would have been grotesque, to define a non-linear social welfare function on *von Neumann–Morgenstern utilities*.[14] Even the axioms for additive separability of the social welfare function over individuals was explicitly criticised (Sen (1973), pp. 39 – 41).

Second, whether we use utilitarianism or not is an important moral issue,[15] and is not disposable by carefully defining individual utilities in such a way that the only operation they are good for is addition. An axiomatic justification of utilitarianism would have more content to it if it started off at a place somewhat more distant from the ultimate destination.[16]

3. AXIOMATIZATION OF THE LEXICOGRAPHIC MAXIMIN RULE

The Rawlsian (1958, 1971) 'maximin' rule ranks social states in terms of the welfare of the worst-off individual in that state. This rule can violate even the Pareto principle. The lexicographic version of the maximin rule (Rawls, 1971; Sen, 1970) does not. This rule, which for brevity, and not out of disrespect, I shall call 'leximin', can be formalized in the following way. Let the worst-off person in state x be called $1(x)$, the second worst-off $2(x)$, and in general the j-th worst-off $j(x)$. When there are ties, rank the tied persons in *any* strict order. For an n-person community, for any x, y in X:

(i) xPy if and only if there exists some r: $1 \leqslant r \leqslant n$ such that

$$(x, i(x))\tilde{I}(y, i(y)) \quad \text{for all} \quad i: 1 \leqslant i < r,$$

and

$$(x, r(x))\tilde{P}(y, r(y));$$

(ii) xIy if and only if $(x, i(x))\tilde{I}(y, i(y))$ for all i: $1 \leqslant i \leqslant n$.

Leximin has been recently illuminatingly analysed in axiomatic terms by Hammond (1975), d'Aspremont and Gevers (1975), and Strasnick (1975). The axiomatization presented here is on similar lines but it differs in some important respects. In particular, the strategy adopted here is first to propose axioms such that the lexicographic maximin rule emerges for 2-person communities, and then to ensure by additional axioms that the lexicographic maximin rule holding for 2-person communities should guarantee the same for n-person communities.

There are, it seems to me, two advantages in this procedure. First, in the 2-person case the axioms are easier to assess and the proof of the theorem is extremely brief. It is my belief that the rationale of the leximin comes out best in this case, and it is worth noting that. Second, this procedure permits the isolation of what appears to me to be the least acceptable feature of leximin, which emerges in the move from 2-person leximin to n-person leximin. The issues raised are discussed in Section 4.

Consider first a 2-person community with persons 1 and 2. The following axioms are defined for a GSWF (generalized social welfare function): $R = f(\tilde{R})$, where R is the social ordering over X and $\tilde{R}$ the extended ordering over the product of X and H.

U (*Unrestricted domain*): Any logically possible $\tilde{R}$ is in the domain of f.

I (*Independence of irrelevant alternatives*): If the restrictions of $\tilde{R}$ and $\tilde{R}'$ on any pair in X are the same, then the restrictions of $f(\tilde{R})$ and $f(\tilde{R}')$ on that pair are also the same.

J (*Grading principle of justice*): For any x, y in X, if for some one-to-one correspondence μ from (1, 2) to (1, 2): $(x, 1)\tilde{R}(y, \mu(1))$ and $(x, 2)\tilde{R}(y, \mu(2))$, then xRy. If, furthermore, one of the two $\tilde{R}$'s is a $\tilde{P}$, then xPy.

T (*Two-person ordinal equity*): For any x, y in X, if one person, say 1, prefers x to y, and the other prefers y to x, and if person 1 is worse off than 2 both in x and in y, then xRy.

U and **I** are standard parts of the Arrow framework applied to extended orderings for a 2-person community. **J** is proposed by Suppes (1966). **T** corresponds to Hammond's Equity Axiom **E** in the two-person case, without the separability requirement built into it in the n-person case (for $n > 2$). It corresponds to **E** in the same way as TEP corresponds to STEP.

(T.2) *For a 2-person community, given at least three social states in X, leximin is the only generalized social welfare function satisfying* **U, I, J** *and* **T**.

Proof. Since it is easily checked that leximin satisfies **U, I, J** and **T**, we need concentrate only on the converse. Suppose **U, I, J** and **T** are satisfied, but not leximin. Leximin can be violated on one of three alternative ways. For some x, y in X:

(I) $(x, 1(x))\tilde{P}(y, 1(y))$, but not xPy.
(II) $(x, 1(x))\tilde{I}(y, 1(y))$ and $(x, 2(x))\tilde{P}(y, 2(y))$, but not xPy.
(III) $(x, i(x))\tilde{I}(y, i(y))$ for $i = 1, 2$, but not xIy.

Since (II) and (III) contradict **J** directly, we need be concerned only with (I). Suppose (I) holds.

If $(x, 2(x))\tilde{R}(y, 2(y))$, then xPy by **J**. Hence it must be the case that $(y, 2(y))\tilde{P}(x, 2(x))$. A fortiori, $(y, 2(y))P(x, 1(x))$. Consider now $\tilde{R}'$ reflecting the following strict descending order involving x, y and a third social state z: $(y, 2(y)), (x, 2(x)), (x, 1(x)), (z, 2(z)), (z, 1(z)), (y, 1(y))$. By **T** and **J**, $zR'y$, where $R' = f(\tilde{R}')$, and by **J**, $xP'z$. Hence $xP'y$. But then by **I**, xPy. So (I) is impossible. This establishes (T.2). ∎

Consider now a family of GSWFs, one for each subset of the community H. In what follows only those for pairs and for H will be used.

In addition to the axioms for 2-person communities, two axioms with a wider scope are now introduced. Let the social states x and y be called 'rank-equivalent' for $\tilde{R}$ if everyone's relative welfare rank is the same in x as in y, i.e. $i(x) = i(y)$ for all i.

B (*Binary build-up*): For any $\tilde{R}$, for any two rank-equivalent social states, for a set π of pairs of individuals in the community H such that $\bigcup \pi = H$, if xRy (respectively, xPy) *for each pair in* π, then xRy (respectively, xPy) for H.

J* (*Extended grading principle*): If $\tilde{R}'$ is obtained from $\tilde{R}$ by replacing i by $\mu(i)$ in all positions (x, i) for some x and all i, where $\mu(\cdot)$ is a one-to-one correspondence from H to H, then $f(\tilde{R}) = f(\tilde{R}')$ for H.

J* is an extension of **J** and is in the same spirit. Notice that it is not satisfied by many conditions, e.g. the method of majority decision; the majority method does not satisfy **J** either. **J*** stipulates essential use of interpersonal comparison information in an anonymous way, e.g. taking note of $(x, i)\tilde{R}(y, i)$ but in the same way as $(x, i)\tilde{R}(y, k)$.

(T.3) *Given at least three social states, if for each pair of persons in* H, *there is a 2-person GSWF satisfying* **U**, **I**, **J** *and* **T**, *then the only GSWF for* H *satisfying* **U**, **J*** *and* **B** *is leximin.*

Proof. It is clear from (T.2) that the GSWF for each pair of individuals is leximin. If the community H has only two members, then (T.3) is trivial. In general for any community H, leximin clearly satisfies **U** and **J***. It remains to be established that it must satisfy **B** also, and then to establish the converse proposition.

Suppose the GSWF is leximin, but **B** is violated. This is possible only if x and y are rank-equivalent, and

(I) xRy for all pairs in π, but not xRy for H.
(II) xPy for all pairs in π, but not xPy for H.

Consider (I) first. Since each R is an ordering, yPx must hold for H. Given the leximin nature of the GSWF for H, this is possible only if there is some rank r such that: $(y, r(y))\tilde{P}(x, r(x))$, and $(y, i(y))\tilde{I}(x, i(x))$, for all $i < r$. Given rank-equivalence, $i(x) = i(y) = i$, say, for all i. Thus: $(y, r)\tilde{P}(x, r)$, and $(y, i)\tilde{I}(x, i)$, for all $i < r$. Since r must belong to at least one pair included in π, for that pair, by leximin, yPx. So the supposition (I) leads to contradiction.

Next consider (II). If not xPy for H, then either yPx, which leads to the same problem as (I), or xIy, which is now considered. For leximin this implies, given rank equivalence, $(x, i)\tilde{I}(y, i)$ for all i. Clearly then xPy is impossible for any pair contained in π, thus contradicting (II).

Now the converse. Let the stated axioms hold. To establish that the GSWF for the community H must be leximin, we have to show that:

(III) If $(x, i(x))\tilde{I}(y, i(y))$ for all i, then xIy for H.

(IV) If there is some r such that $(x, r(x))\tilde{P}(y, r(y))$, and
for all $i < r$: $(x, i(x))\tilde{I}(y, i(y))$, then xPy for H.

Let the antecedent in (III) hold. Take the one-to-one correspondence μ such that $i(x) = \mu(i(y))$ for all i, and let this transformation applied to the y-invariant elements convert $\tilde{R}$ to $\tilde{R}'$. Note that x and y are rank-equivalent for $\tilde{R}'$. Note also that $f(\tilde{R}) = f(\tilde{R}')$ for all subset of H by $\mathbf{J}^*$. Consider now any set π of pairs of persons in H such that $\bigcup \pi = H$. Leximin guarantees $xI'y$ for all such pairs with $R' = f(\tilde{R}')$. By Binary build-up $\mathbf{B}$: $xI'y$ for H. By $\mathbf{J}^*$: xIy.

Finally, let the antecedent of (IV) hold. Consider μ and $\tilde{R}'$ as defined in the last paragraph with x and y rank-equivalent for $\tilde{R}'$. Consider now the set π of pairs $(r(x), i)$ for all $i \neq r(x)$. Since the GSWF for each pair is Leximin, clearly $xP'y$ for all pairs in π. Furthermore $\bigcup \pi = H$. Hence $xP'y$ for H by Binary build-up. By $\mathbf{J}^*$: xPy, which completes the proof. ∎

4. RAWLSIAN AXIOMS: A DISCRIMINATING ASSESSMENT

The axiom structure used in the last section to derive the Rawlsian leximin rule was not chosen to provide an axiomatic 'justification' of the rule. Rawls himself did not seek such a justification (see especially his 'Concluding Remarks on Justification' Rawls (1971, pp. 577–587)), and was much more concerned with being able 'to see more clearly the chief structural features' of the approach chosen by him (p. viii). In (T.2) and (T.3) axioms have been chosen with a view to distinguishing between different aspects of Rawlsian ethics, which would permit a discriminating evaluation.

Before examining the axioms one by one, it is important to clarify the type of aggregation that is involved in the exercise as a whole. Social choice problems can be broadly divided into the aggregation of personal 'interests' and that of 'judgements' as to what is good for society, and as I have tried to argue elsewhere (Sen, 1975), the 'theory of social choice' seems to have suffered from a persistent failure to make clear which particular problem is being tackled. It seems reasonable to take leximin as a proposed solution to the interest-aggregation exercise. The contrast between giving priority to the welfare ranking of the worst off person as opposed to the welfare ranking of the person who 'gains more' (as under utilitarianism) is a contrast between two alternative approaches to dealing with interest conflicts. The problem of aggregating people's different judgements on what should be done (e.g. aggregating different 'views' on the 'right' public policy), while central to Arrow's (1951) analysis of social choice, is not a problem to which leximin can be sensibly addressed.

It is perhaps easiest to think of a generalized social welfare function GSWF as an exercise by a person of deriving ethical judgements from his assessment of everyone's interests implicit in the particular $\tilde{R}$ in terms of which he does the exercise. (This is the sense in which Harsanyi (1955) also uses 'social preference': "When I speak of preferences 'from a social standpoint', often abbreviated to social preferences and the like, I always mean preferences based on a given individual's value judgements concerning social welfare" (p. 310).) The exercise can be institutional also, e.g. taking a person with a lower money income to be invariably worse off as a 'stylized' assumption in a poverty programme (see Atkinson, 1969). These exercises are done with one given $\tilde{R}$ in each case. The problem of basing a 'social judgement' on the n-tuple of 'extended orderings' $\{\tilde{R}_i\}$ – one for each person – is a different issue raising problems of its own (see Sen, 1970, Theorems 9*2, 9*2.1, and 9*3, pp. 154–6, and Kelly, 1975, 1975a).

The fact that a GSWF is defined as a function of $\tilde{R}$, an extended *ordering*, without any information on preference intensities, is of some importance, since this rules out the possibility of varying the ethical judgements with cardinalization. Formally the axiom in question is 'unrestricted domain' U, since if cardinalization made any difference in any particular case, R will not be a *function* of $\bar{R}$ in that case, and such an $\bar{R}$ will not be an element in the domain of $f(\cdot)$. However, even if R were not defined as a function of $\bar{R}$, and the possibility of using intensities of welfare differences were kept open, no essential difference will be made in the axiom structure used in the derivation of leximin. In the 2-person case in (T.2), axioms **J** (Grading principle of justice) and **T** (Two-person ordinal equity) along with **I** (independence of irrelevant alternatives) do 'lock' the social preferences leaving no room for cardinal intensities to exert themselves. In (T.3) there is formally a bit of room which is, however, easily absorbed by a slight variation of the axioms. For the responsibility of elimination of intensity considerations we must critically examine axioms other than **U**.

Axiom **J** (Grading principle of justice) is, however, quite harmless in this respect since it operates on utilizing dominance. Indeed, the preference relation generated by **J** is not merely a subrelation of the Rawlsian leximin relation, it is a subrelation also of the utilitarian preference relation (see Sen, 1970, pp. 159–160).[17] **J** incorporates the Pareto relation but also all similar dominance relations obtained through interpersonal permutations.[18]

The eschewal of intensities of welfare differences as relevant considerations is, however, an important aspect of **T** (Two-person ordinal equity). There is no dominance here, and person i's preference for x over y is made to override j's for y over x if i is worse off in each of the two states without any reference to the relative magnitudes of i's gain and j's loss. This was one reason for our hesitation with TEP also (Section 1) and the same applies to **T**. Both give priority to reducing welfare inequality in the ordinal sense without any concern for 'totals' and for comparing welfare 'differences'. If cardinal welfare comparisons were ruled out *either* because of ordinality of individual welfare, *or* because of level comparability, **T** like TEP would be a lot more persuasive. In so far as **T** is a crucial aspect of the Rawlsian approach, this point about 'informational contrast' is of great importance.[19] The hazier our notion of welfare differences, the less the bite of this criticism of the Rawlsian rules.

The usual criticisms of Arrow's use of the independence of irrelevant alternatives (see Sen, 1970, pp. 39, 89–92; and Hansson, 1973) also apply to the use of **I** for a GSWF. It rules our postulating cardinal measures of intensity based on rank positions (as in 'positional rules' discussed by Gärdenfors (1973), Fine and Fine (1974), and others). This, as it were, puts the last nail in the coffin of using welfare difference intensities.[20]

Turning now to (T.3), axioms **J*** and **B** have to be considered. **J*** uses the interpersonal permutation approach pioneered by Suppes (1966). While **J** uses it for 'dominance' only, **J*** uses this more generally, to the extent of not discriminating between two extended orderings where positions of individuals for some social state are switched around. The objections that apply to the usual 'anonymity' postulates apply here too, and it is particularly serious when considerations of personal liberty are involved (see Sen, 1970, Chapter 6, 1975b; Kelly, 1975).[21]

Binary build-up *B* is in some ways the least persuasive of the axioms used. It permits a lexicographic pattern of dictatorship of *positions* (being least well off) as opposed to persons (as in Arrow's, 1951 theorem). The worst off person rules the roost not merely in a 2-person community, but in a community of any size no matter how many person's interests go against his.[22]

Leximin can be derived without using **B** (see Hammond, 1975; d'Aspremont and Gevers, 1975; and Strasnick, 1975), and using instead conditions that look less narrow in their focus (e.g. d'Aspremont and Gevers' 'elimination of indifferent individuals'). However, leximin must satisfy **B**, as we establish in (T.3). And no matter how we 'derive' leximin, **B** is an integral part of the Rawlsian set-up. This seems to bring out a rather disagreeable feature of leximin. In a 2-person community in the absence of information on welfare difference intensities, it might seem reasonable to argue that the worse off person's preference should have priority over the other's, but does this really make sense in a billion-person community even if everyone else's interests go against that of this one person? The transition from 2-person leximin to *n*-person leximin (making use of Binary build-up) is a long one.

It may be interesting to observe how Binary build-up creeps into axioms that look rather mild. Consider Hammond's Equity axiom **E**. It differs from the 2-person equity axiom **T** used here in being extendable to *n*-member communities also if all others are *indifferent* between *x* and y. This may look innocuous enough, but in the presence of **U** and **I**, this 'separability' assumption is quite overpowering. The 'elimination of indifferent individuals', as d'Aspremont and Gevers (1975) call it, is not merely (as it happens) spine-chilling in the choice of words, but also quite disturbing in its real implications. The condition is defined below in a somewhat different form (to permit ready comparability with Hammond's **E**).

EL (*Elimination of the influence of indifferent individuals*): For any $\tilde{R}$, for any x, y, if $(x, i)\tilde{I}(y, i)$ for all i in some subset G of H, and if f^H and f^{H-G} are the GSWF's for the communities H and $(H-G)$ respectively, then $xf^H(\tilde{R})y$ if and only if $xf^{H-G}(\tilde{R})y$.

Notice that our **T** (Two-person ordinal equity) and **EL** together imply Hammond's Equity axiom **E**. What may, however, not be obvious is that for the class of GSWF satisfying unrestricted domain (**U**) and independence (**I**), **T** and **EL** together eliminate the influence not merely of indifferent individuals but also of non-indifferent ones, leading to a single-minded concern with one person. The same effect is achieved by Hammond's **E** itself in the presence of the other axioms.

SFE(n)(*Single-focus equity for n-member communities*): If for an n-member community, for any x and y, $\tilde{R}$ involves the strict extended order: (y, n), (x, n), $(y, n-1)$, $(x, n-1)$, ..., $(y, 2)$, $(x, 2)$, $(x, 1)$, $(y, 1)$, then xRy.

SFE(2) is equivalent to two-person equity **T** (and trivially to Hammond's **E** also), and may not be thought to be exceptionally objectionable (especially in the absence of preference intensity information). But **SFE**(n) for relatively large n is very extreme indeed, since everyone other than 1 is better off under y than under x, and still xRy.

(T.4) *Given at least three social states, if* **U**, **I** *and* **EL** *hold for the GSWF for each subset of the community H, then* **T** *implies* **SFE**(k) *for GSWF for each subset of the community including the one for the entire community H (i.e. $k \leq n$).*

Proof Suppose **SFE**(m) holds for some $m < n$. We first show that **SFE**($m+1$) holds. Consider the following extended order (with indifference written as $=$) for the triple x, y and some z: $(y, m+1)$, $(x, m+1) = (z, m+1)$, $(y, m) = (z, m)$, (x, m), $(y, m-1)$, $(z, m-1)$,

$(x, m-1), ..., (y, 2), (z, 2), (x, 2), (x, 1), (z, 1), (y, 1)$. By **U**, this is admissible. By **SFE**(m), for the m-member community $(1, ..., m)$, xRz. By **EL**, for the $(m+1)$-member community $(1, ..., m+1)$, xRz also. Again, by **SFE**(m), for the m-member community $(1, ..., m-1, m+1)$, zRy. By **EL**, for the $(m+1)$-member community $(1, ..., m+1)$, zRy also. By the transitivity of R, xRy. By independence **I**, this must be due solely to the restriction of $\tilde{R}$ over the pair (x, y), and this establishes **SFE**$(m+1)$.

The proof is completed by noting that **SFE**(2) holds since it is equivalent to Two-person equity **T**, and then obtaining **SFE**(k) for all $k \leq n$ by induction. ■

While the Rawlsian leximin can be established from axioms that look more appealing, it must end up having the extreme narrowness of focus that is represented by **SFE**(n) for large n. This in itself is obvious enough, since leximin clearly does satisfy **SFE**(n). What (T.4) does is to show precisely how it comes about that such apparently broad-focussed conditions together produce such a narrow-focused property.

It should be observed that the force of the Rawlsian approach as a critique of utilitarian ethics stands despite the limitation of **SFE**(n). **SFE**(2) is equivalent to **T** – an appealing requirement – and while Rawlsian leximin satisfies it, utilitarianism may not. As large values of n are considered, **SFE**(n) becomes less appealing and so does – naturally – Rawlsian leximin, but the criticism of utilitarianism is not thereby wiped out. In this paper a 'warts-and-all' view of Rawlsian leximin has been taken, choosing a set of axioms with the focus on transparency rather than immediate appeal. This 'warts-and-all' ax-iomatization does not, however, give any reason for disagreeing with Rawls' (1971) own conclusions about his theory: (i) "it is not a fully satisfactory theory", and (ii) "it offers ... an alternative to the utilitarian view which has for so long held the pre-eminent place in our moral philosophy" (p. 586). Rawls was, of course, referring to his theory in its broad form including his contractual notion of fairness and justice, but the observations seem to apply specifically to leximin as well.

London School of Economics

NOTES

* Thanks are due to Peter Hammond and Kevin Roberts for helpful comments and criticisms.

[1] See Hansson (1975).

[2] Cf. the concept of 'ordinal intensity' in comparisons of preference intensity used by Blau (1975) and Sen (1975b).

[3] See Sen (1970), Chapters 9 and 9*, for a discussion of the concept of extended ordering.

[4] Note that this criterion permits $(x, \rho(i)) \bar{I} (x, \rho(i))$, in contrast with the examples noted above. But obviously if $\bar{I}$ holds, then there is no inequality in x at all, and this must be less than whatever inequality there is in y.

[5] See Dasgupta et al. (1973), or Sen (1973), pp. 53–6.

[6] Note that STEP subsumes Hammond's (1975) axiom of 'Equity' (E), which extends and generalizes Sen's (1973) 'weak equity axiom' (WEA). Preferring inequality reduction irrespective of any consideration of the 'total' (as in STEP) has the effect of giving priority to the person who is going to be worse off anyway (as in these equity axioms).

[7] Note, however, that the characteristic of 'separability' is shared by STEP with many other criteria, e.g. utilitarianism, or the lexicographic maximin. Cf. the condition of 'elimination of indifferent individuals' (EI) of d'Aspremont and Gevers (1975).

[8] In the special case of 'pure distribution' problem however, there will be no conflict between the welfare sum and the equality of the welfare distribution if everyone shares the same concave welfare function on individual income. This is, however, a very special case, and the condition can be somewhat relaxed without introducing a conflict; on this see Hammond (1975a).

[9] See Sen (1970, 1973), d'Aspremont and Gevers (1975), and Hammond (1975).

[10] However, even with *partial* unit comparability there will be a quasi-ordering of the total (see Sen, 1970; Fine, 1975; Blackorby, 1975). With level comparability only, this quasi-ordering will shrink to the weak n-person version of Suppes' 'grading principle of justice', which will never contradict θ or θ^*.

[11] Contrast Nozick (1973). See also Williams (1973, pp. 77–93).

[12] Indeed the interpretation of Harsanyi's (1955) own theorems on social choice is seriously hampered by his silence on the precise comparability assumption (on which and related issues, see Pattanaik (1968)).

[13] See, for example, Krantz, Luce, Suppes and Tversky (1971).

[14] The only page reference Harsanyi gives on this point, which he discusses so extensively, is to p.18 of Sen (1973). I see nothing there that justifies Harsanyi's presumption that I had non-linear designs on utilities in the *von Neumann–Morgenstern representation*, let alone a 2-parameter non-linear design on them.

[15] For an illuminating debate on this see Smart and Williams (1973).

[16] For some extremely interesting recent contributions in this direction, see d'Aspremont and Gevers (1975), Hammond (1975a), and Maskin (1975), even though more work may still need to be done in terms of starting off from individual welfare functions which are not necessarily confined precisely to the class of positive affine transformations.

[17] Blackorby and Donaldson (1975) demonstrate that with cardinal interpersonal comparability the convex hull of the 'at least as good as' set according to the grading

principle is a subset of the intersection of the utilitarian and leximin 'at least as good as' sets, and in the 2-person case exactly equals the intersection.

[18] Hammond's axioms for leximin include the symmetric part of the grading relation (his **S**) as well as the asymmetric part in the case coinciding with the Pareto strict preference (his **P***), but the remainder of **J** follows from his remaining axioms **U, I, P*** and **S** (as he notes in Theorem 5.1).

[19] See Sen (1970, 1973), and d'Aspremont and Gevers (1975).

[20] Furthermore, combined with the 'separability' assumption implicit in Binary build-up **R** (or in Hammond's **E**, or in d'Aspremont and Gevers' **EI**, to be defined below), independence of irrelevant alternatives can be very demanding from the point of view of inter-pair consistency (see (T.4) below). In this its role here is not dissimilar to that in the class of possibility theorems on social welfare functions without interpersonel comparisons.

[21] In this sense it seems a bit misleading to call Rawls' theory a 'liberal theory of justice' (see, for example, Barry (1973), which is a very helpful contribution otherwise).

[22] It should be remarked that utilitarianism does not satisfy Binary build-up **B**. For example it is possible that in a 3-person community with 1 preferring x to y and the others y to x that $W_1(x) - W_1(y) > W_i(y) - W_i(x)$, for $i = 2, 3$, but $W_1(x) - W_1(y) < \sum_{i=2,3} [W_i(y) - W_i(x)]$. So xPy for the 2-person communities $(1, 2)$ and $(1, 3)$, but not for their union $(1, 2, 3)$. However, utilitarianism satisfies another – and in some ways weaker – binary built-up condition B^*, viz, if xRy for a set of pairs which *partition* the community (with no person belonging to more than one pair), then xRy for the community (and similarly with P). Strasnick's (1975) condition of 'unanimity' incorporates B^* for any partition of the community (not necessarily into pairs), and is more reasonable and much less demanding than **B**, in this sense.

BIBLIOGRAPHY

Arrow, K. J.: 1951, *Social Choice and Individual Values*, Wiley, New York; 2nd edition, 1963.

Arrow, K. J.: 1973, 'Some Ordinalist-Utilitarian Notes on Rawls' Theory of Justice', *Journal of Philosophy* **70**, 245–280.

Atkinson, A. B.: 1969, *Poverty in Britain and the Reform of Social Security*, Cambridge University Press.

Atkinson, A. B.: 'On the Measurement of Inequality'. *Journal of Economic Theory* **2**, 244–263.

Barry, B.: 1973, *The Liberal Theory of Justice*, Clarendon Press, Oxford.

Blackorby, C.: 1975, 'Degrees of Cardinality and Aggregate Partial Orderings', *Econometrica* **43**, 845–852.

Blackorby, C. and Donaldson, D.: 1975, 'Utility vs Equity: Some Plausible Quasi-orderings', mimeographed, forthcoming *Journal of Public Economics*.

Blau, J. H.: 1975, 'Liberal Values and Independence', *Review of Economic Studies* **42**, 395–401.

Daniels, N. (ed.): 1975, *Reading Rawls*, Basil Blackwell, Oxford.

Dasgupta, P., Sen, A., and Starret, D.: 1973, 'Notes on the Measurement of Inequality of Incomes', *Journal of Economic Theory* **6**, 180–7, d'Aspremont, C. and Gevers, L.: 1975, 'Equity and the Informational Basis of Collective Choice', presented at the Third World Econometric Congress, mimeographed, forthcoming *Review of Economic Studies*.

Diamond, P.: 1967, 'Cardinal Welfare, Individualistic Ethics and Interpersonal Comparisons of Utility: A Comment', *Journal of Political Economy* **61**, 765–6.

Fine, B.: 1975, 'A Note on 'Interpersonal Comparison and Partial Comparability'' *Econometrica* **43**, 169–172.

Fine, B. and Fine, K.: 1974, 'Social Choice and Individual Ranking', *Review of Economic Studies* **42**, 303–322, and 459–475.

Gärdenfors, P.: 1973, 'Positionalist Voting Functions', *Theory and Decision* **4**, 1–24.

Hammond, P. J.: 1975, 'Equity, Arrow's Conditions and Rawl's Difference Principle', mimeographed; forthcoming *Econometrica*.

Hammond, P. J.: 1975a, 'Dual Interpersonal Comparisons of Utility and the Welfare Economics of Income Distribution', mimeographed, Essex University.

Hansson, B.: 1973, 'The Independence Condition in the Theory of Social Choice', *Theory and Decision* **4**, 25–49.

Hansson, B.: 1975, this volume, p. 303.

Hardy, G., Littlewood, J., and Polya, G.: 1934, *Inequalities*, Cambridge University Press.

Harsanyi, J. C.: 1955, 'Cardinal Welfare, Individualistic Ethics and Interpersonal Comparisons of Utility', *Journal of Political Economy* **63**.

Harsanyi, J. C.: 1975, 'Nonlinear Social Welfare Functions', *Theory and Decision* **6**, 311–332.

Kelly, J. S.: 1975, 'The Impossibility of a Just Liberal', mimeographed; forthcoming *Economica*.

Kelly, J. S.: 1975a, 'Arrow's Theorem, Grading Principle of Justice and the Axiom of Identity', mimeographed, University of Minnesota.

Kolm, S. Ch.: 1966, 'The Optimum Production of Social Justice', paper presented in the Biarritz Conference on Public Economics; published in J. Margolis (ed.), *Public Economics*, Macmillan, London.

Krantz, D. H., Luce, R. D., Suppes, P., and Tversky, A.: 1971, *Foundations of Measurement*, Academic Press, New York.

Maskin, E.: 1975, 'A Theorem on Utilitarianism', mimeographed, Cambridge University.

Nozick, R.: 1973, *Anarchy, State and Utopia*, Basil Blackwell, Oxford.

Pattanaik, P. K.: 1968, 'Risk, Impersonality and the Social Welfare Function', *Journal of Political Economy* **76**.

Pattanaik, P. K.: 1971, *Voting and Collective Choice*, Cambridge University Press.

Phelps, E. S., ed.: 1973, *Economic Justice*, Penguin, Harmondsworth.

Rawls, J.: 1958, 'Justice as Fairness', *Philosophical Review* **67**, 164–194.

Rawls, J.: 1971, *A Theory of Justice*, Harvard University Press, Cambridge, Mass., and Clarendon Press, Oxford.

Rothschild, M., and Stiglitz, J. E.: 'Some Further Results on the Measurement of Inequality', *Journal of Economic Theory* **6**, 188–204.

Sen, A. K.: 1970, *Collective Choice and Social Welfare*, Holden-Day, San Francisco, and Oliver and Boyd, Edinburgh.

Sen, A. K.: 1973, *On Economic Inequality*, Clarendon Press, Oxford, and Norton, New York.

A. SEN

Sen, A. K.: 1975, 'Social Choice Theory: A Re-examination'. text of lecture at the Third World Econometric Congress; forthcoming *Econometrica.*

Sen, A. K.: 1975a, *'Interpersonal Comparisons of Welfare'*, mimeographed; forthcoming in a festschrift for Tibor Scitoysky.

Sen, A. K.: 1975b, 'Liberty, Unanimity and Rights', mimeographed, London School of Economics; forthcoming *Economica.*

Smart, J. J. C. and Williams, B.: 1973, *Utilitarianism: For and Against*, Cambridge University Press.

Strasnick, S.: 1975, 'Arrow's Paradox and Beyond', mimeographed, Harvard University.

Suppes, P.: 1966, 'Some Formal Models of Grading Principles', *Synthese* **6;** reprinted in P. Suppes *Studies in the Methodology and Foundations of Science*, Dordrecht, 1969.

Williams, B.: 1973, 'A Critique of Utilitarianism', in Smart and Williams (1973).

JOHN C. HARSANYI

NONLINEAR SOCIAL WELFARE FUNCTIONS: A REJOINDER TO PROFESSOR SEN

I shall comment only on Section 2 of Amartya Sen's contribution, where he discusses a recent paper of mine (Harsanyi, 1975). In this paper, I argued for the *utilitarian* view that our social welfare function ought to be defined as a *linear* combination–and, indeed, as the arithmetic mean–of the different individuals' von Neumann-Morgenstern (vNM) utility functions. One basis of my arguments was the following mathematical theorem (proved in Harsanyi, 1955). If (a) the individual members of society follow the Bayesian rationality axioms in their behavior; and *if* (b) our moral choices between alternative social policies likewise follow these rationality axioms; and *if* (c) we are always morally indifferent between two social policies when we know that all individuals in our society would be indifferent between the effects of these two policies; *then* our social welfare function will be, as a matter of mathematical necessity, a *linear* combination of all individuals' vNM utility functions. In the same paper, I also criticized Sen's views favoring *nonlinear* social welfare functions.

In his present paper, Sen claims that my arguments have been inconclusive for the following reasons. (1) At least one of the Bayesian rationality axioms, the sure-thing principle (which Sen calls the strong independence axiom) has no legitimate application to choices among alternative social policies. Here, Sen makes it clear that he endorses Diamond's arguments against this axiom. (2) Even if both individual choices and social-policy choices do obey the Bayesian rationality axioms, it still does not follow that we *have* to define our social welfare function as a linear combination of individual utilities (or of 'individual welfares', as Sen puts it). Sen admits that, under these assumptions, the quantity maximized by our choices among alternative social policies

Butts and Hintikka (eds.), Foundational Problems in the Special Sciences, 293–296.

will be in fact a linear combination of all individuals' vNM utility functions. *But we need not call this particular quantity our social welfare function.* Likewise, even though, under these assumptions all individuals will have well-defined vNM utility functions, *we are free to use indicators other than their vNM utility functions* to measure their individual utility (or individual welfare) levels.

I shall now comment on Sen's arguments.

(1) In my paper, I pointed out some curious implications of Diamond's views. Since Sen still endorses these views, I can only assume that he does not mind these implications. As I pointed out in my paper, Diamond's views would imply that the most extreme hereditary social and economic inequalities would become morally *less objectionable* if newborn babies were regularly interchanged among different families by a big government-conducted lottery—or even if we could get firm assurance from the most reliable theologians that babies were allocated to specific families by a big heavenly lottery in the first place, even without such governmental intervention. Does Sen consider this implication to be really acceptable?

(2) I find the second argument extremely strange. It proposes that, if both individual choices and social-policy choices did follow the Bayesian rationality axioms, then we should *act* as good utilitarians, by always choosing the social policy maximizing a given specific linear combination of all individuals' vNM utility functions. But, at the same time, we should use a terminology which amounts to carefully *disguising* the fact that we are utilitarians. We should refuse to *call* this linear combination of vNM utilities our social welfare function, even though this is the quantity we are maximizing when we choose among different social policies. Likewise, we should refuse to *call* an individual's vNM utility function our measure for his personal welfare, even though, in the mathematical expression used to evaluate alternative social policies, we would go on representing this individual's interests by his vNM utility function. No doubt, we *could*, if we wanted to, proceed in this very pecular manner. But it is hard to see what we could gain by following Sen in this rather elaborate and cumbersome camouflage operation.

It is even harder to see what the admitted possibility of such a make-believe scenario is supposed to prove. Indeed, Sen's scenario concedes my whole point. It concedes that, in the assumed situation, in spite of all the make-believe we *could not* without inconsistency really avoid following a utilitarian approach, i.e. could not avoid maximizing a certain linear combination of the different individuals' vNM utility functions by our social policy choices. We could not really avoid following the utilitarian approach – all we could do is to use a terminology carefully chosen to *hide* this fact.

Sen is complaining that I incorrectly attributed to him advocacy of a *two-parameter* social welfare function which would involve only two arguments, viz. the *mean* of all individuals' utilities, and some measure of *inequality* (or of *dispersion*) among these utilities. I am sorry if I misrepresented Sen's views in this matter. But, of course, my arguments against Sen's position are quite *independent* of the actual number of parameters entering into his social welfare function. A nonlinear social welfare function involving the mean of individual utilities and *any number* of utility-inequality measures (for instance, measures indicating the degrees of inequality at different parts of the utility distribution, etc.) would be still open to the very same *logical* objections. My criticism of the utility-dispersion argument for lotteries (Harsanyi, 1975, pp. 320–321) would still apply; and so would my contention that a social welfare function nonlinear in the individual utilities could be justified only by making the absurd assumption of a decreasing marginal utility for *utility* (*op. cit.*, pp. 321–322).

The grave *moral* objections to Sen's position are likewise quite independent of the number of utility-inequality measures he proposes to use in his social welfare function. Any social welfare function nonlinear in the individual utilities will remain morally completely unacceptable, regardless of its specific mathematical form, because it is bound to give rise to *unfair discrimination* between different individuals by assigning grossly unequal social priorities to one person's and another person's equally urgent human needs. Such discrimination will be morally unacceptable regardless of whether it is undertaken because some individuals already enjoy rather high utility levels, or because they belong to some very unpopular social groups, or for any other reason whatever.

AMARTYA SEN

NON-LINEAR SOCIAL WELFARE FUNCTIONS:
A REPLY TO PROFESSOR HARSANYI

A preliminary point first. John Harsanyi's (1977) rejoinder is concerned exclusively with a relatively small part of my lecture (1977), viz., that dealing with Harsanyi's critique of non-linear social welfare functions. All the formal results, e.g., (T.1) – (T.4), and much of the informal discussion in my lecture were concerned with other issues. While I find Harsanyi's critique of non-linear social welfare functions interesting and important, and welcome this opportunity of discussing further why I disagree with him, I would certainly like to avoid generating the impression that Harsanyi and I are battling on the main substance of my lecture.

Now to Harsanyi's arguments. His first point is that since I endorse Peter Diamond's (1967) rejection of the ethical invulnerability of the 'strong independence axiom' applied to social choice, it follows that I must accept certain 'curious implications of Diamond's views' as seen by Harsanyi. For example, I must accept that "the most extreme hereditary social and economic inequalities would become morally *less objectionable* if newborn babies were regularly interchanged among different families by a big government-conducted lottery" (Harsanyi, 1977, p. 294). I must confess to being not persuaded by this example, which I shall call – for the sake of brevity – Hell ('Harsanyi's exchange-baby lottery legend'). To reject the strong independence axiom, it is sufficient to show that there *exist* circumstances in which the implications of the axiom should be negated; there is no obligation to negate the consequences of that axiom in *every* conceivable circumstance. What is required is a denial of a universal rule, and that can be based on an existential assertion. In contrast, an existential pointer to an example (e.g., Hell) satisfying the axiom for a particular circumstance does not establish the rule. For the type of 'fairness' that is required in the conception of an 'original position', the 'strong independence axiom' does appear to be rather objectionable; Diamond's example was

Butts and Hintikka (eds.), *Foundational Problems in the Special Sciences*, 297–302.

appropriate for that case. The fact that we may well decide not to violate the strong independence axiom in the Hell case, does not re-establish the axiom as a rule with a universal domain.

Harsanyi's summary of my position regarding the strong independence axiom, viz, that it "has *no* legitimate application to choice among alternative social policies" (Harsanyi, 1977, p. 293, italics added), is not entirely accurate. If the axiom had *some* legitimate applications in specified circumstances but had to be violated in other circumstances, then too Harsanyi's axiomatic defence of utilitarianism would collapse, rendering illegitimate his conclusion that social welfare "must be defined as the arithmetic mean of the utility levels of all individuals in the society" (Harsanyi, 1975, p. 312).

Incidentally, it should be noted that the Hell example is rather cleverly chosen since it brings into play the innate revulsion that many people would feel at the thought of redistributing babies by lottery even when this implies merely a permutation of the individual utility vector among the persons involved. In this context it is probably worth pointing out that this revulsion too is difficult to accommodate within the utilitarian framework. Utilitarianism will not recommend the definite rejection of the Hell scheme leading to a permutation of individual utilities, but merely an *indifference* as to whether we should have it or not. Thus the Hell example with Harsanyi's characterization of it may prove 'too much'. Indeed, I would be inclined to argue that the welfare-oriented 'consequentialist' approaches (including utilitarianism, among others) with linear *or* non-linear social welfare functions, cannot really catch the nuances of this innate revulsion at the idea of redistributing babies by lottery.[1]

Harsanyi's second point is concerned with the case in which the strong independence axiom is accepted and so is the entire von Neumann-Morgenstern axiom set for *social* choices involving risk. I had argued that even then utilitarianism might not be acceptable except in a trivial sense. Harsanyi seems to ignore most of the arguments presented, and misinterprets the one he chooses to comment on.

To start with *individual* utilities need not always have cardinal representations and need not be cardinally comparable between persons. Harsanyi does not consider this case, even though it is clear that with mere ordinality and level comparability the Rawlsian rule works completely and so do many other rules, but not utilitarianism. The axiomatics considered in my lecture were based on assuming no more than this (see, for example, Theorems (T.1) – (T.4)). Since Harsanyi puts all such non-utilitarian aggregation procedures under the broad hat of 'non-linearity' – somewhat misleadingly, I think – it is not absurd to raise this question of measurability and comparability in contrasting the merits of such non-utilitarian rules with utilitarianism.

When individual behaviour satisfies the von Neumann-Morgenstern axioms, there will, of course, exist a cardinal representation of the individual utility function; these values we may call V-values. These V-values for social choice, over other representations, is a different representation may, however, co-exist with the von Neumann-Morgenstern representation. Though V-values will *predict* the behaviour of individuals under risk, the question of superior relevance of V-values for social choice, over other representations, is a different issue altogether. I had made the mild claim that "when someone talks about social welfare being a non-linear function of individual welfares, the reference need not necessarily be to the V-values at all" and "there is no obligation to talk about V-values only whenever one is talking about individual or social welfare" (Sen, 1977, p. 277). For reasons that are somewhat obscure, Harsanyi interprets this to mean that I was recommending that "we should use a terminology which amounts to carefully *disguising* the fact that we are utilitarians" and "we should refuse to *call* an individual's vNM utility function our measure for his personal welfare" (Harsanyi, 1977, p. 294). Far from it. The fact is that as with other cardinal representations, there are elements of arbitrariness in the von Neumann-Morgenstern system, which is geared specifically to *predicting* behaviour under uncertainty, and the question is not one of 'disguising' anything but of noting that when people talk about social welfare being a non-linear function of individual welfares, they *in fact* typically refer to representations other than those emanating from the von Neumann-Morgenstern system.[2]

Harsanyi has two essentially different justifications for using the linear form for social welfare functions. One is based on his theorem that if individual and social welfares both satisfy the von Neumann-Morgenstern axioms, and if Pareto indifference is reflected in social indifference, then social welfare (measured in V-values) must be a linear combination of individual welfares (measured in V-values). This I shall call the 'consistency justification'. The other is based on his conception of an *as if* position of primordial equality, where people have to choose between social states without knowing their own positions in them, *assuming* that they have an equal probability of being anyone. If in this *as if* position, people are assumed to behave according to the von Neumann–Morgenstern axioms, then they must be assumed to end up maximizing expected utilities from the different lotteries, and since the 'prizes' are being particular individuals, this would amount to maximizing the mean utility of all individuals taken together (measured in V-values). This I shall call the '*as if* justification'. Both the approaches are justly famous. Indeed, Harsanyi has the right to feel somewhat aggrieved that his conception of *as if* uncertainty has sometimes been overlooked in the excitement that followed the publication of Rawls' *A Theory of Justice*, with its conception of the 'original position'.

The 'consistency justification' incorporates more structure and involves difficulties about individual welfare representations, discussed above. Further, it does not yield utilitarianism as such – only linearity. The '*as if* justification' is more direct. It is, in fact, 'inescapable' provided we accept the premises, viz, *as if* equi-probability and vNM behaviour, and try to relate V-values of social welfare as seen by a *given* person to V-values from individual positions as seen by the *same person* under *as if* uncertainty. The utility of the lottery must of necessity equal the probability-weighted utility sum from the 'prizes' of being different individuals.

The 'as if' model is, of course, extremely interesting and important from the point of view of setting up an ethical requirement of 'impersonality', but the linear form it yields is essentially trivial. $W = \sum_i (1/n) W_i$ follows from the fact that W is the utility of the lottery and W_i's are the prizes with equal probability. The attitudes to risk have already gone into the measures W_i. One would have to misunderstand the structure totally to suggest that under the von Neumann-Morgenstern axioms the utility of the lottery be anything different from the expected utility of it. The result is quite non-controversial, for it is an "inexpensive pleasure" (Sen, 1977, p. 279). Harsanyi might well enjoy that *in this sense* "Sen's scenario concedes [his] whole point" and that "we could not really avoid following the utilitarian approach" (Harsanyi, 1977, p. 295), but its non-controversiality arises essentially from its non-assertion.

I must, therefore, stick to my conclusion in the lecture that "an axiomatic justification of utilitarianism will have more content to it if it started off at a place somewhat more distant from the ultimate destination" (p. 279). Fortunately, rather more revealing axiomatizations of utilitarianism have recently been proposed by several authors, e.g., d'Aspremont and Gevers (1975), Hammond (1975), Maskin (1975), Deschamps and Gevers (1976), permitting us to debate the underlying issues, rather than making utilitarianism 'impossible to avoid' in a trivial sense. Harsanyi's (1955) contribution in proposing his *as if* model was extremely valuable for what it taught us about the requirement of 'impersonality' in ethical judgements. As someone whose understanding of these issues was deeply influenced by Harsanyi's approach, I feel sad that Harsanyi should continue to believe that his contribution lay in providing an axiomatic justification of utilitarianism with real content.

London School of Economics

NOTES

[1] For a critique of consequentialism, see Williams (1973).
[2] Alternative approaches to measurements and comparison of individual welfares are analysed in my 'Interpersonal Comparisons of Welfare', paper prepared for a festschrift for Tibor Scitovsky.

BIBLIOGRAPHY

d'Aspremont, C. and Gevers, L.: 1975, 'Equity and the Informational Basis of Collective Choice', presented at the Third World Econometric Congress, Toronto; forthcoming *Review of Economic Studies.*

Deschamp, R. and Gevers, L.: 1976, 'Leximin and Utilitarian Rules', mimeographed, Faculté des Sciences Economiques et Sociales, Namur, 1976.

Diamond, P.: 1967, 'Cardinal Welfare, Individualistic Ethics, and Interpersonal Comparisons of Utility: A Comment', *Journal of Political Economy* **75,** 765–6.

Hammond, P. J.: 1975, 'Dual Interpersonal Comparisons of Utility and the Welfare Economics of Income Distribution' mimeographed, Essex University; forthcoming *Journal of Public Economics.*

Harsanyi, J. C.: 1955, 'Cardinal Welfare, Individualistic Ethics and Interpersonal Comparisons of Utility', *Journal of Political Economy* **63**, 309–321.

Harsanyi, J. C.: 1975, 'Nonlinear Social Welfare Functions', *Theory and Decision* **6**, 311–332.

Harsanyi, J. C.: 1977, this volume, p. 293.

Maskin, E.: 1975, 'A Theorem on Utilitarianism', mimeographed, Cambridge University; forthcoming *Review of Economic Studies*.

Sen, A. K.: 1977, this volume, p. 271. International Congress of Logic, Philosophy and Methodology of Science, London, Ontario, August 29, 1975; also forthcoming in *Theory and Decision*.

Williams, B.: 1975, 'A Critique of Utilitarianism' in J. J. C. Smart and B. Williams (eds.) *Utilitarianism: For and Against*, Cambridge University Press.

BENGT HANSSON

THE MEASUREMENT OF SOCIAL INEQUALITY

The first question that comes into one's mind about the measurement of the inequality in an income distribution is a simple-minded one: why should such a measure differ from measures of other kinds of inequalities? Statisticians have a collection of such measures and among them the mean deviation, the standard deviation and the mean difference (or Gini coefficient), either absolute or relativised to the arithmetical mean, seem to be reasonable candidates. Is it not the simplest solution to use one of these?

One reason for not doing so was given by Hugh Dalton in his early article 'The measurement of the inequality of incomes' (*The Economic Journal* **30** (1920), 348–361). He argues that "the economist is primarily interested, not in the distribution of income as such, but in the effects of the distribution of income upon the distribution and total amount of economic welfare". Since this welfare is derived from income we must somehow take into account how income and welfare are related, i.e. we must assume some kind of welfare function.

It is quite clear that what he had in mind was a function measuring the welfare an *individual* derives from income, i.e. what is also called the utility of money. Assuming that all individuals have the same utility function for money, that this function is increasing and concave and that their utilities can be added without problems, the aggregate welfare will be at its maximum when all individuals have equal incomes. Dalton, regarding "the resulting loss of potential economic welfare" as the essential drawback of an unequal distribution of income, therefore suggests as a measure of inequality the amount by which the aggregate welfare of the actual distribution falls short of this maximum, divided by the welfare of the actual distribution. If x_i is the income of individual i, n the number of individuals, and u the utility

Butts and Hintikka (eds.), *Foundational Problems in the Special Sciences*, 303–312.

function, Dalton's proposed measure is

$$
(1) \qquad \frac{u\left(\frac{1}{n}\sum x_i\right) - \frac{1}{n}\sum u(x_i)}{\frac{1}{n}\sum u(x_i)}
$$

or equivalently

$$
(2) \qquad \frac{u\left(\frac{1}{n}\sum x_i\right)}{\frac{1}{n}\sum u(x_i)} - 1,
$$

where the subtraction of the one is inessential for comparative purposes.

In a more recent article ('On the measurement of inequality', *Journal of Economic Theory* **2** (1970), 244–263), Anthony Atkinson has pointed out that this measure is not invariant for affine transformations of the function u. In fact, it is easily seen that if $u' = u + k$, both the numerator and the denominator in (2) increase by k, thereby affecting the quotient. Since social welfare is generally considered not to be measurable on a scale with an absolute zero, all such transformations should be equally justified and should yield the same amount of inequality. Atkinson also gives the remedy: instead of looking at the relationship between two amounts of welfare – the actual mean and the maximal mean – we should look at the relationship between the corresponding monetary values, i.e. we should apply the inverse of the function u to both the numerator and the denominator thus transforming Dalton's measure into

$$
(3) \qquad \frac{\frac{1}{n}\sum x_i}{u^{-1}\left(\frac{1}{n}\sum u(x_i)\right)} - 1.
$$

This advantage has to be traded for some loss of intuitive intelligibility: Dalton's original measure could be interpreted as how much we could increase aggregate welfare by distributing income more equally, whereas the transformed version (3) has no equally natural reading.

It should be mentioned that, although Atkinson makes use of the same two numbers as Dalton, he advocates that they be combined in another way. If D is the expression (3), Atkinson prefers $1 - (1/D + 1)$, which of course is ordinally equivalent to D. Thus, his defining formula is

$$(4) \qquad 1 - \frac{u^{-1}\left(\frac{1}{n}\sum u(x_i)\right)}{\frac{1}{n}\sum x_i}.$$

I will not discuss the relative merits of Dalton's and Atkinson's measures here and now. What I will have to say will be equally applicable to both of them because it has to do with their common assumption: that the relationship between actual aggregate welfare and maximal aggregate welfare determines the degree of inequality.

I will question this assumption and my arguments will be both of a measure-theoretical and of a practical kind.

Even if one takes as one's starting-point that the study of inequality of income is worth pursuing because of the social welfare the income generates and that one therefore must take into account how this welfare is related to income, it does in no way immediately follow that a measure of inequality should be construed like Dalton's or Atkinson's or even that "the objection to great inequality of incomes is the resulting loss of potential economic welfare" (Dalton). Let me first give a few arguments to the effect that the measures I have mentioned are less than ideal and let me then try to explain why they are not conceptually linked to inequality at all, even if they accidentally have some relationship to this concept.

A very central problem for the theory of measurement is how to ensure that a certain measure really measures the property one is interested in, how it is conceptually linked to that property. Many times this is done by requiring the measure to agree with some empirical relation or operation. In the simplest cases one uses an ordering relation: if one mineral scratches another and not *vice versa* the first one is said to be harder than the second one and a minimum requirement for a measure of hardness is then that it shall assign a higher number to the harder mineral. Similarly, any measure of length must be required to give to a composite object consisting of two straight rods joined together at a straight angle a numerical value which is the sum of the values given to the two rods separately.

Such requirement must always be thought of as minimal and if several disparate measures all satisfy them, we must consider it a good methodological rule to continue to look for stronger such conceptual links.

Now, what such conceptual links do we have in the case of inequality of income? One such test is what Dalton called the principle of transfers and it is the requirement that any transfer of income from a richer individual to a poorer one diminishes inequality (provided it is not large enough to reverse the relative positions of the individuals). Most proposed measures pass this test, including those mentioned here. Some other tests have also been proposed, but they all deal with the effects of modifying the distribution of income.

But the essential feature of a measure that takes social welfare into account is that it is a function of *two* variables: the distribution of income and the function that relates the welfare to the income. In order to have as varied a set of tests as possible, we should thus try to find out how the measure should respond to changes in that functional relationship. My proposal is the following: whenever the curve relating welfare to income flattens out so that from some point on the new curve is wholly below the old one, but still increasing and concave (like ABD in Figure 1), then the measure of inequality shall decrease for the same distribution of nominal income. This *principle of reduced welfare from excess income* is related to the principle of transfers in the following way: when the curve changes the welfare of some of those who are better off decreases and even if no actual transfer takes place, there is a general shrinking of difference, much like when a series of transfers takes place.

In order to see that both Dalton's and Atkinson's measures violate this principle, let us take a simple situation with only two individuals,

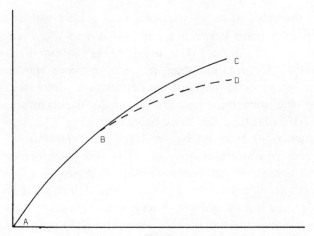

Fig. 1.

as in Figure 2. The two nominal incomes are x_1 and x_2, their mean μ, the two welfares $u(x_1)$ and $u(x_2)$ and the mean of the welfares m. Dalton's formula (2) asks us to take the ratio between $u(\mu)$ and m while Atkinson prefers to consider the corresponding points on the x-axis: μ and $u^{-1}(m)$.

Now, let us see what happens if the curve u changes to u'. In order

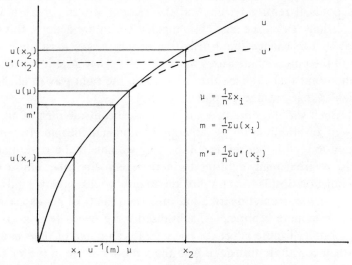

Fig. 2.

to change the value of as few variables as possible I will let u' agree with u up to a point between μ and x_2. μ and $\mu(\mu)$ remain unchanged, but $u'(x_2)$ is less than $u(x_2)$, so m' is less than m and, consequently, $u'^{-1}(m')$ less than $u^{-1}(m)$. In the formula (2) the numerator remains constant, but the denominator decreases, thereby *in*creasing the numerical measure of inequality, contrary to our principle. Trivially, the same thing happens to (3).

The formula (4) fares no better. Here the denominator remains constant and the numerator decreases. The ratio therefore decreases, but since it is negated the numerical value increases.

The principle of reduced welfare from excess income is important not only because it is an example of such a conceptual link between the numerical measure and the intuitive concept as philosophers of science insist on, but also for practical reasons. If we ask why we are interested in measuring inequality, we are likely to get answers about political values. Social equality may or may not be considered a primary value, but almost everyone agrees that it has at least some value, and therefore that they have some interest in measuring it. The questions they seek the answers to can be like: are there any systematic differences between e.g. underdeveloped and advanced countries? what is the long term trend in a particular country? what effect does a certain political reform have? And the reasons we are interested in such questions is that the answers can guide our future actions. If it is a political goal to change the distribution of income we must know the effects of the potential measures we could take. Therefore it is particularly important that the measure responds in the right way to changes in variables that are under political control.

And which variables *are* then under political control? First of all, by powers of taxation: available amount of nominal income. But it is important to realise that this is not the only variable that a government controls. A great many reforms to increase equality take aim at the utility function instead: extra tax on luxury goods, many kinds of rationing, governmental distribution of certain kinds of goods according to non-monetary principles, subsidised basic needs (e.g. no sales tax on food) – all these affect the use people can make of their money rather than constitute transfers. And the kind of change in the welfare curve that the principle of reduced welfare is about corresponds to some kind of tax on luxury goods – the richest get less for their money. And since governments frequently take such measures in order to influence the degree of inequality, it is important that any measure of that inequality correctly responds to this type of change.

It is not difficult to see the reason why these measures have only an accidental correlation with inequality: they are essentially utilitarian. Dalton motivates his interest in equality thus: "The objection to great inequality of incomes is the resulting loss of potential economic welfare". He then notes that, under certain assumptions about the relation between welfare and money, "if a given income is to be distributed among a given number of persons, it is evident that the economic welfare will be a maximum, when all incomes are equal". But from this, it certainly is a mistake to conclude "it follows that the inequality of any given distribution may conveniently be defined as the ratio of the total economic welfare attainable under an equal distribution to the total economic welfare attained under the given distribution".

The relationship that has been established between equality and attained total welfare is much too weak to warrant any such conclusion – much weaker than e.g. an ordinal correspondence. All we know is that equality and total attained welfare reach their maxima at the same time. But this justifies in no way the assumption that the deviation from the maximum in one of the variables can be used as a measure of the same deviation for the other variable.

Furthermore, even the coincidence of the maxima depends on the particular assumptions that Dalton makes. I have already mentioned the importance of not keeping the welfare function constant. Let us also have a look at the assumption that all individuals have the same welfare function. This is a perfectly practical assumption to make, since without it we could never apply our measure to any actual statistics. But it is also clear that it is *only* for practical reasons that we make this assumption, for it certainly *is* the case that different individuals derive different amounts of welfare from the same amount of income. The handicapped e.g. have for purely physical reasons a much narrower set of options available to them to use a given amount of money and they must therefore be considered to derive less welfare from it. But even if it is practically impossible to take this into account, we must still require that if one can derive any qualitative conclusions from the introduction of individual welfare functions, then they must agree with our intuitions about inequality.

But when we do this the utilitarian character of Dalton's and Atkinson's measures is brought out even more clearly. If a given amount of income is to be distributed, the denominator in (4) is constant, so the minimising of inequality according to measure (4) is the maximising of $\sum u(x_i)$ (since u^{-1} is an increasing function). If we generalise to the case of different individual welfare functions we must replace this expression by $\sum u_i(x_i)$. But this expression will not be maximised when all x_i's are the same. If individual 2 has a higher marginal welfare at the common income x than individual 1, then a transfer from 1 to 2 would increase aggregate welfare, but would certainly decrease equality. Furthermore, as can be seen from Figure 3, this may lead not only to less equality in terms of nominal income but also in terms of welfare. If both individuals start with x monetary units, a transfer of δ from 1 to 2 will increase $\sum u_i(x_i)$ because of 2's greater marginal welfare. But since 2 is also absolutely better off than 1 it will also increase the welfare gap from $u_2(x) - u_1(x)$ (D in Figure 3) to $u_2(x + \delta) - u_1(x - \delta)$ (D' in Figure 3). For certain shapes of the curves a

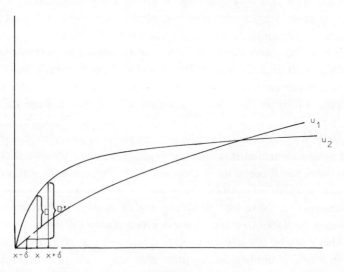

Fig. 3.

Dalton- or Atkinson-like measure will even require all the income to be given to one single individual for the degree of 'inequality' to be minimal.

What, then, would an adequate measure be like? I do not think that there is any need for a high degree of sophistication, but only for a small amount of conceptual clarity. If it is not the distribution of income as such, but the distribution of welfare we are interested in, then of course we should measure the inequality of welfare and not the inequality of income. That the welfare depends on the income does not entail that the inequality of the welfare distribution depends on the inequality of the income distribution. To borrow an analogy from Dalton: if a crop depends on the rainfall, we may well be interested in the rainfall because it affects the crop, but if we want to estimate the farmer's income it is the variation in the crop that matters.

And to the welfare we can apply the dispersion measure of our choice. The standard deviation, e.g. is a reasonable measure and better still is the coefficient of variation (standard deviation over arithmetical mean), because it is independent of the particular unit chosen. The explicit formula for the measure would then be

$$(5) \qquad \frac{\sqrt{V}}{\frac{1}{n}\sum u(x_i)} ,$$

where V is defined by:

$$(6) \qquad V = \frac{1}{n}\sum \left(u(x_i) - \frac{1}{n}\sum u(x_i) \right)^2 .$$

This measure, in all its simplicity, satisfies all reasonable requirements: the principle of transfers and other principles about changes in the monetary distribution present no problem; the principle of reduced welfare from excess income is satisfied and the generalisation to individual welfare curves is straightforward.

A clear distinction between the inequality in the distribution of the pertinent variable, economic welfare, and the inequality in the distribution of the underlying generator of this welfare, income, also helps to keep in mind the important distinction between the degree of inequality and the evaluation of that inequality. Even if a low degree of inequality is generally considered to be a value in itself, this view in no way necessitates that inequality is the *only* factor that determines the collective, holistic welfare of a given society with its income distribution. Certainly, the absolute level of welfare is of great importance too. The introduction of welfare considerations into the measurement of inequality may have led some to assume that the measure of inequality could be directly used as a measure of collective welfare too. Rather, it should be regarded as one characteristic of an income distribution which is one of the factors that determine the collective welfare. The relative importance of this factor is, of course, open to discussion. Probably, social equality is like the absolute zero of temperature: we can get pretty far towards it by relatively simple means, but the last tiny step will require efforts out of proportion and great sacrifices in absolute welfare.

University of Lund

VIII

RATIONALITY
IN SOCIAL SCIENCES

JOHN C. HARSANYI

ADVANCES IN UNDERSTANDING
RATIONAL BEHAVIOR

ABSTRACT. It will be argued that economic theory, decision theory and some recent work in game theory, make important contributions to a deeper understanding of the concept of rational behavior. The paper starts with a discussion of the common-sense notion of rational behavior. Then, the rationality concepts of classical economics and of Bayesian decision theory are described. Finally, some (mostly fairly recent) advances in game theory are briefly discussed, such as probabilistic models for games with incomplete information; the role of equilibrium points in non-cooperative games, and the standard game-theoretical approach to the prisoner's dilemma problem; the concept of perfect equilibrium points; the use of non-cooperative bargaining models in analyzing cooperative games; and the Harsanyi-Selten solution concept for non-cooperative games.

1. INTRODUCTION[1]

The concept of rational behavior (or of practical rationality) is of considerable philosophical interest. It plays an important role in moral and political philosophy, while the related concept of theoretical rationality is connected with many deep problems in logic, epistemology, and the philosophy of science. Both practical and theoretical rationality are important concepts in psychology and in the study of artificial intelligence. Furthermore, rational-behavior models are widely used in economics and, to an increasing extent, also in other social sciences. This fact is all the more remarkable since rationality is a normative concept and, therefore, it has been claimed (incorrectly, as I shall argue) that it is out of place in non-normative, empirically oriented studies of social behavior.

Given the important role that the concept of rationality plays in philosophy and in a number of other disciplines, I have thought it may be of some interest to this interdisciplinary audience if I report on some work in decision theory and in game theory that holds out the prospect of replacing our common-sense notion of rational behavior by a much more general, much more precise, and conceptually very much richer notion of rationality. I feel that successful development of an analytically clear, informative, and intuitively satisfactory concept of rationality would have significant philosophical implications.

Butts and Hintikka (eds.), Foundational Problems in the Special Sciences, 315–343.
Copyright © 1977 by D. Reidel Publishing Company, Dordrecht-Holland. All Rights Reserved.

I shall first discuss the common-sense notion of rational behavior. Then, I shall briefly describe the rationality concepts of classical economic theory and of Bayesian decision theory. Finally, I shall report on some, mostly very recent, results in game theory, which, I believe, are practically unknown to non-specialists. Of course, within the space available, I cannot do more than draw a sketchy and very incomplete picture of the relevant work – a picture no doubt strongly colored by my own theoretical views.

2. THE MEANS-ENDS CONCEPT OF RATIONAL BEHAVIOR

In everyday life, when we speak of 'rational behavior', in most cases we are thinking of behavior involving a choice of the best *means* available for achieving a given *end.* This implies that, already at a common-sense level, rationality is a *normative* concept: it points to what we *should* do in order to attain a given end or objective. But, even at a common-sense level, this concept of rationality does have important *positive* (non-normative) applications: it is used for *explanation*, for *prediction*, and even for mere *description*, of human behavior.

Indeed, the assumption that a given person has acted or will act rationally, often has very considerable explanatory and predictive power, because it may imply that we can explain or predict a large number of possibly very complicated facts about his behavior in terms of a small number of rather simple hypotheses about his goals or objectives.

For example, suppose a given historian comes to the conclusion that Napoleon acted fairly rationally in a certain period. This will have the implication that Napoleon's actions admit of explanation in terms of his political and military objectives – possibly in terms of a rather limited number of such objectives – and that other, often less easily accessible, psychological variables need not be used to any major extent to account for his behavior. On the other hand, if our historian finds that Napoleon's behavior was not very rational, then this will imply that no set of reasonably well-defined policy objectives could be found that would explain Napoleon's behavior, and that any explanation of it must make use of some 'deeper' motivational factors and, ultimately, of some much more specific assumptions about the psychological mechanisms underlying human behavior.

Yet, we do make use of the notion of rationality also in cases where we are not interested in an explanation or prediction of human behavior, but are merely interested in providing an adequate *description* of it. For instance, any historical narrative of Napoleon's political and military decisions will be seriously incomplete, even at a descriptive level, if it contains no discussion of the rationality or irrationality or these decisions. Thus, it will not be enough to report that, in a particular battle, Napoleon attacked the enemy's right wing. Rather, we also want to know whether, under the existing conditions, this attack was a sensible (or perhaps even a brilliant) tactical move or not.

Philosophers, and social scientists outside the economics profession, have often expressed puzzlement about the successful use of the normative concept of rational behavior in positive economics – and, more recently, also in other social sciences[2] – for explanation and prediction, and even for mere description, of human behavior. But there is really nothing surprising about this. All it means is that human behavior is mostly *goal-directed*, often in a fairly consistent manner, in many important classes of social situations. For example, suppose that people are *mainly* after money in business life (even if they do also have a number of other objectives), or are mainly after election or re-election to public office in democratic politics, or are mainly after social status in many social activities, or are mainly after national self-interest in international affairs – at least that these statements are true as a matter of reasonable first approximation. Then, of course, it is in no way surprising if we can explain and sometimes even predict, and can also meaningfully describe, their behavior in terms of the assumption that they are after money, or after public office, or after social status, or after national self-interest.

Even if the subject matter of our investigation were not human behavior, but rather the behavior of goal-pursuing robots, a model of 'rational' (i.e. goal-pursuing) robot behavior would be a very valuable analytical tool. Of course, just as in the case of human beings, such a rationalistic model would only work with highly 'rational' (i.e. with very well-functioning) robots. To explain the behavior of a robot with a faulty steering mechanism, we would need a more complicated model, based on fairly detailed assumptions about the robot's internal structure and operation.

To be sure, while we could at least conceive of a perfectly well-constructed goal-pursuing robot, completely consistent and completely single-minded in working for his pre-established goal, human beings are seldom that consistent. In some situations, they will be deflected from their objectives by Freudian-type emotional factors, while in others they will fail to pursue any well-defined objectives altogether. Moreover, even if they do aim at well-defined objectives, their limitations in computing (information-processing) ability may prevent them from discovering the most effective strategies (or any reasonably effective strategies) for achieving these objectives. (Of course, any robot of less than infinite structural complexity will be subject to similar computational limitations. But he may not be subject to anything resembling emotional problems.)

Obviously, this means that in *some* situations models of rational behavior will not be very useful in analyzing human behavior – except perhaps after substantial modification (e.g. along the lines suggested by Simon's (1960) theory of limited rationality). Clearly, it is an empirical question what types of social situations lend themselves, and to what extent, to analysis in terms of rational-behavior models. But recent work in political science, in international relations, and in sociology, has shown that a much wider range of social situations seems to admit of rationalistic analysis than most observers would have thought even ten years ago.[3]

3.   THE RATIONAL-BEHAVIOR MODEL OF ECONOMIC THEORY

Even at a common-sense level, the means-ends model is not the only model of rational behavior we use. Another, though perhaps less important, model envisages rational behavior as choosing an object (or a person) satisfying certain stipulated formal (possibly non-causal) *criteria*. For instance, if my aim is to climb the highest mountain in California, then it will be rational for me to climb Mount Whitney, and it will be irrational for me to climb any other mountain. But we would not normally say that climbing Mount Whitney is a *means* to climbing the highest mountain in California, because my climbing of Mount Whitney does not causally *lead* to a climbing of the highest mountain. Rather, it already *is* a climbing of the highest mountain. It is a rational action in the sense of being an action (and, indeed, the only action) satisfying the stipulated criterion.[4]

Thus, it would be rather artificial to subsume criterion-satisfying behavior under the means-ends model of rationality. It is more natural to do the converse, and argue that looking for a means to a given end is a special case of looking for an object satisfying a particular criterion, viz. the criterion of being causally effective in attaining a given end.

This implies that the means-ends concept of rational behavior is too narrow because it fails to cover criterion-satisfying behavior. An even more important limitation of this concept lies in the fact that it restricts rational behavior to a choice among alternative *means* to a given end, and fails to include a rational choice among alternative *ends*. Therefore, it cannot explain why a given person may shift from one end to another.

To overcome this limitation, already 19th and early 20th century economists introduced a broader concept of rationality which defines rational behavior as a choice among alternative ends, on the basis of a given set of *preferences* and a given set of *opportunities* (i.e. a given set of available alternatives). If I am choosing a given end (or a given set of mutually compatible ends, which can be described as a unique composite end), then typically I have to give up many alternative ends. Giving up these alternative ends is the *opportunity cost* of pursuing this particular end. Thus, under this model, rational behavior consists in choosing one specific end, after careful consideration and in full awareness of the opportunity costs of this choice.

This model will often enable us to explain why a given individual has changed over from pursuing one objective to pursuing another, even if his basic preferences have remained the same. The explanation will lie in the fact that the opportunity costs of various possible objectives (i.e. the advantages and the disadvantages associated with them) have changed, or at least the information he has about these opportunity costs has done so.

For example, a given person may seek admission to a particular university, but then may change his mind and fail to attend. He may do so because the tuition fees or other monetary costs have increased; or because the studies he would have to undertake have turned out to be harder or less interesting than he thought they would be; or because he has received unfavorable information about the hoped-for economic advantages of a university degree, etc. All these explanations are compatible with the assumption that, during the whole period, his basic preferences remained the same, and that only the situation (i.e. his opportunity costs), or his information about the situation, have changed.[5]

It is easy to verify that the preferences-opportunities model includes both the means-ends model and the criterion-satisfaction model as special cases.

An important result of economic theory has been to show that, if a given person's preferences satisfy certain consistency and continuity axioms, then these preferences will admit of representation by a well-defined (and, indeed, continuous) utility function. (For proof, see Debreu, 1959, pp. 55–59.) Accordingly, for such a person, rational behavior – as defined by the preferences-opportunities model – will be equivalent to *utility-maximization* (utility-maximization theorem).

4. BAYESIAN DECISION THEORY

Classical economic theory was largely restricted to analyzing human behavior under *certainty*, i.e. under conditions where the decision maker can uniquely predict the outcome of any action he may take (or where one can assume this to be the case at least as a matter of first approximation). It has been left to modern decision theory to extend this analysis to human behavior under risk and under uncertainty.

Both risk and uncertainty refer to situations where the decision maker cannot always uniquely predict the outcomes of his action. But, in the case of *risk*, he will know at least the objective probabilities associated with all possible outcomes. In contrast, in the case of *uncertainty*, even some or all of these objective probabilities will be unknown to him (or may even be undefined altogether).

The utility-maximization model provides a satisfactory characterization of rational behavior under certainty, but fails to do so under risk and under uncertainty. This is so because it is not sufficient to assume that any given lottery (whether it is a 'risky' lottery involving known probabilities, or is an 'uncertain' lottery involving unknown probabilities) will have a well-defined numerical utility to the decision maker. Rather, we need a theory specifying *what value* this utility will have, and how it will depend on the utilities associated with the various prizes. This is exactly what decision theory is trying to specify.

The main conclusion of decision theory is this. If the decision maker's behavior satisfies certain consistency and continuity axioms (a larger number of axioms than we needed to establish the utility-maximization theorem in the case of certainty), then his behavior will be equivalent to *maximizing his expected utility*, i.e. to maximizing the mathematical expectation of his cardinal utility function. In the case of *risk*, this expected utility can be defined in terms of the relevant objective probabilities (which, by assumption, will be known to the decision maker). On the other hand, in the case of *uncertainty*, this expected utility must be defined in terms of the decision maker's own subjective probabilities whenever the relevant objective probabilities are unknown to him (expected-utility maximization theorem).[6]

This result leads to the *Bayesian* approach to decision theory, which proposes to define rational behavior under risk and under uncertainty as expected-utility maximization.[7]

Besides the axioms used already in the case of certainty, we need one additional consistency axiom to establish the expected-utility maximization theorem in the cases of risk and of uncertainty. This axiom is the *sure-thing principle*, which can be stated as follows. "Let X be a bet[8] that would yield a given prize x to the decision maker if a specified event E took place (e.g. if a particular horse won the next race). Let Y be a bet that would yield him another prize y, which he *prefers* over x, if this event E took place. There are no other differences between the two bets. Then, the decision maker will consider bet Y to be *at least as desirable* as bet X." (Actually, unless he assigns zero probability to event E, he will no doubt positively *prefer* bet Y, which would yield the more attractive prize y, if event E took place. But we do not need this slightly stronger assumption.)

In my opinion, it is hard to envisage any rational decision maker who would knowingly violate the sure-thing principle (or who would violate the – somewhat more technical – continuity axiom we need to prove the expected-utility maximization theorem). This fact is, of course, a strong argument in favor of the Bayesian definition of rational behavior. Another important argument lies in the fact that all alternative definitions of rational behavior (and, in particular, all definitions based on the once fashionable maximin principle and on various other related principles) can be shown to lead to highly irrational decisions in many practically important situations. (See Radner and Marschak, 1954; also Harsanyi, 1975a.)

In the case of risk, acceptance of Bayesian theory is now virtually unanimous. In the case of uncertainty, the Bayesian approach is still somewhat controversial, though the last two decades have produced a clear trend toward its growing acceptance by expert opinion. Admittedly, support for Bayesian theory is weaker among scholars working in other fields than it is among decision theorists and game theorists. (I must add that some of the criticism directed against the Bayesian approach has been rather uninformed, and has shown a clear lack of familiarity with the relevant literature.)

5. A GENERAL THEORY OF RATIONAL BEHAVIOR

Whatever the merits of the rationality concept of Bayesian decision theory may be, it is still in need of further generalization, because it does not adequately cover rational behavior in *game situations,* i.e. in situations where the outcome depends on the behavior of two or more rational individuals who may have partly or wholly divergent interests. Game situations may be considered to represent a special case of uncertainty, since in general none of the players will be able to predict the outcome, or even the probabilities associated with different possible outcomes. This is so because he will not be able to predict the strategies of the other players, or even the probabilities associated with their various possible strategies. (To be sure, as we shall argue, at least in principle, game-theoretical analysis does enable each player to discover the solution of the game and, therefore, to predict the strategies of the other players, provided that the latter will act in a rational manner. But the point is that, prior to such a game-theoretical analysis, he will be unable to make such predictions.)

Game theory defines rational behavior in game situations by defining solution concepts for various classes of games.

Since the term 'decision theory' is usually restricted to the theory of rational behavior under risk and under uncertainty, I shall use the term *utility theory* to describe the broader theory which includes both decision theory and the theory of rational behavior under certainty (as established by classical economic theory).

Besides utility theory and game theory, I propose to consider *ethics*, also, as a branch of the general theory of rational behavior, since ethical theory can be based on axioms which represent specializations of some of the axioms used in decision theory (Harsanyi, 1955).

Thus, under the approach here proposed, the *general theory of rational behavior* consists of three branches:

(1) *Utility theory*, which is the theory of *individual* rational behavior under certainty, under risk, and under uncertainty. Its main result is that, in these three cases, rational behavior consists in *utility maximization* or *expected-utility maximization.*

(2) *Game theory*, which is the theory of rational behavior by *two or more* interacting rational individuals, each of them determined to maximize his own interests, whether selfish or unselfish, as specified by his own utility function (payoff function). (Though some or all players may very well assign high utilities to clearly altruistic objectives, this need not prevent a conflict of interest between them since they may possibly assign high utilities to quite *different*, and perhaps strongly conflicting, altruistic objectives.)

(3) *Ethics*, which is the theory of rational moral value judgments, i.e. of rational judgments of preference based on impartial and impersonal criteria. I have tried to show that rational moral value judgments will involve *maximizing* the *average utility level* of all individuals in society. (See Harsanyi, 1953, 1955, 1958, 1975a, and 1975b.)

Whereas game theory is a theory of possibly conflicting (but not necessarily selfish) *individual* interests, ethics can be regarded as a theory of the *common* interests (or of the general welfare) of society as a whole.

6. GAMES WITH INCOMPLETE INFORMATION

We speak of a game with *complete information* if the players have full information about all *parameters* defining the game, i.e. about all variables fully determined *before* the beginning of the game. These variables include the players' payoff functions (utility functions), the strategical possibilities available to each player, and the amount of information each player has about all of these variables. We speak of a game with *incomplete information* if some or all of the players have less than full information about these parameters defining the game.

This distinction must not be confused with another, somewhat similar distinction. We speak of a game with *perfect information* if the players always have full information about the *moves* already made in the game, including the *personal moves* made by the individual players and the *chance moves* decided by chance. Thus, perfect information means full information about all game events that took part *after* the beginning of the game. We speak of a game with *imperfect information* if some or all players have less than full information about the moves already made in the game.

It was a major limitation of classical game theory that it could not handle games with *incomplete* information (though it did deal both with games involving perfect and imperfect information). For, many of the most important real-life game situations are games with incomplete information: the players may have only limited knowledge of each other's payoff functions (i.e. of each other's real objectives within the game), and may also know very little about the strategies as well as the information available to the other players.

In the last few years we have discovered how to overcome this limitation. How this is done can be best shown in an example. Suppose we want to analyze arms control negotiations between the United States and the Soviet Union. The difficulty is that neither side really knows the other side's true intentions and technological capabilities. (They may have reasonably good intelligence estimates about each other's weapon systems in actual use, but may know very little about any new military inventions not yet used in actual weapon production.) Now we can employ the following model. The American player, called A, and the Russian player, called R, both can occur in the form of a number of different possible 'types'. For instance, the Russian player could be really R_1, a fellow with very peaceful intentions but with access to very formidable new weapon technologies; and with the expectation that the American player will also have peaceful intentions, yet a ready access to important new technologies. Or, the Russian player could be R_2, who is exactly like R_1, except that he expects the American player to have rather aggressive intentions. Or, the Russian player could be R_3, who shows still another possible combination of all these variables, etc.

Likewise, the American player could be of type A_1 or A_2 or A_3, etc., each of them having a different combination of policy objectives, of access to new technologies, and of expectations about Russian policy objectives and about Russian access to new technologies. (We could, of course, easily add still further variables to this list.)

The game is played as follows. At the beginning of the game, nature conducts a lottery to decide which particular types of the American player and of the Russian player (one type of each) will actually participate in the game. Each possible combination (A_i, R_j) of an American player type and of a Russian player type have a pre-assigned probability p_{ij} of being selected. When a particular pair (A_i, R_j) has been chosen, they will actually play the game. Each player will know his own type but will be ignorant of his opponent's actual type. But, for any given type of his opponent, he will be able to assign a numerical probability to the possibility that his opponent is of this particular type, because each player will know the probability matrix $P = (p_{ij})$.

What this model does is to reduce the original game with *incomplete* information, G, to an artificially constructed game with *complete* information, G^*. The *incomplete* information the players had in G about the basic parameters of the game is represented in the new game G^* as *imperfect* information about a certain chance move at the beginning of the game (viz. the one which determines the types of the players). As the resulting new game G^* is a game with complete (even if with imperfect) information, it is fully accessible to the usual methods of game-theoretical analysis.

The model just described is not the most general model we use in the analysis of games with incomplete information. It makes the assumption that all players' expectations about each other's basic characteristics (or, technically speaking, all players' subjective probability distributions over all possible types of all other players) are sufficiently consistent to be expressible in terms of one basic probability matrix $P = (p_{ij})$. We call this the assumption of *mutually consistent expectations*. In many applications, this is a natural assumption to make and, whenever this is the case, it greatly simplifies the analysis of the game. (See Harsanyi, 1967–68.)

There are, however, cases where this assumption seems to be inappropriate. As Reinhard Selten has pointed out (in private communication – cf. Harsanyi, 1967–68, pp. 496–497), even in such cases, the game will admit of analysis in terms of an appropriate probabilistic model, though a more complicated one than would be needed on the assumption of consistent expectations.

7. NON-COOPERATIVE GAMES AND EQUILIBRIUM POINTS: THE PRISONER'S DILEMMA PROBLEM

We have to distinguish between *cooperative* games, where the players can make fully binding and enforceable commitments (fully binding promises, agreements, and threats, which absolutely *have* to be implemented if the stipulated conditions arise), and *non-cooperative* games, where this is not the case. In real life, what makes commitments fully binding is usually a law-enforcing authority. But in some cases prestige considerations (a fear of losing face) may have similar effects.

Nash (1950 and 1951), who first proposed this distinction, defined cooperative games as games with enforceable commitments *and* with free communication between the players. He defined non-cooperative games as games without enforceable commitments *and* without communication. These were somewhat misleading definitions. Presence or absence of free communication is only of secondary importance. The crucial issue is the possibility or impossibility of binding and enforceable agreements. (For example, in the prisoner's dilemma case, as I shall argue below, the cooperative solution will be unavailable to the players if no enforceable agreements can be made. This will be true regardless of whether the players can talk to each other or not.)

In a cooperative game, the players can agree on any possible combination of strategies since they can be sure that any such agreement would be kept. In contrast, in a non-cooperative game, only self-enforcing agreements are worth making because only self-enforcing agreements have any real chance of implementation.

A self-enforcing agreement is called an equilibrium point. A more exact definition can be stated as follows. A given strategy of a certain player is called a *best reply* to the other players' strategies if it maximizes this player's payoff so long as the other players' strategies are kept constant. A given combination of strategies (containing exactly one strategy for each player) is called an *equilibrium point* if every player's strategy is a best reply to all other players' strategies. The concept of an equilibrium point, also, is due to Nash (1950, 1951).

For example, suppose that the following two-person game is played as a non-cooperative game, so that no enforceable agreements can be made:

	B_1	B_2
A_1	2, 2	0, 3
A_2	3, 0	1, 1

This type of game is called a prisoner's dilemma. (For an explanation of this name, see Luce and Raiffa, 1957, pp. 94–95.)

In this game, the strategy pair (A_2, B_2) is an equilibrium point, because player 1's best reply to B_2 is A_2, whereas player 2's best reply to A_2 is B_2. Indeed, the game has no other equilibrium point. If the two players use their equilibrium strategies A_2 and B_2, then they will obtain the payoffs (1, 1).

Obviously, both players would be better off if they could use the strategies A_1 and B_1, which would yield them the payoffs (2, 2). But these two strategies do not form an equilibrium point. Even if the two players explicitly *agreed* to use A_1 and B_1, they would not do so, and would *know* they would not do so. Even if we assumed for a moment that the two players did expect the strategy pair (A_1, B_1) to be the outcome of the game, *this very expectation would make them use another strategy pair* (A_2, B_2) instead. For instance, if player 1 expected player 2 to use strategy B_1, he himself would not use A_1 but would rather use A_2, since A_2 would be his best reply to player 2's expected strategy B_1. Likewise, if player 2 expected player 1 to use A_1, he himself would not use B_1 but would rather use B_2, since B_2 would be his best reply to player 1's expected strategy A_1.

Of course, if the game were played as a cooperative game, then agreements would be fully enforceable, and the two players would have no difficulty in agreeing to use strategies A_1 and B_1 so as to obtain the higher payoffs $(2, 2)$. Once they agreed on this, they could be absolutely sure that this agreement would be fully observed.

Thus, we must conclude that, if the game is played as a non-cooperative game, then the outcome will be the equilibrium point (A_2, B_2), which is often called the *non-cooperative solution*. On the other hand, if the game is played as a cooperative game, then the outcome will be the non-equilibrium point strategy pair (A_1, B_1), which is called the *cooperative solution.*

More generally, the solution of a non-cooperative game must always be an equilibrium point. In other words, each player's solution strategy must be a best reply to the other players' solution strategies. This is so because the solution, by definition, must be a strategy combination that the players can rationally *use*, and that they can also rationally *expect* one another to use. But, if any given player's solution strategy were *not* his best reply to the other players' solution strategies, then the very expectation that the other players would use their solution strategies would make it rational for this player *not* to use his solution strategy (but rather to use a strategy that was a best reply to the solution strategies he expected the other players to use). Hence, the alleged 'solution' would not satisfy our definition of a solution.

This argument does not apply to a cooperative game, where each player can irrevocably commit himself to using a given strategy even if the latter is *not* a best reply to the other players' strategies. But it does apply to any non-cooperative game, where such a commitment would have no force.

This conclusion is accepted by almost all game theorists. It is, however, rejected by some distinguished scholars from other disciplines, because it seems to justify certain forms of socially undesirable non-cooperative behavior in real-life conflict situations. Their sentiments underlying this theoretical position are easy to understand and deserve our respect, but I cannot say the same thing about their logic. I find it rather hard to comprehend how anybody can deny that there is a fundamental difference between social situations where agreements are strictly enforceable and social situations where this is not the case; or how anybody can deny that, in situations where agreements are wholly unenforceable, the participants may often have every reason to distrust each other's willingness (and sometimes even to distrust each other's very ability) to keep agreements, in particular if there are strong incentives to violate these agreements.

To be sure, it is quite possible that, in a situation that *looks like* a prisoner's dilemma game, the players will be able to achieve the cooperative solution. Usually this will happen because the players are decent persons and therefore attach considerable disutility to using a non-cooperative strategy like A_2 or B_2 when the other player uses a cooperative strategy like A_1 or B_1. Of course, if the players take this attitude, then this will change the payoff matrix of the game. For instance, suppose that both players assign a disutility of 2 units to such an outcome. This will reduce the utility payoff that player 1 associates with the outcome (A_2, B_1) to $3 - 2 = 1$. Likewise, it will also reduce the utility payoff that player 2 associates with the outcome (A_1, B_2) to $3 - 2 = 1$. (If the players assigned a special disutility to violating an agreement, and then actually agreed to use the strategy pair (A_1, B_1), this would have similar effects on the payoff matrix of the game.) Consequently, the game will now have the following payoff matrix:

	B_1	B_2
A_1	2, 2	0, 1
A_2	1, 0	1, 1

This new game, of course, is no longer a prisoner's dilemma since now *both* (A_1, B_1) *and* (A_2, B_2) are equilibrium points. Hence, even if the game remains formally a non-cooperative game without enforceable agreements, the players will now have no difficulty in reaching the outcome (A_1, B_1), which we used to call the cooperative solution, so as to obtain the payoffs $(2, 2)$. This conclusion, of course, is fully consistent with our theory because now (A_1, B_1) *is* an equilibrium point.

This example shows that we must clearly distinguish between two different problems. One is the problem of whether a game that *looks like* a prisoner's dilemma *is* in fact a prisoner's dilemma: does the proposed payoff matrix of the game (which would make the game a prisoner's dilemma) correctly express the players' true payoff functions, in accordance with their real preferences and their real strategy objectives within the game? This is *not* a game-theoretical question, because game theory regards the players' payoff functions as *given*. It is, rather, an empirical question about the players' psychological make-up. The other question *is* a game-theoretical question: it is the question of how to define the solution of the game, once the payoff matrix has been correctly specified. A good deal of confusion can be avoided if these two questions are kept strictly apart.

As a practical matter, social situations not permitting enforceable agreements often have a socially very undesirable incentive structure, and may give rise to many very painful human problems. But these problems cannot be solved by arguing that people should act as if agreements were enforceable, even though they are not; or that people should trust each other, even though they have very good reasons to withhold this trust. The solution, if there is one, can only lie in actually providing effective incentives to keep agreements (or in persuading people to assign high utility to keeping agreements, even in the absence of external incentives). What we have to do, if it can be done, is to *change* non-cooperative games into cooperative games by making agreements enforceable, rather than pretend that we live in a make-believe world, where we can take non-cooperative games as they are, and then analyze them simply as if they were cooperative games, if we so desire.

I have discussed at some length the principle that the solution of a non-cooperative game must be an equilibrium point, because this principle will play an important role in the game-theoretical investigations I am going to report on.

8.   PERFECT EQUILIBRIUM POINTS

After Nash's discovery of the concept of equilibrium points in 1950, for many years game theorists were convinced that the only rationality requirement in a non-cooperative game was that the players' strategies should form an equilibrium point. But in 1965 Reinhard Selten proposed counterexamples to show that even equilibrium points might involve irrational behavior (Selten, 1965). He has suggested that only a special class of equilibrium points, which he called *perfect* equilibrium points, represent truly rational behavior in a non-cooperative game.

Since the difference between perfect and imperfect equilibrium points is obscured in the normal-form representation,[9] let us consider the following two-person non-cooperative game, given in extensive form (game-tree form):

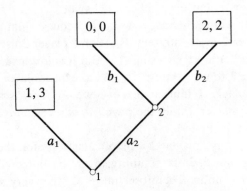

In this game, the first move belongs to player 1. He can choose between moves a_1 and a_2. If he chooses a_1, then the game will end with the payoffs $(1, 3)$ to the two players, without player 2 having any move at all. On the other hand, if player 1 chooses move a_2, then player 2 has a choice between moves b_1 and b_2. If he chooses the former, then the game will end with the payoffs $(0, 0)$; while if he chooses the latter, then the game will end with the payoffs $(2, 2)$. The normal form of this game is as follows:

$$
\begin{array}{c|c|c|}
 & B_1 & B_2 \\
\hline
A_1 & 1,3 & 1,3 \\
\hline
A_2 & 0,0 & 2,2 \\
\hline
\end{array}
$$

The players' strategies have the following interpretation. Strategy A_1 (or A_2) means that player 1 will choose move a_1 (or a_2) at the beginning of the game. On the other hand, strategy B_1 (or B_2) means that player 2 will choose move b_1 (or b_2) *if player 1 chooses move a_2* (while if player 1 chooses move a_1, then player 2 will do nothing). Player 2's strategies can be described only in terms of these *conditional* statements since he will have a move only if player 1 chooses move a_2.

A look at the normal form will reveal that the game has two pure-strategy equilibrium points, *viz.* $E_1 = (A_1, B_1)$ and $E_2 = (A_2, B_2)$. E_2 is a perfectly reasonable equilibrium point. But, as I propose to show, E_1 is not: it involves irrational behavior, and irrational expectations by the players about each other's behavior.

In fact, player 1 will use strategy A_1 (as E_1 requires him to do) only if he expects player 2 to use strategy B_1. (For if player 2 used B_2, then player 1 would be better off by using A_2.) But it is *irrational* for player 1 to expect player 2 to use strategy B_1, i.e. to expect player 2 to make move b_1 should player 1 himself make move a_2. This is so because move b_1 will yield player 2 only the payoff 0, whereas move b_2 would yield him the payoff 2.

To put it differently, player 2 will obviously prefer the outcome (A_1, B_1), yielding him 3 units of utility, over the outcome (A_2, B_2), yielding him only 2 units. Therefore, player 2 may very well try to induce player 1 to use strategy A_1, i.e. to make move a_1: for instance, he may threaten to use strategy B_1, i.e. to punish player 1 by making move b_1, should player 1 counter his wishes by making move a_2. But the point is that this would *not* be a credible threat because, by making move b_1, player 2 would not only punish player 1 but rather would just as much punish himself. This is so because move b_1 would reduce *both* of their payoffs to 0 (while the alternative move b_2 would give both of them payoffs of 2 units).

To be sure, if player 2 could irrevocably *commit* himself to punish player 1 in this way, and could do this *before* player 1 had made his move, then it would be rational for player 2 to make such a commitment in order to deter player 1 from making move a_2. But, in actual fact, player 2 cannot make such a commitment because this is a non-cooperative game. On the other hand, if player 2 is *not* compelled by such a prior commitment to punish player 1, then he will have no incentive to do so since, once player 1 has made his move, player 2 cannot gain anything by punishing him at the cost of reducing his own payoff at the same time.

To conclude, $E_1 = (A_1, B_1)$ is an irrational equilibrium point because it is based on the unreasonable assumption that player 2 would punish player 1 if the latter made move a_2 – even though this punishing move would reduce not only player 1's payoff but also player 2's own payoff. Following Selten's proposal, we shall call such unreasonable equilibrium points *imperfect* equilibrium points. In contrast, equilibrium points like $E_2 = (A_2, B_2)$, which are not open to such objections, will be called *perfect* equilibrium points.

The question naturally arises how it is possible that an equilibrium point should use a highly irrational strategy like B_1 as an equilibrium strategy at all. The answer lies in the fact that, as long as the two players follow their equilibrium strategies A_1 and B_1, player 2 will never come in a position *where he would have to make the irrational move b_1* prescribed by strategy B_1. For, strategy B_1 would require him to make move b_1 only if player 1 made move a_2. But this contingency will never arise because player 1 follows strategy A_1 (which requires him to make move a_1 rather than a_2).[10]

In other words, strategy B_1 would require player 2 to make move b_1 only if the game reached the point marked by 2 on our game tree[11] (since this is the point where he had to choose between moves b_1 and b_2). But, so long as the players follow the strategies A_1 and B_1, this point will never be reached by the game.

This fact suggests a mathematical procedure for eliminating imperfect equilibrium points from the game. All we have to do is to assume that, whenever any player tries to make a specific move, he will have a very small but positive probability ε of making a 'mistake', which will divert him into making another move than he wanted to make, so that *every* possible move will occur with some positive probability. The resulting game will be called a *perturbed game*. As a result of the players' assumed 'mistakes', in a perturbed game every point of the game tree will always be reached with a positive probability whenever the game is played. It can be shown that, if the game is perturbed in this way, only the perfect equilibrium points of the original game will remain equilibrium points in the perturbed game, whereas the imperfect equilibrium points will lose the status of equilibrium points. (More exactly, we can find the perfect equilibrium points of the original game if we take the equilibrium points of the perturbed game, and then let the mistake probabilities ε go to zero.)

Thus, in our example, suppose that, if player 1 tries to use strategy A_1, then he will be able to implement the intended move a_1 only with probability $(1-\varepsilon)$, and will be forced to make the unintended move a_2 with the remaining small probability ε. Consequently, it will not be costless any more for player 2 to use strategy B_1 when player 1 uses A_1. This is so because now player 1 will make move a_2 with a positive probability and, therefore, player 2 will have to counter this by making the costly move b_1, likewise with a positive probability. As a result, strategy B_1 will no longer be a best reply to A_1, and (A_1, B_1) will no longer be an equilibrium point.

The difference between perfect and imperfect equilibrium points can be easily recognized in the extensive form of a game but is often hidden in the normal form. This implies that, contrary to a view that used to be the commonly accepted view by game theorists, the normal form of the game in general fails to provide all the information we need for an adequate game-theoretical analysis of the game, and we may have to go back to the extensive form to recover some of the missing information.

On the other hand, if the normal form often contains too little information, the extensive form usually contains far too much, including many unnecessary details about the chance moves and about the time sequence in which individual moves have to be made. For this reason, Reinhard Selten and I have defined an intermediate game form, called the *agent normal form*, which omits the unnecessary details but retains the essential information about the game. (We obtain the agent normal form if we replace each player by as many 'agents' as the number of his information sets in the game, and then construct a normal form with these agents as the players.) For a more extensive and more rigorous discussion of perfect equilibrium points and of the agent normal form, see Selten (1975).

9. NON-COOPERATIVE BARGAINING MODELS FOR COOPERATIVE GAMES

Ever since 1944 (the year when von Neumann and Morgenstern first published the *Theory of Games and Economic Behavior*), most research in game theory has been devoted either to a study of the mathematical properties of saddle points in *two-person zero-sum* games, or to a construction and study of solution concepts for *cooperative* games. Many very interesting cooperative solution concepts were proposed. But this work on cooperative games showed little theoretical unity: taken as a group, the different solution concepts that were suggested shared few common theoretical assumptions, and no clear criteria emerged to decide under what conditions one particular solution concept was to be used and under what conditions another.

Part of the problem is that the authors of the different solution concepts have seldom made it sufficiently clear what institutional arrangements (negotiation rules) each particular solution concept is meant to assume about the bargaining process among the players, through which these players are supposed to reach an agreement about the final outcome of the game. Yet, it is well known that the very same cooperative game may have quite different outcomes, depending on the actual negotiation rules governing this bargaining process in any particular case. The nature of the agreements likely to arise will be often quite sensitive to such factors as who can talk to whom and, in particular, who can talk to whom *first*, ahead of other people; the degree to which negotiations are kept public, or can be conducted in private by smaller groups if the participants so desire; the conditions that decide whether any agreement remains open to repeal and to possible re-negotiation, or is made final and irrevocable; the possibility or impossibility of unilaterally making binding promises and/or threats, etc.

As a simple example, consider the following three-person cooperative game (called a three-person majority game). Each player, acting alone, can only achieve a zero payoff. Any coalition of two players can obtain a joint payoff of $100. The three-person coalition of all three players can likewise obtain a joint payoff of $100. Obviously, in this game, if pairs of players can meet separately, then the two players who manage to meet first are very likely to form a two-person coalition, and to divide the $100 in a ratio 50:50 between them. In contrast, if the negotiation rules disallow pairwise meetings, and if the negotiation time permitted is too short for forming any two-person coalition during the three-person negotiating session, then the likely outcome is a three-person coalition, with payoffs $33\frac{1}{3}:33\frac{1}{3}:33\frac{1}{3}$. Finally, under most other negotiation rules, both two-person and three-person coalitions will arise from time to time, and the probability of either outcome will depend on the extent to which these rules tend to help or hinder two-person agreements.

Another limitation of most cooperative solution concepts is this. Their application is basically restricted to fully cooperative games, and does not extend to that very wide range of real-life game situations which have a status intermediate between fully cooperative games and fully non-cooperative games – such as social situations where some kinds of agreements are enforceable while others are not, or where different agreements may be enforceable to different extents and with different probabilities; or where enforceable agreements are possible among some particular players but are impossible among other players; or where enforceable agreements cannot be concluded at some stages of the game but can be concluded at other stages, etc. In many contexts, it is particularly regrettable that most of these cooperative solution concepts are inapplicable to games possessing a strongly sequential structure, making the emergence of agreements a very gradual process, later agreements being built on earlier agreements and extending the former in various ways.

Yet, John Nash, when he introduced the very concepts of cooperative and of non-cooperative games, also suggested what, in my opinion, is a possible remedy to these deficiencies in the theory of cooperative games (Nash, 1951, p. 295). He suggested that an analysis of any cooperative game should start with constructing a precisely defined formal bargaining model (bargaining game) to represent the bargaining process among the players. Of course, this bargaining model must provide a mathematical representation, in the abstract language appropriate to such models, for the negotation rules we want to assume, whether on empirical or on theoretical grounds, to govern this bargaining process. Then, according to Nash's proposal, this bargaining model should be analyzed as a *non-cooperative* game, by a careful study of its equilibrium points.

Nash's suggestion is based on the assumption that close cooperation among the players in a cooperative game usually requires a prior agreement about the payoffs, which, in most cases, can be achieved only by bargaining among the players. But this bargaining itself must have the nature of a non-cooperative game, unless we want to assume that the players will agree in an even earlier subsidiary bargaining game on how they will act in the main bargaining game – which would be not only a rather implausible assumption but would also lead to an infinite regress.

Nash's proposal, if it can be successfully implemented, will enable us to unify the whole theory of cooperative games, because it provides a uniform method of analysis for all cooperative games. Of course, even under Nash's proposal, it will remain true that any given cooperative game may have a number of different solutions, depending on the details of the bargaining process assumed to occur among the players. But, as we have argued, this is how it should be, since in real life different bargaining methods do lead to different outcomes. Yet, the game-theoretical analysis of this bargaining process can be based on the same theoretical principles in all cases.[12]

Indeed, Nash's approach will result in a unification of *both* the theory of cooperative games *and* of the theory of non-cooperative games, because it essentially reduces the problem of solving a cooperative game to the problem of solving a non-cooperative bargaining game. Moreover, it can be easily extended to games which have any kind of intermediate status between fully cooperative games and fully non-cooperative games (including games of a sequential nature, mentioned before).

10. A BAYESIAN SOLUTION CONCEPT
FOR NON-COOPERATIVE GAMES

Nash's proposal, however, runs into a very basic difficulty – the same difficulty, which, up to very recently, also prevented the emergence of any really useful theory of non-cooperative games. This difficulty lies in the fact that almost any interesting non-cooperative game – including almost any interesting non-cooperative bargaining game – will have a great many, and often infinitely many, very different equilibrium points. (This remains true even if we restrict ourselves to perfect equilibrium points.) This means that, if all we can say is that the outcome of the game will be an equilibrium point (or even that it will be a perfect equilibrium point), then we are saying little more than that almost anything can happen in the game.

For instance, consider the very simplest kind of two-person bargaining game, in which the two players have to divide $100. If they cannot agree on how to divide it, then both of them will receive zero payoffs. This game can be analyzed by means of the following formal bargaining model. Both players name a number between 0 and 100. Let the numbers named by players 1 and 2 be x_1 and x_2. (Intuitively, these numbers represent the two players' payoff demands.) If $x_1 + x_2 \leq 100$, then player 1 will obtain $\$x_1$ and player 2 will obtain $\$x_2$. On the other hand, if $x_1 + x_2 > 0$, then both players will get $0.

If we assume that money can be divided in any possible fractional amount, then this game has infinitely many equilibrium points, since any pair (x_1, x_2) of payoff demands is an equilibrium point so long as $x_1 + x_2 = 100$. (Of course, by the rules we have assumed for the game, we must also have $0 \leq x_i \leq 100$ for $i = 1, 2$.) But even if we assumed that money can be divided only in amounts corresponding to whole numbers of dollars, the game will still have 101 equilibrium points. (Of these, 99 equilibrium points will even be perfect equilibrium points. Only the two 'extreme' equilibrium points giving one player $100 and giving the other player $0 turn out to be imperfect.) The situation will be even worse if we study more interesting, and therefore inevitably more complicated, bargaining games.

In view of these facts, several years ago Reinhard Selten and I decided to look into the possibility of defining a new solution concept for non-cooperative games, which will always select *one* particular equilibrium point as the solution for the game. This research project proved to be much more difficult than we had anticipated. But in 1974 we did find such a solution concept which seems to satisfy all intuitive and mathematical requirements. Conceptually, it amounts to an extension of the Bayesian approach, so successful in the analysis of one-person decision situations, to an analysis of non-cooperative games. The definition of this solution concept is based on the disturbed agent normal form of the game (see Section 8 above).

Let me introduce the following notation. We shall assume that a given player i $(i = 1, 2, ..., n)$ has K_i different pure strategies. Therefore, a mixed strategy s_i of player i will be a probability vector of the form $s_i = (s_i^1, s_i^2, ..., s_i^{K_i})$, where s_i^k $(k = 1, 2, ..., K_i)$ is the probability that this mixed strategy s_i assigns to the kth pure strategy of player i.

A strategy combination of all n players will be denoted as $s = (s_1, s_2, ..., s_n)$. Let $\overline{s}_i$ denote the strategy combination we obtain if we omit player i's strategy s_i from the strategy combination s. Thus, $\overline{s}_i$ is a strategy combination of the $(n - 1)$ players *other* than player i. We can write $\overline{s}_i = (s_1, ..., s_{i-1}, s_{i+1}, ..., s_n)$.

Our solution is defined in two steps. As a first step, we construct a *prior probability distribution* p_i, over the pure strategies of each player i $(i = 1, 2, ..., n)$. The second step involves a mathematical procedure which selects one specific equilibrium point $s^* = (s_1^*, s_2^*, ..., s_n^*)$ as the solution of the game, on the basis of these n prior probability distributions $p_1, p_2, ..., p_n$.

Each prior probability distribution p_i over the pure strategies of a given player i has the form of a probability vector $p_i = (p_i^1, p_i^2, ..., p_i^{K_i})$, where each component p_i^k $(k = 1, 2, ..., K_i)$ is the initial *subjective* probability that every other player j $(j \neq i)$ is assumed to assign to the possibility that player i will use his kth pure strategy in the game. Consequently, the prior probability distribution p_i is a probability vector of the same mathematical form as is any mixed strategy s_i of player i. But, of course, p_i has a very different game-theoretical interpretation. Whereas a mixed strategy s_i expresses the *objective* probabilities s_i^k that player i *himself* chooses to associate with his various pure strategies as a matter of his own strategical decision, the prior probability distribution p_i expresses the *subjective* probabilities p_i^k

that the *other* players are assumed to associate with player i's various pure strategies, simply because they do not know in advance which particular strategy player i is going to use.

The numerical prior probability p_i^k our theory assigns to a given pure strategy of each player i is meant to express the theoretical probability that a rational individual, placed in player i's position, will actually use this particular pure strategy in the game. More specifically, p_i^k is meant to express the theoretical probability that player i will find himself in a situation where his best reply is to use this particular pure strategy. (This theoretical probability p_i^k, of course, is not directly given, but can only be obtained from a suitable probabilistic model about the players' behavior in the game.)

For convenience, I shall write the n-vector consisting of the n prior probability distributions as $p = (p_1, p_2, ..., p_n)$. I shall write the $(n-1)$-vector consisting of the $(n-1)$ prior probability distributions associated with the $(n-1)$ players other than player i as $\overline{p}_i = (p_1, ..., p_{i-1}, p_{i+1}, ..., p_n)$. Thus, $\overline{p}_i$ is the $(n-1)$-vector we obtain if we omit the ith component p_i from the n-vector p.

The second step in defining our solution involves a mathematical procedure, based on Bayesian ideas, for selecting one particular equilibrium point s^* as solution, when the vector p of all prior probability distributions is given. The simplest Bayesian model would be to assume that each player i would use a strategy s_i that was his best reply to the prior probability distribution vector $\overline{p}_i$ he would associate with the other $(n-1)$ players' pure strategies, and then to define the solution as the strategy combination $s = (s_1, s_2, ..., s_n)$. But this simple-minded approach is unworkable because in general this best-reply strategy combination s will not be an equilibrium point of the game.

Accordingly, our theory uses a mathematical procedure, called the *tracing procedure*, which takes this best-reply strategy combination s as a starting point, but then systematically modifies this strategy combination in a continuous manner, until it is finally transformed into an equilibrium point s^*, which will serve as the solution of the game.

This mathematical procedure is meant to model the psychological process, to be called the *solution process*, by which the players' expectations converge to a specific equilibrium point as the solution of the game. At the beginning, the players' initial expectations about the other players' strategies will correspond to the subjective probability distributions (prior distributions) p_i, and their initial reaction to these expectations will be an inclination to use their best-reply strategies s_i. Then, they will gradually modify their expectations and their tentative strategy choices until in the end *both* their expectations and their strategy choices will converge to the equilibrium strategies s_i^* corresponding to the solution s^*. (See Harsanyi, 1975.)

Of course, within the confines of this paper, I could do no more than sketch the barest outlines of our solution concept for non-cooperative games and, more generally, could draw only a very incomplete picture of some recent work in game theory. But I feel my paper has achieved its purpose if it has induced a few people from other disciplines to take some interest in the results of Bayesian decision theory and in some recent developments in game theory, and in the implications both of these may have for a deeper understanding of the nature of rational behavior.

University of California, Berkeley

NOTES

[1] The author wishes to express his thanks to the National Science Foundation for supporting this research by Grant GS-3222 to the Center for Research in Management Science, University of California, Berkeley.

[2] For references, see Harsanyi (1969, p. 517, footnote 9).

[3] See footnote 2.

[4] The concept of criterion-satisfying behavior is probably not very important in everyday life. But it is very important in ethics (see Harsanyi, 1958).

[5] Of course, in many cases, when a person has changed his goals, the most natural explanation will be that his preferences themselves have changed. In such cases, the model of rational behavior will be inapplicable, or at least will have to be supplemented by other explanatory theories, e.g. by learning theory, etc.

[6] A very simple proof of this theorem for *risk* is given by Luce and Raiffa (1957, pp. 23–31). But note that their Assumptions 1 and 5 could be easily stated as one axiom, whereas Assumption 6 could be omitted because it follows from the other axioms. (Of course, the use of extra axioms was intentional and made it possible for the authors to simplify their statement of the proof.) Note also that their substitutability axiom (Assumption 4) could be replaced by a form of the sure-thing principle (see below). A simple proof of the theorem for *uncertainty* is found in Anscombe and Aumann (1963).

[7] The term 'Bayesian approach' is often restricted to the proposal of using expected-utility maximization as a definition of rational behavior in the case of *uncertainty*, where expected utility must be computed in terms of *subjective* probabilities.

[8] The terms 'bet' and 'lottery' will be used interchangeably.

[9] For a non-technical explanation of the terms 'normal form' and 'extensive form', see Luce and Raiffa (1957, Chapter 3).

[10] From a logical point of view, strategy B_1 does satisfy the formal criteria for an equilibrium strategy because, in applying these criteria, the conditional statement defining strategy B_1 ('player 2 would make move b_1 if player 1 made move a_2') is interpreted as *material implication*. In contrast, B_1 fails to satisfy our informal criteria for a 'rational' strategy because, in applying these latter criteria, the same conditional statement is automatically interpreted as a *subjunctive conditional*.

[11] We say that a given point of the game tree is *reached* by the game if it either represents the starting position in the game or is reached by a branch representing an actual move by a player or by chance. Thus, in our example, the point marked by 1 is always reached whereas the point marked by 2 is reached only if player 1 chooses to make move a_2 (rather than move a_1).

[12] While Nash's proposal has many advantages, it certainly does not provide an easy routine method for solving cooperative games because, except in the very simplest cases, finding a suitable formal bargaining model for any given game – just as a modeling of any other complicated dynamic process – may be a very difficult task, requiring a good deal of insight and ingenuity, and subject to no mechanical rules.

BIBLIOGRAPHY

Anscombe, F. J. and Aumann, R. J.: 1963, 'A Definition of Subjective Probability', *Annals of Mathematical Statistics* **34**, 199–205.

Debreu, G.: 1959, *Theory of Value*, John Wiley & Sons, New York.

Harsanyi, J. C.: 1953, 'Cardinal Utility in Welfare Economics and in the Theory of Risk-taking', *Journal of Political Economy* **61**, 434–435.

Harsanyi, J. C.: 1955, 'Cardinal Welfare, Individualistic Ethics, and Interpersonal Comparisons of Utility', *Journal of Political Economy* **63**, 309–321.

Harsanyi, J. C.: 1958, 'Ethics in Terms of Hypothetical Imperatives', *Mind* **47**, 305–316.

Harsanyi, J. C.: 1967–68, 'Games with Incomplete Information Played by 'Bayesian' Players', *Management Science* **14**, 159–182, 320–334, and 486–502.

Harsanyi, J. C.: 1969, 'Rational-Choice Models of Political Behavior vs. Functionalist and Conformist Theories', *World Politics* **21**, 513–538.

Harsanyi, J. C.: 1975, 'The Tracing Procedure: A Bayesian Approach to Defining a Solution for *n*-Person Non-cooperative Games', *International Journal of Game Theory* **4**, 61–94.

Harsanyi, J. C.: 1975a, 'Can the Maximin Principle Serve as a Basis for Morality? A Critique of John Rawls's Theory', *American Political Science Review* **59**, 594–606.

Harsanyi, J. C.: 1975b, 'Nonlinear Social Welfare Functions', *Theory and Decision* **7**, 61–82.

Luce, R. D. and Raiffa, H.: 1957, *Games and Decisions*, John Wiley & Sons, New York.

Nash, J. F.: 1950, 'Equilibrium Points in n-Person Games', *Proceedings of the National Academy of Sciences, U.S.A.* **36**, 48–49.

Nash, J. F.: 1951, 'Non-cooperative Games', *Annals of Mathematics* **54**, 286–295.

Radner, R. and Marschak, J.: 1954, 'Note on Some Proposed Decision Criteria', in R. M. Thrall *et al.*, *Decision Processes*, John Wiley & Sons, New York, 1954, pp. 61–68.

Selten, R.: 1965, 'Spieltheoretische Behandlung eines Oligopolmodells mit Nachfrageträgheit', *Zeitschrift für die gesamte Staatswissenschaft* **121**, 301–324 and 667–689.

Selten, R.: 1975, 'Reexamination of the Perfectness Concept for Equilibrium Points in Extensive Games', *International Journal of Game Theory* **4**, 25–55.

Simon, H. A.: 1960, *The New Science of Management Decision*, Harper & Brothers, New York.

von Neumann, J. and Morgenstern, O.: 1944, *Theory of Games and Economic Behavior*, Princeton University Press, Princeton, N.J.

J. W. N. WATKINS

TOWARDS A UNIFIED DECISION THEORY: A NON-BAYESIAN APPROACH

1. THE PROBLEM

In his 'Advances in Understanding Rational Behavior' John Harsanyi (1977) raises, and answers, some very interesting problems. The one with which I will be mainly concerned is the problem of *unifying* decision theory. (He prefers the term 'utility theory' for a general theory that covers rational decision making under conditions of certainty, risk, and uncertainty.) Before turning to that I will make one quick suggestion concerning another problem that he raises: how can the *normative* idea of rationality play a central role in a *positive* science such as economics?

People often act in a sub-optimal way. Now there are situations which are rather undemanding and tolerate a good deal of sub-optimal behaviour. A monopolist who sets his price at a sub-optimal level may still lead a very comfortable life. But there are other more demanding situations; and some of these exert a strong corrective feedback on an agent who is not acting rationally. A seller in a highly competitive market who sets *his* price at a sub-optimal level will soon find himself facing bankruptcy unless he revises his behaviour.[1] In situations of the latter kind behaviour is likely to converge towards the rational norm. Rationalistic models are more likely to yield good predictions when applied to such situations. When applied to situations of the other kind they may yield good advice, but advice with which people do not conform very closely in their empirical behaviour.

So much for the empirical applicability of the idea of rationality. In this paper I will henceforth be concerned with rationality primarily in its normative aspect.

In his paper Harsanyi (relying on some recent results obtained by Selten and himself) gives a bold sketch of a comprehensive solution of the problem of unifying decision theory. It involves some far-reaching and systematic extensions of Bayesian ideas to game theory.

Butts and Hintikka (eds.), Foundational Problems in the Special Sciences, 345–379.
Copyright © 1977 by D. Reidel Publishing Company, Dordrecht-Holland. All Rights Reserved.

For me, both Harsanyi's main problem and his programme for dealing with it present an exciting challenge: the problem because of its obvious interest and importance, and his programme because, greatly as I admire his execution of it, I find myself in disagreement with its underlying ideas and therefore under an obligation to try to adumbrate, however tentatively and amateurishly, an alternative approach.

As I see it, Harsanyi's problem arises in the following way. At the present time we have a plethora of theories (or models) each designed with an eye to some particular kind of decision problem. For instance: economic theory provides indifference-curve models for consumers' behaviour and marginal revenue-cost equation models for sellers' behaviour, the latter proliferating into special models for monopoly, bilateral monopoly, and oligopoly; for a single individual acting under conditions of risk, there is the principle of maximising expected utility; for two (or a few) competing individuals, there is, of course, game theory; but this proliferates into a variety of models each designed to fit some main type of game-situation. The above all involve some kind of maximisation; but there is also a conception of rational conduct which dispenses with this idea, namely Simon's conception of 'satisficing'.[2]

Now one might – especially if one had instrumentalist leanings – take a 'tool-bag' view of this multiplicity: what we have here, one might say, is not a set of rival theories but only a set of different tools, each appropriate to its special task(s). That view might be tenable if there were either no overlap between the respective domains of application of each of these models of rationality, or if there were no inconsistency where they do overlap. But there is both overlap and inconsistency. Harsanyi has exposed some large conflicts between the view that he espouses and a view (associated with the minimax principle) which he vigorously opposes. And I will myself bring out some conflicts between his view and one that I shall espouse.

Now I do not want, especially at this early stage, to get involved in a meta-level discussion concerning the status of theories of rational decision making. But it would seem that, whatever view one adopts, the existence of such conflicts must surely be regarded as a sign that something is wrong somewhere. The main alternatives seem to be these.

(1) A theory of rational decision making is a formal system whose axioms constitute an implicit *definition* of rationality and whose theorems are logical implications of that definition.

(2) A theory of rational decision making yields substantive rules which, however, may be applicable only to restricted classes of decision makers. (For instance, some kind of minimax rule may be presented as suitable for 'security-minded' agents.)

(3) A theory of rational decision making yields substantive rules that are applicable to any decision maker *provided* that full information is available concerning his preferences and temperament.

(4) A theory of rational decision making yields empirical hypotheses to the effect that people, when they are acting in ways that are generally regarded as wise and prudent, behave very much *as if* they were consciously complying with the theory.[3]

According to (1), conflicts should no more arise between alternative 'theories' than they do between, say, Euclidean and non-Euclidean geometries. According to both (2) and (3), if two rationality rules conflict, then at least one of them must be wrong; however, according to (2) but not to (3), the trouble may be rectified by restricting one or both to a narrower class of decision maker. According to (4), if two theories of rationality have significantly divergent implications, at least one of them must be false.

Notice that (3) and (4) are compatible. And (1) can easily grow into (3) and/or (4) if the axiom system is supposed to capture (be a rational reconstruction of) some central though informal idea of rational behaviour.

To put the problem picturesquely: what we have at present is a baronial system of semi-autarkic theories whose territories sometimes overlap; and where they do overlap, they sometimes conflict. Admittedly, when such a conflict comes to light a boundary commission might restore peace by redrawing the domains of application of the theories involved. But apart from its *ad hoc* nature, this would give no assurance that further conflicts will not come to light in the future. Surely it would be more satisfactory if we had one sovereign theory of rational decision making which answers in a consistent way all the questions that existing theories answer in various ways and which, perhaps, also answers some questions which present theories cannot answer.

How might we tackle the programme of developing a unified theory of rational action? Well, we might begin with some very simple and, as it were, paradigmatic idea of rationality which I will refer to as R. (For instance, R might say that if an agent A has a choice between two strategies, a_1 and a_2, and if the total payoff of a_1 will certainly be x while that of a_2 will certainly be y, and if he prefers x to y, then A should choose a_1.) Within its range of application R should be ungainsayable; but its range of application will be very restricted. The problem now is to extend it, step by step, in a cogent way and without loss of precision, so that it applies to ever more kinds of situation. If R', R'', ... were such a sequence of successive generalisations of the original R, then the relation of a later theory to its predecessor should, ideally, be this: the earlier theory should be a limiting case of its successor in the sense that the latter has an adjustable parameter which assumes a fixed value (usually one or zero) in the former. Obvious examples of this would be: a theory of rational decision making under conditions of risk which yields as a limiting case a theory of rational decision making under conditions of certainty when all probability values are put at either one or zero; or an n-person game theory which yields a one person 'game against nature' theory when n is put at one. (More sophisticated examples are given by Harsanyi; see below.) Ideally, such a sequence should terminate with a theory, say R^*, which does not seem open to further generalisation and which solves, in principle, all solvable decision problems. (I will suggest in Section 3 that there are some decision problems that are not rationally solvable.)

2. HARSANYI'S PROGRAMME

Let us now look into the guiding ideas in Harsanyi's programme for a unified, or general, theory of rational decision making (or utility theory, as he prefers to call it).

In his paper he presents a sequence, of the kind indicated above, of theories of ever-increasing generality. The first one – as it were, the R in his R, R', R'', ... sequence – is the idea that under conditions of certainty (i.e. when the agent knows what would be the result of each of the choices open to him) one should maximise utility. His R' states that under conditions of risk (i.e. when the numerical probabilities of all the possible results of all the choices open to him are objectively given) the agent should maximise expected utility, the expected utility of a possible outcome being its utility multiplied by its probability. Here, Harsanyi's *Bayesianism* makes its first appearance. But it emerges more fully with his R'': this says that under conditions of uncertainty, where objective probabilities are *not* given for (at least some of) the possible outcomes, the agent should assign (explicit or tacit) *subjective* probabilities to them, and then proceed as if under conditions of risk.

The idea of expected utility is central to Harsanyi's approach. As well as allowing us to regard certainties as a limiting kind of risk, namely one where the probabilities are 1 or 0, it also, more interestingly, allows us to deal with risks as if we were dealing with certainties. For as well as allowing us to give cardinal values (on an interval scale) to the utilities of different payoffs for an agent who complies with the axioms of expected utility theory, it allows us to give a cardinal utility value to a given *chance* of getting a specified payoff. Thus if we present our agent with a choice, not between the prizes a and b and c, but between lottery tickets a' and b' and c' which give him, respectively, certain chances of getting a or b or c, then according to the theory of expected utility each of these lottery tickets will have a definite value for him and he will be able to choose between a' and b' and c' just as if he were choosing between a and b and c. A combination of a utility-value and a risk-factor can be replaced by just a utility-value (namely, the expected utility of the combination): we can, as it were, take the riskiness out of risks. (This idea will come under attack in Section 7.)

The novel and impressive thing about Harsanyi's programme is that he has gone on to incorporate the whole territory of game theory into his Bayesian empire, ending with an R^* which yields, at least in principle, a solution for every decision-theoretical problem. His first step is to treat game situations, not as being in a category of their own, but as 'a special case of uncertainty' (p. 322). This means that one of the first steps in a player's solution of a game-theoretical problem is to assign probabilities to each of the (pure) strategies available to each of the other players, his aim being to arrive eventually at an assignment of probabilities to each of his own (pure) strategies resulting in a (mixed) strategy that constitutes his *best reply* to the (mixed) strategies, taken collectively, of all the other players.

In the course of proceeding to his R^* Harsanyi carries out several important reductions and generalisations. A striking example is his reduction of a game with incomplete information to the status of a special case of a game with complete information (pp. 324–326). Orthodox game theory usually assumes that the players are transparent to each other at least in the sense that each knows what strategies are available to the others and what payoffs they ascribe to the possible outcomes of the game. Indeed, this usually has to be assumed if the game is not to become indeterminate through lack of a well-defined payoff matrix. But in real life this assumption will often be unrealistic: a player will often have very incomplete information about the possible strategies and actual utilities of his opponent(s). As I understand him, Harsanyi deals with this along the following lines. You first split your real but more or less shadowy and enigmatic opponent into an (exhaustive?) disjunction of possible opponents, each of whom has a well-defined set of strategies and well-defined payoff values. You then assign to each of these possible opponents a numerical probability according to your estimate of the relative likelihood that he is your real opponent. Vis-à-vis each possible opponent you calculate the expected utility of that strategy of yours which constitutes your best reply to him; you then multiply this by the probability that he is your real opponent. Finally, you calculate which of your strategies has the highest expected utility vis-à-vis *all* your possible opponents, and hence vis-à-vis the man himself.

Another of Harsanyi's reductions consists of turning a cooperative game into a special case of a non-cooperative game. But rather than dwell on that I will turn at once to his *tracing procedure* for arriving at a solution for any n-person non-cooperative game (hence for *any* game).[4] This begins with each player i making initial probability-distributions over the strategies of each of the other $n-1$ players. These probabilities are not arbitrary but are determined by a mathematical function of the parameters of the game. (Thus if a certain player j has two strategies, say s_{j_1} and s_{j_2}, such that s_{j_1} would be a best reply to more situations in this game than would s_{j_2}, then player i

will assign a higher probability to s_{j_1} than to s_{j_2}.) In the light of his initial probability assignments to the other players' strategies, player i now calculates what assignment of probabilities to his own (pure) strategies would yield a best reply to them. This done, he now revises his probability distributions over the other players' strategies so as to let them take into account *his* best reply, as presently estimated. This done, he then revises his own best reply ... and so on. Harsanyi calls this the solution process, which is mathematically modelled by his tracing procedure. The latter defines a unique solution for the game, a solution to which all the players' probability distributions will converge as they get continuously modified.

Harsanyi's achievement might be summed up thus. He has (bar a few details still to be filled in) actually constructed an $R, R', R'', ..., R^*$ sequence such that, no matter under what conditions he is acting, the decision maker is in principle able to attach to each of the strategies open to him a lottery ticket (it may be enormously long) which assigns a probability-cum-utility value to each of the possible outcomes of that strategy, the probabilities summing to one. And this means that each strategy can, in principle, be characterised by its expected utility. And this in turn means that, no matter how complex and fraught with uncertainty his situation may be, the decision maker can proceed as if under conditions of certainty. Finally, Harsanyi extends his Bayesian-ism to ethics, where he reinterprets utilitarianism as 'maximising the average utility level of all individuals in society' (p. 323). But I will not follow him into ethical territory in the present paper.

I regard Harsanyi's achievement as a triumph for Bayesianism. But I disagree with Bayesianism. And since I prefer not to engage in nagging, piecemeal criticisms, but to make my criticisms from an alternative standpoint, I will, in Sections 4, 5, and 6, present the first few steps in an alternative $R, R', R'', ...$ sequence. I apologise for my termerity in undertaking this: I am a philosopher, not a decision-theorist. But as I mentioned, I have felt that I ought to try to respond to the challenge of Harsanyi's paper. I hasten to add that I am nowhere near being able to complete my alternative sequence with an R^* and that there will be little originality in the steps of it that I will present. In the meanwhile I will now make a disclaimer of a different kind. My motive for including the next section is to secure consistency with an earlier claim of mine, and the reader in a hurry is advised to skip it.

3. RATIONALLY INDETERMINATE DECISION PROBLEMS

I have argued elsewhere that there are decision-problems that are quasi-paradoxical and for which there is no determinate rational solution.[5] I want to say a word about this now and then set it aside.

One way of generating such rationally indeterminate problem situations is this. We take two problem-situations, P_1 and P_2, such that: (1) both allow just two alternative types of strategy, say S_1 and S_2; (2) S_1 is the rational solution for P_1 and S_2 is the rational solution for P_2; (3) by continuously varying the parameters of P_1 we can transform it into P_2. In such a case there must presumably be some intermediate problem-situation, say P^*, which has a strong resemblance both to P_1 and to P_2, and for which it is undecidable whether S_1 or S_2 is the right solution. For instance, let P_1 be a straightforward case of the prisoners' dilemma, and let S_1 and S_2 be, respectively, cooperative and non-cooperative strategies. I will give P_1 this payoff matrix:

	b_1	b_2
a_1	20, 20	0, 30
a_2	30, 0	10, 10

(P_1)

Here, a_2 and b_2, the non-cooperative strategies, clearly dominate the strategies a_1 and b_1, and the a_2b_2 outcome seems inevitable if the players act rationally.

Let P_2 have this matrix:

	b_1	b_2
a_1	30, 30	0, 29
a_2	29, 0	0, 0

(P_2)

Here, a_1 and b_1, the cooperative strategies, dominate a_2 and b_2, and the $a_1 b_1$ outcome seems assured.

Let P_3 have this payoff matrix:

	b_1	b_2
a_1	30, 30	0, 30
a_2	30, 0	0, 0

(P_3)

Here, a_1 and b_1 are not dominated by a_2 and b_2 and rational players (provided they are not positively malevolent, which would alter the payoffs) have no reason not to use their cooperative strategies.

As a candidate for an intermediate and equivocal P^* I put the following:

	b_1	b_2
a_1	29, 29	0, 30
a_2	30, 0	1, 1

(P^*)

If we consider the payoffs of P^* merely ordinally, it is a replica of P_1: a_2 and b_2 dominate a_1 and b_1. But if we take into account the cardinal values of its payoffs it is *almost* a replica of P_3: a_2 and b_2 only *marginally* dominate a_1 and b_1. And P_3 differed only marginally from P_2. It seems to me that a rational player in this situation might very well oscillate between an S_2 and an S_1 strategy.

If my general claim here is right it means that we shall never have a completely *comprehensive* theory that could, at least in principle, deliver a determinate rational solution for every kind of decision problem. But this does not tell against the possibility of a *unified* (though incomplete) theory of rational decision making.

4. AN ALTERNATIVE PROGRAMME

I will outline an alternative programme to Harsanyi's in two stages. At first, *gambling* will be excluded. We can actually get quite far without bringing it in. A situation that involves risks need not involve gambles. (If one has to choose between a half chance of x and a quarter chance of y, and if one prefers x to y, one does not *gamble* in choosing the former.) And in situations where we might gamble we often prefer not to. Even a hardened gambler may prefer to get to the casino by a reliable method of transport rather than chance his luck with a cut-price minicab firm notorious for breakdowns, wrong destinations, etc. Shortly before takeoff a fuel mechanism on an airliner is found to be worn. There is only a one-in-a-million chance that it would actually break down during the coming flight; but *if* it did so the plane would crash. To replace it will involve considerable delay and considerable expense to the airline in providing hotel accommodation for the passengers, etc. But the airline will not hesitate: the sheer *possibility* of a crash will over-ride other considerations.

Of course, there are situations that require people in them to gamble. And there are people who like gambling in situations which do not require them to. I will turn to this in Section 6. What will be said there will involve some qualifications to what is said in earlier sections, and I ask the reader to bear that in mind. However, the qualifications will turn out to be less drastic than might have been expected.

I will begin with the second step in my sequence. It will be remembered that Harsanyi's second step – the R' in his $R, R', R'', ..., R^*$ sequence – was the idea of maximising expected utility under conditions of risk. For my R' I put the game-theoretical idea (already alluded to above) that if one has a dominant strategy one should use it. (I shall henceforth ignore the kind of case mentioned in the previous section where a strategy is only marginally dominant.) This idea is well known; but to be on the safe side I will spell it out.

Consider a game with two players, A and B, where A has n alternative strategies, $a_1, a_2, ..., a_n$, and B has m strategies, $b_1, b_2, ..., b_m$. Let the payoff (or utility) to A of the outcome of the intersection of a strategy a_i of his with a strategy b_j of B's be designated by 'u_{ij}'; and let all these payoffs be given in a matrix of the form:

	b_1	b_2		b_m
a_1	u_{11}	u_{12}		u_{1m}
a_2	u_{21}	u_{22}		u_{2m}
a_i	u_{i1}	u_{i2}		u_{im}
a_n	u_{n1}	u_{n2}		u_{nm}

Then a particular strategy a_i is dominant if for every strategy a_k $(k \neq i)$ and for every strategy b_j, u_{ij} is at least as good as u_{kj}. (Where there are two or more such strategies I will adopt the convention of saying that the player has *one* dominant strategy consisting of the disjunction of his undominated strategies.)

Notice that someone with a dominant strategy available to him need not try to predict the other player's moves or assign probabilities to them. He reviews all *possibilities*, regardless of their relative probability, and selects a strategy that deals with all of them better than any other strategy.

It seems to me that this idea has the sort of intuitive convincingness that we demand of at least the first few members of an R, R', R'', ... sequence. When applied to conditions of certainty, this R' yields a suitably trivial R. The 'player' B now becomes 'Nature', whose reactions to A's moves are supposed to be fully predictable; thus the payoff matrix now has only one column and A chooses a strategy whose payoff is at least as good as that of any alternative strategy.

For my R'' I put the following: if you are in a two-person situation and do not have a dominant strategy, *but the other player does*, then make the assumption that he will act rationally and select a strategy that is a best answer to his dominant strategy. I here introduce an idea which will become increasingly important later, namely the idea of a *quasi-dominant strategy*. A quasi-dominant strategy is one which, although it is not dominant when *all* possibilities are taken into account, becomes dominant when a certain restricting assumption is introduced which delimits the possibilities. (Of course, if the idea of a quasi-dominant strategy is not to degenerate, we must be very careful as to what kinds of restricting assumption may legitimately be introduced.)

Later, this R'' will be subject to a proviso when gambling temperaments are introduced. In the meanwhile I may mention that in real life a so-called 'dominant' strategy will usually be only a quasi-dominant strategy because a whole host of more or less remote possibilities – that the other player drops dead from a heart attack, say, or that you inherit £1 m from an unknown cousin in Australia – will have been tacitly excluded. The tidy payoff matrices in game theory textbooks involve the restricting assumption that the game is not disturbed by such external contingencies. (I will return to the question of the rational handling of rather far-out possibilities in Section 6.)

In the meanwhile I will take a sideways look, from the standpoint of my R' and R'', at Harsanyi's paper; for one thing that struck me about it is that Harsanyi is curiously silent about the idea of a dominant strategy, even when this idea is clearly relevant to a game-situation that he is analysing. For instance, he discusses at considerable length (pp. 331–333) a game which, presented in normal form, looks like this:

	b_1	b_2
a_1	1, 3	1, 3
a_2	0, 0	2, 2

I have no complaint with his conclusion that the a_1b_1 and a_2b_2 outcomes are, respectively, an 'imperfect' and a 'perfect' equilibrium point. But he seems to me to make unnecessarily heavy weather of this game. For the plain fact is that b_2 dominates b_1 and a_2 is the best answer to b_2. Hence a_2b_2 is the rational outcome.

The idea of a dominant strategy is negatively relevant to another case that Harsanyi discusses (pp. 329–330). Here, although neither player has a dominant strategy, Harsanyi treats it as if the game had an easy solution.

This game is a modified version of the prisoner's dilemma in which the payoffs have been revised to reflect the fact that the players are decent people who 'attach considerable disutility to using a non-cooperative strategy'. (I was interested in this because, in a discussion with Sen a few years ago,[6] I reached the discouraging conclusion that the introduction even of rather strong moral considerations into the prisoners' dilemma does not help much.) Harsanyi's revised payoffs are as follows:

	b_1	b_2
a_1	2, 2	0, 1
a_2	1, 0	1, 1

Harsanyi comments: 'the players will now have *no difficulty* [my italics] in reaching the outcome (A_1, B_1), which we used to call the cooperative solution, so as to obtain the payoffs (2, 2)' (p. 330).

But is this so? It would be if a_1 and b_1 were dominant strategies; but they are not: to use them involves a *gamble*. Whereas a_2 guarantees player A a payoff of 1, a_2 will give him either 2 or 0, according to whether B chooses b_1 or b_2. Suppose that A would like to play cooperatively but believes B to be a cautious person who prefers to play safe with b_2 and the certainty of 1. Then A may reluctantly choose a_2, despite his desire for the $a_1 b_1$ outcome. This confirms the conclusion I reached in my discussion with Sen, namely that only the addition of *exceptionally* strong assumptions will turn a prisoners' dilemma game into one in which the $a_1 b_1$ outcome is assured.

5. QUASI-DOMINANT STRATEGIES IN COMPETITIVE SITUATIONS

Very often, of course, no dominant strategy is available to any of the players in a game situation. However, as we have already seen, the idea of a dominant strategy can be modified to yield various kinds of quasi-dominant strategies. In this section I want to consider two such kinds of strategy that are appropriate to *competitive* situations. I will confine myself to two-person games in which, whether the games are zero-sum or not, the players are opponents with conflicting interests.

For my R''' I propose that an agent in such a competitive situation should delimit the possibilities by attending just to the *worst* possible outcome attached to each of his strategies. Having in this way associated each strategy with *one* outcome, he now chooses a strategy that is dominant with respect to *these* outcomes.

This is, of course, the famous (in some eyes, notorious) minimax or maximin principle. Harsanyi has vigorously opposed this principle.[7] But he was considering an unrestricted version of it whereas I am here confining it to competitive situations.

The principle says that if neither you nor your opponent has a dominant strategy, you should seek to maximise your minimum gain (minimise your maximum loss). To some ears this has a defeatist ring. Maximise my *minimum* gain? Why not maximise my maximum gain? Well, this principle will usually turn out to be a maximising principle on the assumption that your opponent is no less rational a player than you are. This is particularly clear in the case of zero-sum games. Consider the following simple example (only A's payoffs are shown):

	b_1	b_2
a_1	100	-5
a_2	-1	1

Here, A might be tempted to play a_1 in the hope of winning 100. But the maximum rule rejects this as wishful thinking: the important thing about a_1 is that it may result in -5 for A, whereas the worst that a_2 can result in is -1. Moreover, if B plays a minimax strategy, a_1 *will* result in -5 for A; for if B is to minimise his maximum loss, he must avoid b_1 with its risk of -100 and stick to b_2.

Harsanyi has suggested (in discussion) that the maximin rule can lead to paradoxical reasoning when each player both tries to follow it and assumes that the other will also try to follow it.[8] I think that his objection could be illustrated, vis-à-vis the above matrix, by the following sequence of oscillating decisions. I begin with B.

B: "I daren't choose b_1, so I must choose b_2".

A: "B won't dare choose b_1, so he will choose b_2, in reply to which I must choose a_2".

B: "A will choose a_2 in reply to my b_2. But my best reply to a_2 is b_1. So I will switch to b_1".

A: "B will expect me to choose a_2 in reply to his b_2. But his best reply to a_2 is b_1. So I should switch to a_1 which will give me 100. Hurray!"

B: "But A may anticipate my switch to b_1 and meet it by switching to a_1, which will give me -100. Ugh! So I daren't choose b_1."

A: "B won't dare choose b_1 ..."

And off they go on another round ...

Is this, then, another case of a rationally indeterminate decision problem? When gambling is considered, in Section 6 below, it will turn out that, on a certain (rather unlikely) assumption, A *may* rationally gamble on a_1. However, this very possibility tends, indirectly, to nullify itself; for this possibility means that there is a real possibility that B gets -100 if B gambles on b_1; so it is very likely that B will be deterred from using b_1; so it is very likely that A will get -5 if he gambles on a_1. The fact that a maximin strategy is not rationally obligatory for A will, via its effect on B, make a maximin strategy advisable for A.

Suppose now that the above game, which we were supposing to be played only once, were to be played a large or indefinite number of times by A and B. Then each player could calculate what probability mix of his pure strategies would give him the best statistical expectation on the assumption that the other player likewise uses his best strategy.

This brings me to my R''''. It would not do for A to treat each play of this long-term game as a once-only game. For if, on each occasion, A played a_2 in accord with the maximin rule for a once-only game, B could risk an occasional b_1. Indeed, if B *knew* that A would always play a_2, B could win outright by always playing b_1. A's defence against this possibility is to insert an occasional a_1 into a sequence consisting predominantly of a_2's. Of course, he must not do this in a way that allows B to *predict* when he will use a_1 (for then B could do even better). The safest way for A to make his occasional departures from a_2 unpredictable by B is to make them unpredictable in principle by employing a suitable randomising device. This brings us to the familiar idea of a *mixed strategy:* player A gives to each of the pure strategies available to him a certain probability (which may be zero), the probabilities summing to one; and the actual selection of a pure strategy on each occasion is effected by some suitably adjusted randomising device. If the right mix of probabilities is chosen, the result will be a mixed strategy that is a statistical equivalent of a pure maximin strategy: it will be the optimal reply to an optimal strategy. Should B use a sub-optimal strategy, A can expect to do better. Admittedly, A might have done better still against B's sub-optimal strategy by using another strategy exactly appropriate to *it* rather than to B's optimal strategy. So A's strategy will not be strictly dominant. But it is quasi-dominant when we restrict the possibilities by introducing the assumption that B is a fully rational opponent.

My $R, ..., R''''$ sequence is far from satisfying the exacting desiderata that I proposed for an $R, ..., R^*$ sequence in Section 1. However, as far as it goes it has, I think, a certain unity and coherence. Its underlying idea is that the agent tries, where possible, to attend to all the possibilities facing him and then to find the best way of coping with them all. Where he cannot do this, he makes a simplifying assumption which delimits the possibilities, and then proceeds as before. The assumptions used in generating this sequence are, in reverse order:

R'''':   a competitive game is to be played a large or indefinite number of times against a rational opponent;

R''':   a competitive game is to be played once (or a small fixed number of times) against a rational opponent;

R'':   a game (not necessarily competitive) is to be played with a player who is rational and who has a dominant strategy;

R':   a game is to be played with a player (not necessarily rational) by an agent who has a dominant strategy;

R: a 'game' is to be played against 'nature' whose reactions to the agent's moves are predictable.

Probabilities play only a minor role in this approach as so far sketched: they appear on the scene only with R'''' with its idea of a mixed strategy. But probabilities have not been used to *weight* the significance of the different possibilities: the agent has been supposed to consider each possibility as such.

This concludes the first stage of my programme. As it stands it is seriously incomplete. For it may very well happen that there is *no* admissible simplifying assumption that will so delimit the possibilities as to allow a quasi-dominant strategy to emerge. In such cases the agent is obliged to *gamble* whether he wants to or not. I now turn to the second stage in my programme in which I try to make full provision both for the fact that people do find themselves in gambling situations and for the fact that they approach such situations with very varying gambling attitudes.

6. Gambling

From an early enthusiasm for Shackle's theory I retain the idea that an adequate theory of decision making will allow that people have different gambling temperaments and will not seek to impose a uniform gambling policy on everyone.[9] From the wide variety of gambling types I will, for ease of exposition, select just two, which I will embody in the hypothetical persons of a *Mr Cautious* and a *Mr Bold*. As their names suggest, Mr Cautious is generally averse to, whereas Mr Bold tends to enjoy, risk-taking.

Before considering how these two gentlemen might deal with gambling problems in a rational way, let us first consider how much disturbance their contrasting temperaments may cause to the ideas set out in Sections 4 and 5, where gambling temperaments were ignored.

My R and R' will remain intact: however cautious or bold the decision maker may be, there is no strategy for him to switch to that *could* give him a better payoff than that which his dominant strategy will give him. And I think that my R'''' will also remain essentially intact (at least on the orthodox definition of a *strategy* as something laid down in advance which cannot be altered in the course of the game – a definition which, incidentally, excludes the possibility that one player is gulled by misleading moves by the other into shifting to an inferior strategy). For suppose that A, before embarking on a long-run game with B, expects B to use a sub-optimal strategy and would like to gamble on this expectation by discarding his maximin strategy in favour of one that might more fully exploit the inferiority of B's strategy. But how could A anticipate *in what direction B* may deviate from his optimal norm? Even if B has only two pure strategies, b_1 and b_2, he already has a potentially infinite number of mixed strategies (for he can give b_1 and b_2 respectively any probabilities p and $1-p$ where $0 \leqslant p \leqslant 1$), one of which is optimal and all the rest of which are sub-optimal.

However, my R'' and my R''' do get somewhat disturbed by the existence of different types of gambling temperament.

Consider R'', which advises you to set aside the possibility that the other player irrationally uses a dominated strategy. Suppose that Mr Cautious is in A's position in a situation with this payoff matrix:

	b_1	b_2
a_1	3, 2	−100, 1
a_2	2, 3	0, 2

Here, b_1 dominates b_2 and a_1 is the better reply to b_1. So by R'' A should choose a_1. But Mr Cautious might prefer to insure himself against −100 with a_2.

Conversely, if Mr Bold is in A's position vis-à-vis this matrix:

	b_1	b_2
a_1	2, 3	100, 2
a_2	3, 2	1, 1

he might prefer to gamble on a_1 in the hope that B will use his dominated strategy b_2.

As to R''': it can hardly be disturbed by the existence of Mr Cautious, since it not only fits in with his 'safety-first' attitude but has the added advantage of being a quasi-maximising rule against a skilful opponent. However, if Mr Bold were in A's position vis-à-vis the matrix discussed at some length earlier (pp. 358 f.) namely:

	b_1	b_2
a_1	100	-5
a_2	-1	1

we could expect him at least to *toy* with a_1 with its possibility of 100. Whether he might rationally gamble on a_1 we will consider below (p. 366).

I said earlier that in real life a so-called 'dominant strategy' will really be a quasi-dominant strategy because various more or less remote possibilities will have been cut off.

Harsanyi has exploited this in an attack[10] on Rawls's use of a maximin rule: Suppose that A has a choice between two certainties, x and y, and that he much prefers y to x. However y involves a one-in-a-million chance of z, which is much worse than x. In Harsanyi's example, x is staying in a boring job in New York and y is accepting an exciting job in Chicago which, however, involves an airflight; and z is the event that the plane crashes. An unrestricted maximin rule (which Rawls actually rejected)[11] would seem to require A to choose x.

But it is unfair to spotlight *one* remote contingency associated with *one* of the alternatives, when *both* alternatives are surrounded by *innumerable* contingencies. (Maybe A believes that, in the event of nuclear war, New York is a more likely target than Chicago ...) A Bayesian can hardly claim that the expected utility of *all* the contingencies, however remote, that attend *each* of the agent's possible actions could actually be computed. Whatever theory of decision-making we adopt, we must allow the decision-maker to disregard the vast majority of such eventualities; otherwise he would be paralysed.

Occasionally, however, he may allow a far-out possibility to rise above his threshold of practical awareness and lodge itself in his decision-scheme. *Example.* For your summer holiday you are going to a mountain village in Ruritania. Hotel reservations have been made and tickets bought. And then, a week before you are due to go you read that there is some cholera in Ruritania, though only a few cases have been reported and these in a district far from your village. What do you do?

There are various alternatives. One is that you brush aside the possibility of getting cholera if you go, bracketing it with other possibilities – that your plane will crash or that your village will be destroyed by an earthquake – which you were already disregarding. Another is that you take the possibility seriously and try to *neutralise* it, for instance by having an anticholera jab. But suppose no vaccine is available, or that there is but you doubt its effectiveness. Then you have to decide whether to go ahead and gamble on not getting cholera, or play safe and cancel the holiday. How should you proceed to a decision?

To render the problem more tractable let us introduce some artificial precision into it. Let a_1 and a_2 be, respectively, the decisions to go ahead with, and to cancel, the holiday; and let N_1 and N_2 be, respectively, Nature's 'decisions' that you do, or do not, catch cholera if you go. (We may imagine the choice between N_1 and N_2 to be determined by a randomising device which gives a very small positive probability to N_1.)

In the next section I will be operating with payoffs expressed in *utiles*, which allow for variations in the marginal utility of money. But not much harm will be done if in the present section we use cash-values. Suppose that you put these values on the possible outcomes:

	N_1	N_2
a_1	-1000	10
a_2	0	0

I propose that you should proceed to a decision with the help of the following thought-experiment. First imagine that the option a_2 is not available to you: you have *got* to go ahead with your holiday. Now ask yourself what, given this counter-factual supposition, is the maximum bribe you would just be willing to pay Nature to reduce the probability of N_1 to zero. I will call this your *certainty-equivalent* for the a_1N_1 outcome. Let us designate its value by 'C'.[12] (Unless you first imagine a_2 away, an arbitrary upper limit will be imposed on C: for you would not be willing to insure yourself against cholera at a price that would leave you worse off than if you had cancelled the holiday.) In other words, C is the price at which you are indifferent between *either* going on holiday and risking cholera *or* going on holiday and paying C to insure yourself against cholera. Thus your choice between a_1 and a_2 would not be affected if a_1, instead of having the two possible outcomes,

	N_1	N_2
a_1	-1000	10

had the one certain outcome:

	N_2
a_1	$10-C$

You now reintroduce a_2 and proceed as if facing a choice between these two certainties:

	N_2
a_1	$10-C$
a_2	0

How you decide will, of course, depend on the size of C.

Essentially the same procedure can be used to resolve A's dilemma (see p. 362 above) over whether to comply with R'' when B's dominant strategy is b_1 in a situation which gives these payoffs to A:

	b_1	b_2
a_1	3	-100
a_2	2	0

J. W. N. WATKINS

In the previous case a_2 had only one payoff. In this case, a_2 has two payoffs. So A will have to determine two certainty-equivalents. He first imagines himself confined to a_1 and asks how much, in that case, he would pay to eliminate the risk that B irrationally chooses b_2. I will again designate this insurance premium by 'C'. He then imagines himself confined to a_2 and asks himself the same question. Since his insurance premium will obviously be much smaller in this second case, I will designate it by 'c'. Thus A would be indifferent between the choice that actually confronts him and this choice:

	b_1
a_1	$3 - C$
a_2	$2 - c$

C-equivalents, as I will call them, can also, of course, be given for uncertain prospects of a desirable kind: one asks oneself what is the lowest price at which one would just be willing to sell the prospect. The question arose in Section 4 as to whether A might rationally discard the maximin rule and gamble on a_1 in this zero-sum game:

	b_1	b_2
a_1	100	-5
a_2	-1	1

Well, A should determine a C-value and a c-value such that he is indifferent between that choice and:

	b_2
a_1	$C - 5$
a_2	$1 - c$

A C-equivalent is like an expected utility in that it puts a definite value on an uncertain prospect. However, it is *not* determined by just the utility and probability of the prospect: it is *also* a function of the person's gambling temperament. An anxious person may regard a very slight risk of getting cholera if he goes on holiday as nearly as bad as the certainty that he will get it if he goes. Mr Cautious will generally tend to put considerably higher C-equivalents on bad possibilities than Mr Bold; and Mr Bold will have some tendency to put higher C-equivalents on good possibilities.

I have indicated in a rather bitty way how my R'' and R''' need to be revised now that gambling temperaments have been introduced. Can we attempt a more general revision?

Using our present terminology we could represent R'' and R''' as advising the decision maker to treat specific kinds of possibilities as if they had negligible C-equivalents and could therefore be set aside, and then to adopt a strategy that is dominant with respect to the remaining possibilities. The trouble with this advice was that in some cases the possibilities which the decision maker was advised to set aside had *non*-negligible C-equivalents for him. This suggests that we should proceed in the opposite direction. If he has no strictly dominant strategy, the decision maker should first set aside possibilities which he does in fact treat as if they had negligible C-equivalents. If this allows a quasi-dominant strategy to emerge, he should use it; if not, he finds himself in a gambling situation and should proceed to a decision by making the kind of thought-experiment, involving non-negligible C-equivalents, outlined above.

I admit that it is a theoretical disadvantage of my approach that it introduces a complication which gets ironed out in the Bayesian approach, making decisions a function of three variables instead of two. (My third factor roughly corresponds to Allais's 'fourth element'.)[13] But I hold that this complication corresponds to a separately identifiable factor in the psychology of decision making and that it is better to make explicit provision for it than to handle it surreptitiously by imputing its distorting effects to hidden variations in the other two factors.

This completes my attempt to develop an alternative standpoint from which to criticise Bayesianism. I now proceed to my criticism.

7. AGAINST EXPECTED UTILITY

'In the case of risk, acceptance of Bayesian theory is now virtually unanimous', Harsanyi declares (p. 322). I am in no position to dispute that, having once said that its adherents 'seem to include almost everyone except Shackle and myself'.[14] (At that time, however, I did not, regrettably, know of Allais;[15] and I now notice that Raiffa describes himself as belonging 'to a minority party, the so-called Bayesians'.)[16] Anyway, even if virtual unamimity does prevail, that does not exclude all possibility of error.

Harsanyi also says that some of the criticism directed against Bayesianism by non-experts 'has been rather uninformed'. Well, I am a non-expert who has directed some rather uninformed criticism against Bayesianism;[17] and without expecting to disturb the prevailing virtual unanimity, I am going to have another go.

In what follows I am going to operate (for a reason which will transpire later) with payoffs expressed in *utiles*. This is a well-defined concept within expected utility theory.[18] Relative to a given individual, the utile value of a payoff is its money value corrected for variations in his marginal utility of money. And there is a famous method, due to von Neumann and Morgenstern,[19] for eliciting the shape of an individual's marginal utility curve for money. Take some unit of money appropriate to the financial circumstances of the person in question. Let 0 represent no change in his fortune, and let 1, 2, 3, ... and -1, -2, -3, ... represent, respectively, gains or losses of so many of these monetary units. We now confront him with a choice between the certainty of b and a lottery involving some chance of a and the complementary chance of c, where a, b and c are payoffs in money values such that a is inferior to b and b is inferior to c. (Thus we might start with $a = -1$, $b = 0$, $c = 1$.) We elicit from him a probability value p_0 such that he is indifferent between b and a lottery giving the probability p_0 of getting a and the probability $1 - p_0$ of getting c. Representing the utile value to him of a, b and c by, respectively, $u(a)$, $u(b)$ and $u(c)$, we can now say that the ratio of the interval, on his utile-scale, between $u(a)$ and $u(b)$ to the interval between $u(b)$ and $u(c)$ is $1 - p_0$ to p_0. Repeating this process for other triples of monetary payoffs, we can map money values on to utile values for this individual.

If (which I will not assume in what follows) his marginal utility of money were continuously diminishing, its curve would look like this:

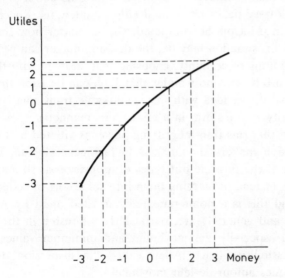

I now assume our individual's marginal utility curve for money to have been ascertained in the above manner. Of course, it is likely to change if his circumstances change at all significantly; but I am going to assume that there has been no significant change in his circumstances since it was ascertained. I am also going to assume that the mere fact that he is presented with a choice between lotteries is not a circumstance which will, by itself, cause his marginal utility curve to change. In short, I am going to assume that his curve is stable during the period in question. Thus if we want to offer him payoffs with certain utile values we have only to read off from this curve their corresponding cash values and offer him those.

As we saw earlier, Harsanyi has, with his R^*, brought off the following achievement: given that a player A can exhaustively specify the strategies $a_1, a_2, ..., a_n$ that are open to him, then even if no objective probabilities are available and even if he has very incomplete information about the other players, he can, in principle, always associate with each of his strategies something like a *lottery ticket* which gives a list – it may be a very long list – of each possible payoff of the strategy together with its relative probability, the probabilities summing to one. Moreover – and this is what I am about to challenge – each such lottery ticket has a *total value*, namely its expected utility, which can in principle be computed: so, no matter how fraught with uncertainty his situation may be, the decision maker can proceed as if under conditions of certainty, choosing that strategy with the highest (expected) utility. It is no wonder that Bayesian decision theorists are loath to give up an idea with this remarkable transmogrifying power ('transmogrify' = 'transform in a miraculous manner').

But I say that this transmogrifying power is vitiated by two factors. First, decision makers should, and in fact generally do, regard the disjunction of the possible outcomes of a decision as so many distinct possibilities instead of merging them into some sort of collective total. Second, and this is a point repeatedly insisted upon by Allias, they should be, and generally are, particularly interested in the spread of the values, especially the maximum and minimum values, of these distinct possibilities (what Allais called "la dispersion des valeurs psychologiques autour de leur moyenne").[20]

This two fold objection can be brought out with the help of numerical examples. Having ascertained A's marginal utility curve for money we offer him a once-only choice between two lotteries, each with six possible outcomes, the result to be decided by the throw of a fair die. Their respective payoffs, in utiles, are:

	1	2	3	4	5	6
a_1	36	24	12	0	−12	−24
a_2	10	8	6	4	2	0

Gamble 1

In this case Bayesianism prescribes that A should choose a_1 because its expected utility for him is 6 while that of a_2 is 5. More generally, Bayesianism prescribes – once the individual has determined his utilities and subjective probabilities – a *uniform gambling policy* which depends only on an overall characteristic and ignores maxima and minima (except, of course, as they affect the total expected utility).

I say that people do have, and are perfectly entitled to have, very different gambling temperaments; and that, for the most part, they neither do, nor should, comply with Bayesian prescriptions. Consider Mr Cautious. He tries to minimise his involvement in gambling situations; and when he does find himself involved in one he tends to prefer the least risky choice. In the case of Gamble 1 above he would assuredly prefer a_2, on which he cannot lose.

Mr Bold might very well incline towards a_1. However, Mr Bold might side with Mr Cautious against Bayesianism in the following gamble:

	1	2	3	4	5	6
a_1	30	30	30	30	30	−24
a_2	30	30	30	30	0	0

Gamble 2

For although Mr Bold tends to be attracted by larger maxima, he may not be much interested in a slightly larger chance of a maximum no larger than that of the less risky lottery.

What I called the transmogrifying power of Bayesianism depends essentially on the assumption that under conditions of uncertainty, when objective probabilities are no longer available, it is rational to plug the gaps with subjective probabilities and to proceed as if under conditions of risk. But it seems to me that my twofold objection gains renewed force when we turn to conditions of uncertainty. Consider Gamble 3 that is like Gamble 1 except that this time the result is to be decided by spinning a disc divided into six numbered segments whose arcs (which may be very unequal) are entirely unknown to A. What subjective probabilities should A assign to the outcomes? Presumably he should rely on the principle of insufficient reason and assign them equal probabilities; in which case the expected utilities of the two lotteries will be the same as before; and Bayesianism will again recommend that A gamble on a_1.

I am not a member of the Anti-Gambling League but I do think that a decision theory goes too far if it says that gambling is the rational thing to do whenever it is marginally justified on expected utility grounds. It seems to me that in the present case Mr Cautious would be even more entitled than before to play safe with a_2. For all he knows, outcome 6 may have an objective probability close to 1. (True, its objective probability may be close to 0; but Mr Cautious is less impressed by that.)

Suppose that once-only gambles like Gambles 1, 2, and 3 were presented, under carefully controlled conditions, to a number of subjects selected on the basis of a reputation for their shrewd handling of decision problems involving risks and uncertainties. I would confidently predict that the majority of choices would be for the lottery with less risk and less expected utility. Maybe I would turn out to be wrong. But suppose I turned out to be right: how might a Bayesian respond to such an experimental result?

Much would depend on which of the four meta-level views (indicated in Section 1) he took concerning the status of Bayesian decision theory. If he took view (1) he might say with a shrug that these people do not comply with the axioms of expected utility, *period*. View (2) is inapplicable, since Bayesian decision theory is supposed to be a quite general theory. If he took view (3) be might take a tough line and roundly declare that his *normative* theory is not overthrown by these *de facto* decisions, which were sub-optimal.[21] If he took view (4) he would, presumably, admit that his hypothesis had been falsified. In their (1948) Friedman and Savage, in order to explain from an expected utility point of view the fact that the same people may be willing to purchase both insurance and lottery tickets, postulated a marginal utility curve for money with 'a rather special shape' (p. 282); they also described it as a 'rather peculiar utility function'. However, having postulated this 'wiggly utility curve' (p. 297), they concluded their paper, in good Popperian fashion, by specifying in advance a number of potential falsifiers of their hypothesis. And in their (1952) they raised the question whether it had actually survived empirical tests, to which question they gave a 'highly qualified affirmative answer' (p. 486). This would presumably turn into a negative answer if the above experiment were carried out and the results were as I predicted.

But there is another possible response, to which, I fear, some Bayesians would be all too ready to resort. It might be claimed that the experimenters must have made mistakes about the subjects' subjective utilities and probabilities; for if someone after due deliberation chooses lottery a_2 in preference to a_1, then that just *means* that a_2 has the higher expected utility for him.

It was in an attempt to forestall this defence that I (a) introduced an assumption concerning the stability, over a short time period, of the man's marginal utility curve, and (b) operated, in the first instance, with gambles where the probabilities are objectively given. But a defender of Bayesianism could always claim that the man's marginal utilities may have fluctuated or that his subjective probabilities did not correspond to the objective probabilities because, say, he feared that the die was loaded against him or that it was not his lucky day.

Were Harsanyi regularly to defend his position against seeming counter-examples in such ways I would be reluctantly obliged to withdraw what I said earlier about the *power* of his R^*. For if the factual hypothesis that people generally regarded as wise and prudent in practical affairs do act as if maximising expected utility is rendered unfalsifiable by a policy of systematically imputing seeming counter-examples to hidden variations among their utilities and subjective probabilities, then the normative idea that a rational person *should* maximise expected utility is rendered vacuous: *whatever* he does will count as maximising expected utility.

Now Harsanyi has a (to my mind rather alarming) tendency to postulate such hidden variations. To take a favourite example of his: suppose we come across a man who has staked $5 on a 1/1000 chance of winning $1000; that does not *look* like expected utility maximisation; but there is a ready explanation for his behaviour: the marginal utility of money for him turns up sharply around $1000 because he badly wants something – medical treatment, a foreign holiday, or whatever – which he cannot get without an extra $1000.[22] To this I say: if we meet with such a case, let us investigate it; if we do we may indeed find something of this kind at work. *But suppose we do not:* what will Harsanyi say then? Will he concede that the man is a pretty reckless gambler who is not maximising expected utility? Or will he say that there must have been some such factor at work despite our failure to find it?

The other variable in expected utility is, of course, subjective probability. About this Harsanyi has declared: 'a rational decision maker simply *cannot help* using subjective probabilities, at least implicitly'.[23]

Well, that goes right against my idea that a rational decision maker may seek to deal as best he can with various possibilities *as such* and without weighting them probabilistically. Harsanyi illustrates his claim with the following example. He offers you a choice between these two bets: either (1) you bet that X will win the next election; you get $100 if he does win and nothing if he loses; or (2) you bet that Y will win; again you get $100 if he wins and nothing if he loses. Suppose you choose (1). Then, according to Harsanyi, 'I can infer that (at least implicitly) you are assigning a probability of $\frac{1}{2}$ or *higher* to Mr. X's winning the next election.' But this inference is invalid. For you may have arrived at your choice in a different, but no less rational, way: perhaps you had no view at all about who *will* win (it may be a new constituency in which no opinion polls have been taken); but you *hoped* that Y will win; so your best way of dealing with the two possibilities was to choose (1); then if Y wins you will be happy, and if he loses you will get $100 in compensation. Subjective probabilities may have played *no* role in your reasoning.

The idea that the decision maker should *aggregate* the utility-cum-probability values of the mutually exclusive possible outcomes of a lottery into one overall value goes right against my idea that he should keep the different possibilities distinct and attend to them separately. What I want to consider now, however, is the psychological feasibility of the former idea.

Suppose that in my right hand I hold a £10 note and in my left hand a ticket giving me a half chance of £20 and a half chance of nothing. If there were a market in lottery tickets I might be able to sell that ticket for something approaching its expected monetary value. But assume that there is no such market: there is nothing I can do with this ticket except hang on to it until noon tomorrow when the result is going to be decided by the spin of a coin.

Then according to Bayesianism, between now and noon tomorrow this piece of paper will have a definite value for me very much as the piece of paper in my right hand has. Indeed, if my marginal utility for money is fairly constant over this range, they will have approximately the same value. At noon tomorrow something will occur that might be called (by analogy with a famous concept in quantum mechanics) the 'reduction of the lottery packet': the ticket will be thrown into one state or the other and it will take on one extreme utility value or the other. But in the meanwhile it is supposed to have an unequivocal middling value for me, roughly equal to that of the piece of paper in my right hand. But it does not. I keep its two possibilities apart. I view it *ambivalently* as worth *either* nothing *or* £20.

Savage would retort that the piece of paper in my right hand has a similar status: 'a cash prize is to a large extent a lottery ticket in that the uncertainty as to what will become of a person if he has a gift of a thousand dollars is not in principle different from the uncertainty about what will become of him if he holds a lottery ticket of considerable actuarial value'.[24] Fair enough. But I could avoid this retort by making the bits of paper, respectively, a theatre ticket for a play I am keen to see and a half chance of this ticket plus another one for another play that I am also keen to see.

I could, it is true, specify a *certainty-equivalent* for the lottery ticket – the lowest price at which I would just be willing to sell it if it were negotiable. But in my case this would be quite a bit less than its expected monetary value. I have a cautious gambling temperament and would pay a sizeable premium to be rid of the uncertainty. (I would actually settle for about £7.)[25] Of course, a Bayesian could say that this only means that my marginal utility of money diminishes rather rapidly in this region; but I can promise him that it is actually pretty constant over the £0 to £20 region.

I have one last criticism of the expected utility approach. It has been nicely formulated by Hansson:

Consider a certain result r and a family of gambles G_p, defined as getting r with probability p and nothing otherwise. Classical utility theory evaluates the utility of G_p as $p \cdot u(r)$. Since $u(r)$ is constant this means that $u(G_p)$ is a linear function of p. In particular, equal increases in p mean equal increases in $u(G_p)$. But do we not evaluate the step from $G_{0.99}$ to G_1 higher than e.g. from $G_{0.5}$ to $G_{0.51}$?[26]

One of Allais's arguments against 'the American School' had been that people do not evaluate risks bordering on certainty as they should according to expected utility theory.[27] For some years now I have been in the habit of asking lecture-audiences whether anyone present would prefer a 99% chance of £100 to a 100% chance of £98. So far, no one has. But Professor Narens has told me that he has been putting similar questions to a wide variety of both normal and institutionalised people, and that he *did* once get the answer that Bayesianism indicates – from an elderly schizophrenic lady.

I said at the outset that we would like one sovereign theory of rational decision making. Yes, but we do not want this theory to be dictatorial; nor do we want it to be vacuous. I have claimed: (1) that if Bayesianism allows that a decision maker's utilities and subjective probabilities might be elicited from him prior to the decision he is about to make, then it turns 'dictatorial' by prescribing a uniform gambling policy which ignores maxima and minima, and the significant difference between a certainty and a near-certainty, except in so far as these are reflected in a total expected utility; (2) that if, on the other hand, Bayesianism says that which of the alternatives before the decision maker has the highest expected utility for him can be ascertained only retrospectively, by observing which he actually chooses, then it turns vacuous.

In opposition to Bayesian decision theory I have proposed that a rational decision maker should, ideally, attend to the full matrix of *separate* possibilities before him. If no dominant strategy is available to him he should delete those possibilities to which he gives negligible certainty-equivalents in case a quasi-dominant strategy emerges. If it does not, he should proceed to reduce the multi-column matrix to a single column one, in the manner indicated in Section 6, by incorporating C-equivalents for the possibilities imagined away into the surviving possibilities, so that he is indifferent between the imaginary decision problem thereby created and the actual decision problem that confronts him; he can then proceed as if under conditions of certainty.

Although this approach seems better to me than the Bayesian one, I must admit that I feel abashed when I compare my amateurish and unfinished attempt at a general decision theory with Harsanyi's precise, rigorous and comprehensive R^*.

*The London School of Economics
and Political Science*

NOTES

[1] See Latsis (1972, pp. 209f) on the 'single exit' nature of the entrepreneur's situation, according to neoclassical economics, under perfect competition.

[2] Satisficing behaviour is behaviour that attains a certain level of aspiration. Thus a firm which is content to maintain a certain level of profits 'satisfices' rather than maximises. See Simon (1957) and (1959).

[3] Friedman and Savage adopted this view which they explained with the help of their famous analogy (1948, p. 298) of an expert billiard player playing his shots *as if* he knew the mathematical formulas governing the movements of the balls.

[4] Harsanyi (1975b). Harsanyi had reported that he had 'succeeded in developing a new approach to game theory which does yield determinate solutions for *all* classes of games, and thereby provides a rational-behavior concept rich enough to serve as a basis for a general individualistic theory of social behavior' in (1968, p. 309), referring there to Harsanyi (1966).

For his theory of games with incomplete information see Harsanyi (1967–8).

[5] Watkins (1970, pp. 197–8 and 202f). My thinking there had been strongly influenced by Rapoport (1966, pp. 141f) and Schelling (1960). The two examples from the literature which I drew upon there also constitute Howard's first and second 'breakdown of rationality' (1971, pp. 8f and 44f). Howard also presented a third 'breakdown' (pp. 168f) involving inducements and counter-inducements.

[6] Sen (1974); Watkins (1974).

[7] Harsanyi (1975a).

[8] This point was briefly made in Luce and Raiffa (1957, p. 69).

[9] Shackle (1949; second edition 1952). Actually, Shackle brought in the entrepreneur's gambling temperament twice over: once implicitly, when 'primary focus-outcomes' are standardised (see note 12 below), and once explicitly when pairs of *standardised* focus-gains/losses are plotted on the agent's gambler indifference-curve map.

[10] Harsanyi (1975a, espec. p. 595).

[11] "Clearly the maximin rule is not, in general, a suitable guide for choices under uncertainty". (Rawls, 1971, p. 153).

[12] This idea of a certainty-equivalent C is quite unoriginal. Friedman and Savage (1948, pp. 289f) spoke of the 'income-equivalent' I^* of a lottery involving incomes I_1 and I_2 (though their I^* is determined by the expected utility of the lottery). My C-equivalent comes close to Hansson's (1975, pp. 181–2) *risk premium* which is a function of the riskiness of the gamble *and* of the individual's propensity to avoid risks.

I originally got this idea from my reading of Shackle's (1952, pp. 18f) 'standardised focus-outcomes' which I interpreted as follows: "It is as if the agent asked himself, 'For how much would I sell this uncertain prospect of gain and how much would I pay to be rid of this uncertain prospect of loss?' " (Watkins, 1957, p. 112.) Actually, this interpretation was strictly incorrect (see Watkins, 1970, p. 194).

[13] Allais (1953, *passim*). Allais's main objections to 'the American school' will be revived in section 7 below.

[14] Watkins (1972, p. 187).

[15] When in the course of preparing this paper I eventually read Allais (1953) I was at once chagrined to discover how many of my (1970) criticisms of the expected utility approach had already been made, and made better, by him, and heartened by this powerful endorsement of them. This time I will try to make proper acknowledgements to him.

In the meanwhile, partly by way of an excuse for my previous neglect, I will make a complaint about the way Allais has been treated by the 'Bayesian' camp. It was perhaps unfortunate that, among his various power criticisms of what he called 'the American school', he gave prominence to an ingenious numerical counter-example (1953, pp. 525–8) to Savage's sure-thing principle. In his (1954) Savage concentrated exclusively on this. (He admitted that he had at first found it intuitively convincing, but claimed that a careful analysis of it shows that his initial intuition had been mistaken.) This seems to have created a widespread impression that Allais's whole criticism hinges on this particular counter-example which, moreover, can be rebutted. For instance, the sole reference to Allais in Luce and Raiffa (1957, p. 25) refers to Savage's handling of it; and Raiffa (1968, pp. 80f) again concentrates exclusively on it.

It is perhaps also unfortunate that Allais's example concerned gambles involving prizes of hundreds of millions of francs, giving it an air of unreality. This allowed Edwards, in the course of his generally excellent, and influential, review of the development of decision theory, to conclude a brief report on Allais with the remark: "However, from a simple-minded psychological point of view these [sophisticated examples of Allais] are irrelevant. It is enough if the theory of choice can predict choices involving familiar amounts of money ..." (1954; reprinted 1967, p. 42). Allais's example would not, of course, have lost its force if the prizes had consisted of 'familiar amounts of money.'

[16] Raiffa (1968, p. 20).

[17] Watkins (1970, pp. 184f).

[18] Friedman and Savage (1948, p. 292); Mosteller and Nogee (1951; reprinted 1967, pp. 126f); Edwards (1954; reprinted 1967, pp. 28f).

[19] Von Neumann and Morgenstern (1947, Section 3 and Appendix).

[20] Allais (1953, p. 513).

[21] Savage inclined to this view in (1954): he held that the theory of expected utility 'makes factual predictions many of which can easily be observed to be *false*', but added that 'it may have value as a *normative* theory' (second edition, 1972, p. 97, my italics).

[22] Friedman and Savage (1948, pp. 298–299) had suggested that increases of income may have increasing marginal utility if they would be large enough to lift the individual *into a higher socio-economic class* and not merely leave him better off within his present class.

[23] Harsanyi (1975a, p. 599, italics in the original).

[24] Savage (1954; 1972, p. 99).

[25] According to Allais (1953, p. 525), experience shows that people regarded as rational will typically prefer 40 francs for sure to a half chance of 100 francs.

[26] Hansson (1975, p. 187).

[27] Allais (1953, pp. 505 and 525f).

BIBLIOGRAPHY

Allais, M.: 1953, 'Le Comportement de l'Homme Rationnel devant le Risque: Critique des Postulats et Axiomes de l'Ecole Americaine', *Econometrica* **21**, 503–546.

Edwards, W.: 'The Theory of Decision Making', *Psychological Bulletin* **51**, 380–417. (Reprinted in Edwards and Tversky (1967), 13–64.)

Edwards, W. and Tversky, A. (eds.): 1967, *Decision Making*, Penguin Books, Harmondsworth.

Friedman, M. and Savage, L. J.: 1948, 'The Utility Analysis of Choices Involving Risk', *The Journal of Political Economy* **56**, 279–304.

Friedman, M. and Savage, L. J.: 1952, 'The Expected-Utility Hypothesis and the Measurability of Utility', *The Journal of Political Economy* **60**, 463–474.

Hansson, B.: 1975, 'The Appropriateness of the Expected Utility Model', *Erkenntnis* **9**, 175–193.

Harsanyi, J.: 1966, 'A General Theory of Rational Behavior in Game Situations', *Econometrica* **34**, 613–634.

Harsanyi, J.: 1967–8, 'Games with Incomplete Information Played by "Bayesian" Players', *Management Science* **14**, 159–182, 320–334, 486–502.

Harsanyi, J.: 1968, 'Individualistic versus Functionalistic Explanations in the Light of Game Theory', in I. Lakatos and A. Musgrave (eds.), *Problems in the Philosophy of Science*, North Holland, Amsterdam, pp. 305–321, 337–348.

Harsanyi, J.: 1975a, 'Can the Maximin Principle Serve as a Basis for Morality? A Critique of John Rawls's Theory', *American Political Science Review* **69**, 594–606.

Harsanyi, J.: 1975b, 'The Tracing Procedure: a Bayesian Approach to Defining a Solution for *n*-Person Noncooperative Games', *International Journal of Game Theory* **4**, 61–94.

Harsanyi, J.: 1977, this volume, pp. 315–343.

Howard, N.: 1971, *Paradoxes of Rationality: Theory of Metagames and Political Behavior*, M.I.T. Press, Cambridge, Mass.

Latsis, S. J.: 1972, 'Situational Determinism in Economics', *The British Journal for the Philosophy of Science* **23**, 207–245.

Luce, R. D. and Raiffa, H.: 1957, *Games and Decisions*, Wiley, New York.

Mosteller, F. and Nogee, P.: 1951, 'An Experimental Measurement of Utility', *Journal of Political Economy* **59**, 371–404; reprinted in Edwards and Tversky (1967), pp. 124–169.

Raiffa, H.: 1968, *Decision Analysis*, Addison-Wesley, Reading, Mass.

Rapoport, A.: 1966, *Two-Person Game Theory*, U. of Michigan Press, Ann Arbor.

Rawls, J.: 1971, *A Theory of Justice*, Clarendon, Oxford.

Savage, L. J.: 1954, *The Foundations of Statistics*, Wiley, New York; second edition, 1972, Dover, New York.

Schelling, T. C.: 1960, *The Strategy of Conflict*, Harvard University Press, Cambridge, Mass.

Sen, A.: 1974, 'Choice, Orderings and Morality', in S. Korner (ed.), *Practical Reason*, Blackwell, Oxford, pp. 54–67 and 78–82.

Shackle, G. L. S.: 1949, *Expectation in Economics*, Cambridge University Press; second edition 1952.

Simon, H. A.: 1957, *Models of Man*, Wiley, New York.

Simon, H. A.: 1959, 'Theories of Decision-Making in Economics and Behavioral Science', *American Economic Review* **49**, 253–283.

von Neumann, J. and Morgenstern, O.: 1947, *Theory of Games and Economic Behavior*, second edition, Princeton University Press.

Watkins, J. W. N.: 1957, 'Decisions and Uncertainty,' in C. F. Carter, G. P. Meredith, and G. L. S. Shackle (eds.), *Uncertainty and Business Decisions*, second edition, Liverpool University Press, pp. 107–121.

Watkins, J. W. N.: 1970, 'Imperfect Rationality', in R. Borger and F. Cioffi (eds.), *Explanation in the Behavioural Sciences*, Cambridge University Press, pp. 167–217, 228–230.

Watkins, J. W. N.: 1972, 'Social Knowledge and the Public Interest', in W. A. Robson (ed.), *Man and the Social Sciences*, Allen & Unwin, London, pp. 173–196.

Watkins, J. W. N.: 1974, 'Self-Interest and Morality', in S. Körner (ed.), *Practical Reason*, Blackwell, Oxford, pp. 67–77; a comment on Sen (1974).

JOHN C. HARSANYI

ON THE RATIONALE OF THE BAYESIAN APPROACH: COMMENTS ON PROFESSOR WATKINS'S PAPER

I

The main thrust of John Watkins's paper (Watkins, 1977) is criticism of the Bayesian approach. The strange thing is that he never even mentions the standard arguments for the Bayesian approach, and certainly makes no attempt to *refute* these arguments. In fact, he writes as if he were quite unfamiliar with the basic literature on Bayesian theory, which contains a thorough discussion of these arguments.

The main argument for the Bayesian approach, of course, lies in the fact that any decision maker following a few very compelling rationality requirements (rationality axioms) will always act, as a matter of mathematical necessity, *as if* he tried to maximize his expected utility – whether he makes any conscious effort to maximize his expected utility or not. (More precisely, his behavior will always define a von Neumann-Morgenstern (vNM) utility function for him, such that his behavior will be equivalent to maximizing his expected utility when the latter is measured in vNM utility units.)

In other words, a Bayesian need not have any special desire to maximize his expected utility *per se*. Rather, he simply wants to act in accordance with a few very important rationality axioms; and he knows that this fact has the inevitable mathematical *implication* of making his behavior equivalent to expected-utility maximization. As long as he obeys these rationality axioms, he simply cannot help acting *as if* he assigned numerical *utilities*, at least implicitly, to alternative possible outcomes of his behavior, and assigned numerical *probabilities*, at least implicitly, to alternative contingencies that may arise, and *as if* he then tried to maximize his expected utility in terms of these utilities and probabilities chosen by him. (In fact, after observing his actions, we can always *compute* the utilities and probabilities implicit in his behavior.)

Butts and Hintikka (eds.), Foundational Problems in the Special Sciences, 381–392.
Copyright © 1977 by D. Reidel Publishing Company, Dordrecht-Holland. All Rights Reserved.

Of course, once we understand that, if our behavior is consistent with the relevant rationality postulates, then it will always involve at least an *implicit* choice of numerical utilities and of numerical probabilities, we may very well decide to choose these utilities and probabilities in a fully *conscious* and *explicit* manner, so that we can make fullest possible use of our conscious intellectual resources, and of the best information we have about ourselves and about the world. But the point is that the basic claim of Bayesian theory does not lie in the suggestion that we *should* make a conscious effort to maximize our expected utility; rather, it lies in the mathematical theorem telling us that *if* we act in accordance with a few very important rationality axioms then we *shall* inevitably maximize our expected utility.

This means that the Bayesian approach stands or falls with the validity or invalidity of its *rationality axioms*. It cannot be refuted at all, as Watkins is trying to do, by arguing against expected-utility maximization as such. It can be refuted only by showing that some or all of its rationality axioms lack logical force. This Watkins has not even attempted to do.

There are three main rationality axioms underlying the Bayesian approach:

1. *Consistent ranking* of lottery tickets: The decision maker can always decide about any two lottery tickets A and B, whether he prefers A over B, or prefers B over A, or is indifferent between them. Moreover, if he prefers A over B, and B over C, then he also prefers A over C. (The last statement remains true if we replace 'prefers' by 'fails to prefer', at both places.)

2. *Continuity.* Suppose the decision maker prefers A over B, and B over C. Then, there exists a probability p such that he will be indifferent between a certainty of obtaining B, and a lottery ticket yielding either A or C, with probabilities p and $(1-p)$, respectively.

3. *Sure-thing principle.*

(a) *As applied to risky situations.* Suppose that L is a lottery ticket yielding the prizes A_1, A_2, ..., A_n with the positive probabilities p_1,

$p_2, ..., p_n$, respectively. Suppose that L' is a lottery ticket exactly like L, except that it substitutes for A_1 another prize A_1' which the decision maker prefers over A_1. Then, the latter will prefer this second lottery ticket L' over the first lottery ticket L.

(b) *As applied to uncertain situations.* Suppose that L_0 is a lottery ticket yielding prize A_1 if event E_1 occurs; yielding prize A_2 if event E_2 occurs;...; and yielding prize A_n if event E_n occurs. Suppose that L_0' is a lottery ticket exactly like L_0, except that it substitutes for A_1 another prize A_1' which the decision maker prefers over A_1. Then, the latter will prefer L_0' over L_0, or at least will be indifferent between the two. (Now, we must admit the possibility that the decision maker will be indifferent between the two lottery tickets, which will happen if he assigns *zero* probability to event E_1. For, in this case, a replacement of A_1 by A_1' will be of course, a matter of indifference to him.)

AXIOM 1 essentially says that the decision maker knows what he wants, and that what he wants is self-consistent. To be sure, not *all* decision makers satisfy this requirement, and even highly rational people may fail to satisfy it on some occasions. (Perhaps nobody can be fully rational all the time, and in all possible situations.) But a normative theory of rationality, such as Bayesian theory, simply cannot help people who do not know what they want. All it can tell them is, "Go home, try to make up your own mind, and come back again when you have managed to do so." In this one respect, I must choose alternative (2) of Watkins's paper (Section 1): Bayesian theory is for those decision makers only who know what they want, and have managed to bring some consistency into whatever they want.

AXIOM 2 considers lottery tickets yielding A with probability p, and yielding C with probability $(1-p)$. It is a natural assumption that the utility of such a lottery ticket to the decision maker will be a *continuous* function of this probability p. (This continuity assumption could fail only if A had *infinitely many times* as much utility as C has, or conversely.) Consequently, as we move from $p = 1$ (where the lottery would yield A with certainty) to $p = 0$ (where it would yield C with certainty), the utility of the lottery ticket will move continuously from the utility level of A to the utility of C. Since B lies at an intermediate utility level between A and C, as we move from $p = 1$ to $p = 0$, we must reach a p value at which the lottery ticket in question will *cross* B's utility level. At this point, the decision maker will be exactly indifferent between obtaining B and obtaining the lottery ticket.

In any case, I am sure that Watkins does not deny the validity of Axiom 2, or of the continuity assumption underlying this axiom – since he himself uses this continuity assumption in his paper (end of Section 3, in comparing the three payoff matrixes).

I am equally confident that Watkins will not reject the *sure-thing principle* (Axiom 3), since it is essentially a restatement, in lottery-ticket language, of the *dominance principle*, which plays such a prominent role in his own theory. The dominance principle says, If one strategy yields a better outcome than another does under *some* conditions, and never yields a worse outcome under *any* conditions, then always choose the first strategy, in preference over the second. On the other hand, the sure-thing principle essentially says, If one lottery ticket yields a better outcome under *some* conditions than another does, and never yields a worse outcome under *any* conditions, then always choose the first lottery ticket. Surely, the two principles express the very same rationality criterion!

Accordingly, I must conclude that Watkins's rejection of the Bayesian approach is due merely to his *failure to realize* that the expected-utility maximization requirement *mathematically follows from a few rationality postulates that he himself fully accepts.* This is all the more surprising because this logical relationship between expected-utility maximization and the rationality axioms in question has been extensively discussed in the standard literature dealing with Bayesian theory, and has been conclusively established there repeatedly by rigorous mathematical proofs.

Equally strange is the fact that the certainty-equivalent method that Watkins himself is advocating (in Section 7) is nothing else but a rather crude, exceedingly vague and unsystematic, and manifestly *inferior version* of the very same expected-utility maximization principle he claims to reject! (Watkins argues that his method is preferable to the Bayesian approach because it makes allowances for differences in people's 'gambling temperaments' while the Bayesian approach does not. But this argument does not hold water. See my Section II, below.)

II

I shall now briefly discuss a few more specific points raised by Watkins. His unhappiness with expected-utility maximization apparently is mainly due to the fact that he feels the utility of a lottery ticket to a rational man must depend, not only on its *expected* (*mean*) *utility* but also on some measure of the *risk* associated with it (for instance, on the distance between the highest and the lowest possible utility outcomes, etc.). He argues we have to introduce this risk factor in order to take proper account of people's different 'gambling temperaments', i.e. of their attitudes toward risk taking. This is, of course, a standard objection to Bayesian theory. It has been answered many times, and the answer is readily available in the literature (see, e.g. Luce and Raiffa, 1957, p. 32).

Fundamentally, the answer is that the decision maker's 'gambling temperament' *has already been allowed for* in defining his von Neumann-Morgenstern (vNM) utility function. Therefore, if the utilities of the various possible outcomes are measured in vNM utility units, then the expected utility of a lottery ticket will already fully reflect the decision maker's positive or negative (or neutral) attitude toward risk. If we now allowed for his 'gambling temperament' for a second time, by introducing some measure of the risk involved into our computations at this point, then we would engage in unnecessary and impermissible *double counting*. Indeed, we would certainly *violate* at least one of the *Bayesian rationality axioms*: this would have to be the case since these axioms are known to logically *entail* equating the utility of a lottery ticket with its expected utility.

Thus, we are coming back to the same point. In order for Watkins's arguments to carry any force, he would have to show that the Bayesian rationality axioms lack validity. One simply cannot reject a mathematical theorem without showing that some of the axioms entailing the theorem are erroneous.

The only part of his paper where Watkins makes even an attempt to rebut this standard Bayesian answer to his gambling-temperament objection is in Section 7. There, he argues that he would not pay more than £7 for an even chance of winning £20 or £0. He claims this shows that he is *not* maximizing his expected utility. He admits that a

Bayesian may try to explain his reluctance to pay more than £7 for this gamble, by postulating that Watkins's "marginal utility for money diminishes rather rapidly in this region." But he adds that this would be the wrong explanation because "I can promise him that [my marginal utility for money] is actually pretty constant over the £0 to £20 region."

What is pretty constant? Is it the marginal utility of Watkins's vNM utility function? Surely not. Watkins seems to forget that his vNM utility function is actually *defined* in terms of his choices among gambles, so that his unwillingness to pay more than £7 is in fact quite conclusive evidence to show that his vNM marginal utility *is* rapidly diminishing in the relevant region – provided that Watkins's behavior conforms to the Bayesian rationality axioms at all. (Of course, if his behavior fails to conform to these axioms, then he simply will not have any vNM utility function altogether, and any question about the mathematical properties of this utility function will become otiose.)

Of course, John Watkins may also be able to define a completely different utility function for himself, based on equating those utility differences which he *feels* to be equal, according to his introspection. But *introspective* utility functions need not have any simple relationship to utility functions inferred from people's actual *behavior* (in the way vNM utility functions are).

This is true even about *ordinal* utility functions (as used in analysing people's behavior under certainty). After Freud, it will not come as a surprise to anybody if he finds that people's introspective utility functions tend to show much more 'noble' and morally much more 'respectable' preferences than their actual behavior (or utility functions inferred from their actual behavior) may do. In the case of *cardinal* utility functions (as used in analysing people's behavior under risk and uncertainty), there is even less reason to expect any simple relationship between introspective and behavioral utility functions. What we know about introspective psychological measurements (often described as psychophysical measurements) in fields other than utility theory seems to suggest that, even in cases where such Freudian differences between people's introspective and behavioral utility functions can be neglected, we probably still cannot expect any simple linear relationship between the two types of utility functions. If these two types of utility functions are connected by any simple mathematical law at all, then this could be just as easily a logarithmic law, or a quadratic or cubic or any other power law, etc., as it could be a linear one.

Thus, even if Watkins's introspective utility function does have a constant slope within the relevant range, it does not follow at all that the same is true about his vNM utility function. His attempt to infer constancy of his vNM marginal utility on the basis of an introspective estimation of his marginal utility for money within this range only betrays a basic misunderstanding of the nature of vNM utility functions, and of the way they are defined.

III

In Section 3 of his paper, Watkins writes as if I had rejected the maximin (or minimax) principle as invalid for two-person zero-sum games and for other strictly competitive games. This is, of course, a complete misunderstanding. No game theorist would even think of rejecting the maximin principle as invalid for strictly competitive games given in *normal form*. But two points must be made:

1. In the special case of strictly competitive games given in *normal form*, the maximin principle is mathematically *equivalent* to the equilibrium-point principle[1]: In these games, any strategy pair used by the two players will be an equilibrium point if and only if both players' strategies are maximin strategies. This means that we simply *do not need* the maximin principle as a separate rationality axiom because any conclusion we could draw from it we can just as easily obtain from the equilibrium-point principle. On the other hand, while the validity of the maximin principle is restricted to strictly competitive games, the equilibrium-point principle equally applies to *all* non-cooperative games and, therefore, can be used as a unifying principle for the whole theory of non-cooperative games, whether strictly competitive or not.

2. As Aumann and Maschler (1972) have shown, even in the case of strictly competitive games, the maximin principle will yield counterintuitive implications if these games are given in *extensive form*, and not in normal form. In contrast, the equilibrium-point principle *always* furnishes reasonable conclusions for strictly competitive (as well as for other non-cooperative) games, whether they are given in extensive form or in normal form.

These two mathematical facts provide rather strong reasons for basing our formal theory of non-cooperative games on the equilibrium-point principle, and *not* on the maximin principle – even if, in the very special case of strictly competitive games given in normal form, the maximin principle would yield equally valid conclusions.

IV

Watkins is defending the use of the maximin principle, not only in strictly competitive games, but also in one-person decision problems. Actually, what he is defending is not the clear and unambiguous formal principle that game theorists and decision theorists call the maximin principle; rather, what he is defending is a much weaker and much vaguer principle, with a wide-open escape clause which is draining away most of its operational content. He does not suggest that we should always strictly obey the maximin principle and should always opt for the policy maximizing our security level – a suggestion which may very well yield disastrous practical recommendations, but which at least always has clear and unambiguous behavioral implications. If we tried to follow this principle, we would always know what to do – even if this meant never taking a plane since planes might crash, never crossing a road since we might be hit by a car, and never getting married because marriages might come to a terrible end, etc.

Instead, Watkins merely suggests that we should always opt for the policy that would maximize our security level *if we disregarded some* – unspecified – '*remote contingencies*'. Thus, according to Watkins, we *may* take planes after all (the chance of a plane crash is rather small in many cases); we *may* cross roads (the chance of an accident is often rather small), etc. Unfortunately, Watkins fails to tell us *what criterion* to use in deciding what contingencies are sufficiently 'remote' to be reasonably disregarded; and, without such a criterion, Watkins's principle has no operational meaning.

The term 'remote' seems to suggest that the criterion that Watkins has in mind is in terms of the small *probabilities* associated with these 'remote' contingencies. But this surely will not do: It may be reasonable to disregard even a ten per cent chance of a *minor* mishap, even though it may be highly imprudent to disregard even a 1000 times smaller chance of a *major* disaster. Thus, the criterion must depend on the sizes of both the *probabilities* and the *utilities* (disutilities) associated with these contingencies. Of course, we could take the *product* of the relevant probability and the relevant utility (disutility) as our criterion. But this would amount to subordinating the maximin principle to an essentially *Bayesian* criterion, and I am sure this is not what Watkins has in mind!

In any case, so long as Watkins does not give us a specific criterion for identifying those 'remote' contingencies which can be safely neglected in applying the maximin principle, he has not suggested any clear alternative either to the classical (unrestricted) maximin principle, or to the Bayesian criterion of expected-utility maximization. Indeed, he has not suggested any well-defined decision rule at all.

V

In Section 7 of my original paper (this volume, p. 326), I have argued that in a prisoner's dilemma game, if the players are decent people, they may attach considerable disutility to using a non-cooperative strategy when the opponent uses a cooperative strategy. If the players do take this attitude, then this may change the payoffs of the game in such a way as to turn the cooperative outcome into an equilibrium point – even though the non-cooperative outcome may still remain likewise an equilibrium point. For instance, I proposed the new payoff matrix may take the form:

	B_1	B_2
A_1	2, 2	0, 1
A_2	1, 0	1, 1

I added that in the new situation the players would have no difficulty in reaching the cooperative outcome [viz., the strategy pair (A_1, B_1), in our example].

Watkins questions this conclusion (near the end of Section 4 of his paper). He feels that in choosing (A_1, B_1) each player would incur a considerable risk. But in actual fact, *no risk would be involved* in agreeing to use the strategies A_1 and B_1 because *such an agreement would be self-enforcing*, without any real risk of default, precisely because now (A_1, B_1) would be an equilibrium point. (That is to say, so long as the two players expected each other to follow this agreement, each player would have a clear incentive to follow it. This is just the opposite to what would happen in the original prisoner's dilemma game: there, the very expectation by either player that his opponent would keep to such an agreement would give the first player a clear incentive to *violate* this agreement. This is so because, in a prisoner's dilemma game in its unmodified form, (A_1, B_1) would *not* be an equilibrium point.)

Of course, if the two players *cannot talk* to each other, then the game will become simply a non-cooperative game with two different equilibrium points in pure strategies.[2] In this case, I agree with Watkins that the players will presumably choose between the two equilibrium points in terms of *risk* considerations (which game theorists sometimes call *risk-dominance* considerations). Equivalently, they can choose between the two equilibrium points in terms of the Harsanyi-Selten solution theory, which formalizes these risk considerations in the form of a general mathematical theory.

Thus, contrary to Watkins's claim in his present paper and in earlier publications,[3] a prisoner's dilemma game will have an easily identifiable unique rational solution,[4] at least in two cases:

(1) If the prisoner's dilemma payoffs are the actual payoffs the players are trying to maximize (i.e. these payoffs are *not* modified by the players' moral attitudes): in this case, the only rational solution will be the *non-cooperative* outcome – whether the players can communicate or not.

(2) If the original prisoner's dilemma payoffs are modified by the players' moral attitudes in the way indicated above, and if the players

are free to communicate with each other: in this case, the only rational solution will be the *cooperative* outcome.

Moreover, again contrary to Watkins's claim, neither of these two conclusions is based on introducing any 'strong assumptions' (let alone, '*extremely* strong assumptions', as he puts it). The only assumption we need is that the players will try to *maximize their payoffs* as stated in the payoff matrix of the game (which, in case (2), of course means the payoff matrix as modified by their moral attitudes). This is surely not a 'strong' assumption. Indeed, it follows from the very definition of payoffs as the quantities the players are trying to maximize in the game.

Looking at it from another point of view, it amounts to no more than assuming that the person who stated the payoff matrix of the game *knew what he was doing;* and that, in particular, he knew that the payoffs must provide a correct specification of the players' assumed preference ranking among alternative possible outcomes – whether this preference ranking is based on pure self-interest, on institutional loyalty, on moral attitudes, or on any other factors.

ACKNOWLEDGEMENT

The author wishes to thank the National Science Foundation for supporting this research through Grant SOC75-16105 to the Center for Research in Management Science, University of California, Berkeley.

University of California, Berkeley

NOTES

[1] By 'equilibrium-point principle' I mean the principle that the solution of any non-cooperative game must be an equilibrium point.
[2] There is also a third equilibrium point in mixed strategies.
[3] See the references listed at the end of his present paper.
[4] In the sense that no elaborate game-theoretical analysis is needed to find the solution.

J. J. LEACH

THE DUAL FUNCTION OF RATIONALITY

1. INTRODUCTION: A SOCIAL RESEARCH PROGRAMME IN SEARCH OF A RATIONALE

I would like to explore some of the brief comments made by Prof. Harsanyi about applying normative notions of rationality for positive purposes in the social sciences. I share some of the usual puzzlement about such use. How is it that rationality theory, as derived from classical economics to mean something like efficient goal-directed behaviour, serves the dual function of prescribing conditionally what behaviour ought to be and also of describing, predicting, and explaining actual behaviour? After all, rationality has traditionally been associated with other normative disciplines, mainly logic, epistemology, and ethics. And no one seriously takes these domains to serve this dual function. Normative they are, but positive they are not.

If anything, psychologists and social scientists use the negative aspects involved in these disciplines to characterize positively much of human behaviour and reasoning. We are usually afflicted with accounts of how inconsistent, how silly, how inconsiderate, how emotionally charged, how perverse – i.e. how irrational – much of our behaviour actually is. Even when, on those rare occasions, we do manage to reason and behave in accord with valid logical or ethical principles, social scientists still explain such acts in non-rational terms. It hardly seems scientifically appropriate to claim that we believe X because x is true, or that we reasoned thus because such is valid, or that we did y because y is the ethically correct thing to do. At best, such acts are seen as cases of 'psycho-logic'; at worst, cases of past conditioned reinforcement. Why then presume that one special sense of rationality, as developed by decision theorists to explicate the common-sense notion of practical reasoning, is unique in serving both normative and positive functions? Unlike Harsanyi and most economists, I find this claim puzzling and philosophically intriguing.

Butts and Hintikka (eds.), Foundational Problems in the Special Sciences, 393–421.
Copyright © 1977 by D. Reidel Publishing Company, Dordrecht-Holland. All Rights Reserved.

Two easy resolutions to the puzzle merit immediate dismissal as blocking further inquiry. First, to avoid the issue by claiming that social behaviour is simply not rational in the positive decision-theoretic sense flies in the face of a mounting array of counter evidence from all the social sciences. To ask instead how normative rationality theory manages to account for such behaviour seems much more appropriate. But then, secondly, to be told by Harsanyi that "there is really nothing surprising about this. All it means is that human behaviour is mostly *goal-directed* · · · " (1977, p. 317) merely sweeps the puzzle and resulting need for philosophic clarification under the empirical fact.

That human social behaviour is largely rational in this sense and can be successfully explained in this framework seems to me both surprising and immensely fruitful. For this suggests that rationality is a fundamental unifying concept of much social scientific theorizing. Just how this theorizing conceptually hangs together to form a unified research programme, a general empirical theory of social action, if you will, then needs considerable scientific clarification. Moreover, if the programme does successfully account for such behaviour, we are still left with the residual philosophic issue of how the dual functions, the normative and positive uses of rationality theory are related.

Although Harsanyi, and an expanding list of others, have devoted much attention to formulating an abstract logical theory of rationality and to developing rational explanations of specific types of social behaviour, little work has been devoted to either of the above tasks: to the appropriate structure of the general research programme or to the accompanying philosophic issues. In what follows I shall, accordingly, comment only briefly and indirectly on the logical foundations of utility and game theory. Nor shall I review the burgeoning literature which directly and explicitly employs rationality theory to account for such social phenomena as power, voting, social status, organization, and international relations. Likewise shall I resist the temptation to criticize literary expropriations of game-theoretic vocabulary for descriptive or dramatic affect. Instead, I shall concentrate on the two above, relatively neglected, tasks.

The major part of my paper will try to elaborate the surprising unifying force of rationality theory for the social sciences. Here I will argue against the picture of social theorizing as consisting of a batch of mini-theories or even descriptive hypotheses, both unrelated and basically non-rational. The contrasting perspective will be presented by defending the thesis that rationality theory provides the rationale, the generative mechanism, for a current and relatively successful research programme in the social sciences. There is indeed a general theory of social action developing among social scientists and at its core is the decision-theoretic notion of rationality.

Moreover, this programme extends from action theory to operant behaviour theory, at least when applied to social learning and social exchange theories, from social and symbolic interaction theories to cognitive balance theories, from role theory to anthropological and sociopathological theories. It includes work on reinforced learning behaviour, social structures and human interaction. When viewed in this context, much of the usual philosophic debate about scientism and humanism subsides.[1] Grafting of theories from both domains around the core notion of rationality make up the research programme. Fit into a general action-theoretic framework, rationality serves as a dual functioning mechanism for both appraising and explaining social behaviour, for bringing people (with all of their wants, preferences, beliefs, social roles, and rituals) back into the social sciences. It systematically integrates a programme of strategic interaction.

In the concluding part, this framework is expanded to raise deeper philosophic issues regarding the nature of justification and the relations between normative and empirical considerations. Here I will try to sketch an argument for a naturalistic theory of justification based on the dual functioning of rationality. Failure to function adequately in either role may be grounds for revision of one's theory of rationality, thus requiring a mutual trade-off or adjustment between the two roles. This gives more serious weight than usually accorded to positive explanatory adequacy as a justificatory condition of normative theories.

A brief suggestion will also be made about the parallel use of rationality as an epistemic notion of justified theory choice, and recent emphasis on the positive history of science. The connection between these seems to provide impetus for a new revival of the currently wilting field misleadingly referred to as the 'sociology of knowledge'. Hopefully what emerges is a picture of an even more general research programme, much too broad to be even sketched on this occasion; one not restricted to social-scientific theories but extended to cover such related philosophic domains as epistemology and ethics as well. In the words, if not the spirit, of an older tradition: "man is by nature a rational animal". In more contemporary jargon, life is a game, an indeterminate set of complex strategic and adaptive gambles fraught with the hazards of risk and uncertainty. So much for almost joining Aristotle, von Neumann, and Pirandello.

2. SCIENTISM AND HUMANISM

Now to the first task of elucidating the surprisingly unified structure of a current social research programme, of a general theory of social action. Let me begin by rehearsing some of the more salient features of the continuing debate among philosophically inclined social scientists regarding scientism and humanism. But to avoid the banalities associated both with special methodologies and with distorted representations of each position, I shall try to formulate the issue in a way that allows of a mutually satisfactory integrative resolution.

By stressing the more cogent theses of each side, I shall argue that they are not merely logically compatible but contributing parts of a coherent theoretical framework. My contention is that rapprochment eludes the debaters because the appropriate common framework seems to lack a rationale, a unifying core theory, a mechanism for generating behaviour and social structure. Such a rationale provides relief from the enduring dilemma. Choosing between scientifically objective experimentally verified theories as one option and subjectively meaningful anthropomorphic principles as the other, as if the issue amounted somehow to a zero-sum game, seems bizarre. Not even the most cut-throat of poker games is so neatly characterized. Most important choice situations turn out to be mixed-motive games with varying degrees of cooperation and competition, of consensus and conflict. Just so in this debate as well. Part of the function of a core theory is to offer guidance as to how the diverse theories are to be grafted together into a coherent programme.

Positivistically oriented social theories usually stress, with some good epistemic reason, the objectively observable and nomological dimensions of behaviour. From this spectator's view a mechanistic model directs attention to social behaviour as a passive conditioned response to external environmental factors, either as discriminative stimuli or as reinforcing consequences; and explicable via functional laws, which may or may not invoke intervening variables of a physiological or motivational sort, to mediate the linkages. Field research only hinders experimental isolation of controlling independent variables, hence historical causation is emphasized. Behaviour appears as a one way influence process which in turn produces social structures of varying degrees of complexity. Reductive empirical analyses, controlled modification of performance by direct experience of consequences, and atheoretical development of functional statistical correlations are the hallmarks of this option. Its main protagonist is of course a radical version of Skinner's operant behaviourism.

Central to the opposed option is the subjectively meaningful, intentional, and purposive dimension of human action. From this agents' perspective an anthropomorphic model, derived from common sense accounts and field work, stresses social action as an active self-directed adaptation to the social pressures of interacting others, and explicable only via social structural roles, rules and rituals. Internalized personal reasons or deliberations are involved in negotiating shared meanings and commonly accepted conventions. These non-causal factors in the agent's social field represent the justificatory restructuring of his situation as a feedback process which severely alters experimentally derived functional relationships. Non-reductive conceptual analyses, interactional self-control of action by responsible agents, and an action-theoretic development of situational logic are the keys of this alternative humanistic option.

When posed in the usual fashion as radically incompatible scientific paradigms of inquiry (Harré and Secord, 1972), each model faces numerous deficiencies making a choice between them intolerable. One seems to lose as much as one gains by qualitative acceptance or rejection of either paradigm. A wiser strategy is to negotiate the differences so as to effect a viable reconciliation. But to so negotiate between the mechanistic – causal and the teleological – rule-governed models requires both a conceptual framework and a generative mechanism capable of accommodating and directing the grafting of the varied special theories into a coherent whole, into a general research programme of social action. An action theory seems to me the most promising framework to house rationality theory.

Though much progress toward fruitful compromise has already occurred among social scientists, major philosophic barriers still persist. On the one hand, the impoverished narrow methodological scruples inherited from an earlier positivist tradition must be replaced by non-reductive introduction of structural and functional concepts. By rejecting methodological individualism, e.g. these latter take on added importance. Operant and related behavioural notions can still provide a base, though not a logical foundation, for a general social theory. On the other hand, the looser metaphysical scruples inherited from a long-standing humanistic tradition needs to be replaced by a causal analysis of reasons, i.e. of wants and beliefs, as determinants of behaviour. Roles, rules, and rituals remain as important factors but not as generative mechanisms. Instead they function as premises in practical reasoning, as social structural constraints of choice behaviour, both in terms of restricting or opening available alternatives and of governing opportunity costs.

3. ACTION THEORY

Yet however well an action-theoretic framework might accommodate insights from each model, the central problem remains. How to devise a generative mechanism capable of serving both practical and causal purposes, as a logic of choice and as an account of behaviour; of providing the dual normative and positive functions? But before suggesting how rationality theory satisfies this core task, let me briefly remind you of some recent developments in action theory. We will then be in a position to appreciate current research centered mainly on social interaction with an eye to proposing a general theory.

Much of the dispute between the mechanical and teleological models centers on the twofold sense of the word 'because' when used in accounts of human behaviour. The logical and causal senses are then represented in the opposed agent and spectator perspectives from which behaviour is either understood as a reasoned practical inference or explained as an external casual process. By exploiting the ambiguity and resisting the obvious move of treating reasons as causes, one is led – by the example of Anscombe, Melden, Hart, Hamlyn, Taylor, Dray, and Winch – in the futile quest of non-casual conceptual explanatory mechanisms connecting beliefs and wants with actions.

Part of the problem of course is that, strictly speaking, reasons are not causes. From the agent's point of view practical deliberation is viewed as a logical relation between the propositional contents of beliefs, wants and actions. A normative theory thus amounts to a codification of 'logical' principles indicating which propositional premises constitute good reasons for what conclusions. At the same time, but now from the spectator's point of view, decision making is seen as an internal casual process connecting the agent's believings, wantings, and doings. A positive theory then consists of a codified set of casual laws indicating the empirically unified linkages between these psychological states and behaviour.

Hence – as Davidson, Peters, Alston, Goldman and others have amply argued – despite the fact that reasons are contents of propositions while the associated causes are the believings and wantings of these contents, it nonetheless remains the case that reasoned accounts of behaviour are causal explanations. But now the normative rules of rational choice are transformed into causal laws by way of a behavioural decision rule of action. The normative logic of the situation becomes the positive account of behaviour.

Psychologists and social scientists of varied stripe have long approached this major plank of an action theoretic platform: Weber, Mead, Kurt Lewin, Tolman, Heider, Festinger, Piaget, Bandura and Pribram, to mention but a few. It seems clear then that reason, à la Kant, is not practical. Rules, norms, principles, propositional content, or logical relations do not move men to act. Hence the futility of the quest for structural, functional, or other such social mechanisms as determinants generating behaviour. Instead reason, in Humean fashion, is the servant of the passions. It is a normative yet natural mechanism for organizing and unifying inclinations, believings, preferences, expectations and desires, all of which in turn are the moving causal forces of behaviour.

Ignore the gross confusions surrounding 'pleasures and pains', and it appears that nature has indeed placed mankind under the governance of rationality in its dual normative and positive functioning. Our beliefs and wants serve as practical reasons in providing guidance for what we ought to do. Likewise they serve as causes of what we in fact do. In both cases the underlying mechanism is a decision theoretic logic of rational choice.

Recent work by action theorists contributes considerable conceptual clarification of many problems still plagueing social scientists. Beyond the reason – cause dispute, they provide cogent logical analyses of actions, intentions, and agency; of abilities, excuses and constraints. They also offer important insight about act individuation, about locating the appropriate units of behaviour, an issue sadly neglected by most behaviourists. More importantly for present purposes, however, an action–theoretic framework accommodates in the most fruitful manner much scientific theorizing stressing cognitive, symbolic, social and interactional aspects of behaviour. In thus conceptually clarifying these theories, it provides the first stage in recognizing the proliferation of many small theories as actually part of a unified general research programme. Rather than enumerate the many examples, let me here just mention a few representative cases.

Following the early suggestions of Mead's social behaviourism, Kurt Lewin postulated a field theory of social interaction which Heider later developed into an attribution and balance theory. Goffman followed with a dramaturgical interpretation of role behaviour and strategic face to face interaction. And recently there emerged more sophisticated studies of attitude change and altruistic equity theory. From a different philosophic view, notably continental phenomenology, came Alfred Schutz's rational theory of social action which ethnomethodologists pursued in elaborating the extensiveness of rule-governed behaviour. Common to these varied approaches are two further action–theoretic themes.

Stress falls, first, on the close relation between systematic theorizing and common sense understanding, between what Heider called 'systematic and naive psychology'. Part of the point was the importance of a unified conceptual framework capable of handling both positive and evaluative practices. Ordinarily we appraise each other's actions in quite the same conceptual terms as we describe and explain our behaviour. And this because what we do is largely a function of how we perceive and evaluate our social environment. Not terribly much sense could be made of recent political catastrophies, e.g. without realizing that the rewarding reinforcements of well-laundered money affects people differentially if they see it as justly deserved political support, or as illegally attempted bribery, or as morally neutral power moves to gain economic control. Even some reductionist behavioural accounts of social action resort to just such normative notions as distributive justice and the rule of reciprocity to supplement the law of effect.

The second common theme derives from the insistence that these factors operate causally, not merely as epiphenomena or as functional correlates. Action explanation then becomes a contemporaneous, not an historical, affair. Employing the fundamental concept of a psychological field or 'life space', consisting of a person and his social environment in interdependent unity, Lewin argued for a situational logic. Thereby actions are explained by the field existing contemporaneously with the behaviour, by the situational factors present at the time. Past events or circumstances do not directly determine present behaviour, though they surely operate in causal chains to produce the present situation. Past conditioning, via levels of deprivation and continuous or intermittent schedules of reinforcement, might well help to determine one's present psychological field, but it is the latter that accounts for action and subsequent structural changes.

4. BEHAVIOURISM AND SOCIAL INTERACTION

Even many operant theorists – freed of Skinner's independent methodological scruples – invoke cognitive, social, symbolic and non-deprivation based preferences as intervening variables to complete the causal chain linking past conditioning and present behaviour. This again suggests a convergence of behavioural and action research, despite underlying philosophic or ideological dissonances.

By casting off the additional theoretical burden of instincts, drives and inner motivational forces, behaviourists settle into the framework of social interaction and interdependent reciprocal reinforcement. People actively construct their own reinforcement contingencies, select out appropriate discriminative stimuli from their environment, and strategically respond to situational interactions of the moment. In Bandura's social learning theory (Bandura, 1971, p. 11), e.g., human functioning relies upon the following interrelated threefold regulatory system: stimulus control of antecedent inducements, reinforcement control of response feedback influences, and the cognitive control of processes guiding and regulating action.

Once we replace radical behaviourism by the subtle richness of a more liberalized operant social behaviour model, I suggest, the case for separating scientism and humanism becomes progressively more untenable. Wiser indeed to link them in a common action-theoretic framework and to develop them as a unified research programme focussed on the notion of social interaction and an underlying theory of rationality.

For (as I shall argue in the context of some detailed examples on another occasion) when one rejects instincts, drives and inner motivational forces as determinants of behaviour in favor of an interactional field approach stressing cognitions, preferences, roles, group norms, social sanctions, and self-directed action – an approach common to humanism and scientism – one thereby creates a major theoretical gap. Perspicuous by its absence in all of these theories is a central unifying generative mechanism. Consequently, the theories and hypotheses invoked to describe and explain various dimensions of interactional social behaviour, though perhaps codifiable within an action framework, still appear splintered and unrelated. Witness the continuing dispute between behaviourists and structuralists. Each unsuccessfully quests for separate, distinct and competitive mechanisms to account for behaviour: various learning principles on one side and numerous structural principles on the other. In good Feyerabendian spirit, such middle-range theories pluralistically proliferate either in despair that there is no method or in hopes that if there is it will emerge via a free competitive market.

My suggestion, by contrast, is that many of these varied interactional accounts constitute parts of a major social research programme with the missing mechanism provided by maximization principles of rationality theory at its core.

In other words, interactional theories replace instincts or drives by rewarding incentives and cognitive expectations, by subjective utilities and beliefs. But they overlook the fact that for these factors to serve the activating and directional functions required by a motivational theory, some method for combining them with action must be postulated. Unless there is some conceptual and empirical unity to a person's beliefs, preferences and actions, some patterned regularity to goal-directed choice behaviour, some logic to the reinforcement value of costly and beneficial consequences, it is doubtful that any social interactional theory could move far beyond the descriptive stage.

One could, of course, deliberately and arbitrarily restrict investigation to atheoretical functional correlations between contingencies of reinforcement and rate of response strength. But a deeper predictive and explanatory account of behaviour requires theoretical development of unifying principles indicating, in determinate fashion, how goals activate and direct action in specific patterns within particular social contexts.

The obvious and simplest candidate is a decision theory which combines beliefs and preferences multiplicatively, via probability and utility functions, to exhibit risky behaviour as a maximization of expected utility. The point is not that on this optimizing model of rationality deviantly aggressive behaviour, say, is unmotivated. Only that it is not determined by innate instinctual forces, nor by frustration produced drives, nor even by special inner motivational forces. Rather, particular cognitive expectations and valued incentives provide the motives to action as a strategic transaction with the social environment. Aggression can then be understood not merely as aversive but also as functionally adaptive behaviour designed to achieve desired outcomes in an efficient manner. Only by conceptualizing aggression, and deviance generally, as a rationally selective response to interacting situational factors can one account for its increasing prevalence in our everyday affairs. Only then do we have an appropriate framework and mechanism for both appraising and explaining the causes, maintenance, and control of such behavior. Deviance, just as much as conformity and other human actions, is then seen as a matter of coping with the problems of living in a complex social structure.

But while this general view of social behavior is shared by diverse social scientists of interactional persuasion, the dual unifying function of rationality theory remains notoriously suspect. Some even employ cost-benefit terminology, use decision matrices as descriptive devices, invoke bargaining coalitions as social structural emergents, and characterize interactional behaviour as strategically adaptive game playing. Yet, uniformly, the underlying mechanisms invoked are restricted to reinforced learning processes or to structural concepts.

Naturally, the main advocates of applying rationality theory to specific types of behaviour have been economists. They range from Downs' maximization study of political parties and voting participation to Olson's analysis of selective incentives in social groups and collective action, from Harsanyi's bargaining theory of power and social status to Schelling's coordinating view of warfare and international relations, from Simon's satisficing study of organizations to Arrow's critique of welfare and amalgamation principles.

Resistors, on the other hand, are counted among psychologists and all the other social sciences. Some of these (to be considered in detail on another occasion) include Bandura's social learning theory of modelling behaviour; the social exchange theory of Thibaut and Kelley, Homans, Blau and Emerson; Goffman's dramaturgical role analysis; Evans-Pritchard's anthropological studies of the Nuer and Azande; and the sociopathological theories of Szasz, Laing and Bateson. Some of this latter group remain behavioural learning theorists; others persist in some form of structuralism: resistors all. This pervasive neglect of rationality theory as a unifying mechanism thus conduces both to the philosophic impasse and to the appearance of unrelated competing mini-theories mentioned above.

To explore all of the important reasons for such suspicion is far beyond the scope of this paper. Nonetheless, some comments are in order, since they relate to the specific role of rationality theory in the current social research programme and to some of the ensuing philosophical implications of its dual functioning.

5. RATIONALITY THEORY: THE LOGIC OF THE SITUATION

Likely the main cause of neglecting rationality theory, aside from the usual aversions of social scientists to all normative notions, is the disarray within the theory itself. For many of the reasons reviewed by

Harsanyi, this afternoon and on other occasions, fundamental disagreement still prevails among decision theorists, even on construction of an adequate logic or postulate set for rational behaviour. His account of the logical development of rationality – from the common sense notion of means-end deliberation to the certainty model of classical economics, to the early decision-theoretic view of risky actions as expected-utility maximization, to the later game-theoretic model of uncertainty along Nash bargaining lines, and finally to his own Bayesian version of non-cooperative bargaining games – provides a lucid initiation to problems engulfing the normative dimension of rationality. By no means, then, does my case for unification rest on achieved settlement in decision theory. Rather it is also to be seen as part of the continuing research programme. As might be expected, positive applications follow similar patterns, but they also raise additional difficulties concerning both the abstract logical theory and the applied social theory. They will be my main concern here.

Methodological formulations of rationality postulates range, in other words, from quite weak and informal philosophic and common sense proposals to social-psychologically informed analyses to the other extreme of mathematically refined formalizations. Popper represents the former and Harsanyi the latter with Simon and Schelling defending the empirically oriented position.

Now I want to maintain that this methodological difference leads to alternative strategies of theory construction, and hence bears heavily on whether one sees current research in social science as splintered or as a unified programme. Moreover, how one develops a general theory of social action based on rationality theory depends mainly on just these methodological committments.

Popper and Harsanyi, for instance, share the conviction of methodological individualism. All social structural concepts, on this view, must be reduced to, or defined in terms of, concepts referring to individuals and their properties: specifically their objectives, beliefs, the information available to them, and their resulting choices. From this heuristic guideline, the resulting general theory of social action becomes largely analytic, either as an indeterminate situational logic or as a proto-economic abstract theory. Both developments are, when pursued alone, objectionable. They provide either no guidance or misguidance to the social scientist. This suggests another reason why the latter tend to neglect rationality theory. It also argues for combining them with the more empirical approach, a position which coheres better with my unification thesis about a research programme.

Consider Popper's position first. From his earliest to his most recent writings he has advocated explaining human actions in terms of the logic of the agent's situation and in accord with a rationality principle. The latter maintains that "Individuals always act in a fashion appropriate to the situation in which they find themselves" (Popper, 1967, p. 14), i.e. appropriate to their aims and situational appraisal. When applied to social anthropological studies of traditional societies, Jarvie and Agassi follow this weak proposal by defining rational action as merely goal-directed or "*conducive* to desired ends". Watkins in turn uses the postulate, now relabelled the "imperfect rationality principle" (1970), to bridge the empirical gap between decision schemes and action, particularly in explaining historical cases of bungled or unsuccessful actions. In effect, then, Popper's rationality principle serves as a heuristic rule of social inquiry. It directs us to be dissatisfied with any explanation of human action which makes the action appear irrational or inappropriate. Sound advice indeed, both scientifically and humanistically.

Nevertheless, Popper leaves so totally unspecified what is to count as appropriate that almost any action might so count. I am not objecting here that Popper's principle is analytic in the sense of being unfalsifiable. For it seems to me no more so than the law of effect or any other fundamental theoretical principle. Any such principle might be rejected or modified but only at the price of revising the entire theoretical framework of which it is a part.

Instead, the problem is twofold. On the one hand, the vagueness of 'appropriate' behaviour results from severing the connection between the positive and normative functions of rationality. Without some normative criterion, i.e. without some determinate situational logic, there is no chance of providing any but the most vague heuristic advice to social inquirers. Popper's approach therefore requires supplementing by some normative postulates and a decision rule specifying what is appropriate to each situation. The response of sociologists and anthropologists to the Jarvie-Agassi version demonstrates just this point (Wilson, 1970).

As if to acknowledge this weakness, Popper has only recently advocated his unduly strong "principle of transference; what ıs true in logic is true in psychology". But again he adds the unexplicated qualifier "provided it is property transferred" (1972, pp. 6, 24). This revised version of the heuristic principle manages merely to reopen, surely not to clarify, the issue about the dual functioning of rationality. In any event, without further specification of criteria of appropriateness, this weak formulation seems destined to an ideographical fate. The appropriate logic of the situation becomes relativized to each individual agent. Appropriateness is determined in each situation by the particular agent's own decision rule. This virtually precludes the nomological task of the social scientist. For there is no regularity, no determinate pattern to deliberate action, no 'logic' of the situation, except that in each case the action is vaguely appropriate according to the agent's own subjective rule. How Popper might reconcile this consequence with his general covering law analysis of scientific explanation raises important problems.

On the other hand, what little guidance Popper's weak construal of the rationality principle does provide to social scientists regarding theory construction seems quite misleading. For the implication of methodological individualism is that individual choice behaviour is an autonomous matter not to be corrupted by related findings either in, say, behavioural learning theory or by theoretical developments regarding social roles, rules, rituals, or interactions. So far as I can determine Popperians have accordingly said very little, beyond general philosophic critiques of Freud and Marx, about theory construction in the social sciences. What a general theory of social action would look like on this approach remains an open question.

6. PROTO-ECONOMIC RATIONALITY

From Harsanyi's opposed position, however, the answer is clear. Such a general theory would explicate situational logic along proto-economic lines. It consists of a normative analytic theory in the sense of an abstract axiomatized calculus. It provides formalized codification of the connection between beliefs, wants and actions in terms of well-ordered preference structures or utility functions, along with a set of rationality postulates governing increasingly complex choice situations of risk and uncertainty. It thereby offers some relief from the 'imperfect rationality' of Popper's indeterminate situational logic by demonstrating, if only in abstract form, determinate solutions to complex bargaining game situations.

The basic postulates are 'rational' in the sense of specifying both the optimal demands of the pursuit of self-interest and the consistent expectations such players can entertain about each other's behaviour, and also in the sense of yielding a uniquely defined decision rule for the players. The latter represents a unique rationally negotiated compromise in terms of each player's self-interest without use of any moral criteria. In particular, Harsanyi generalizes the earlier Nash-Zeuthen two-person bargaining model solution by showing that the only decision rule satisfying the basic postulates for n-person cooperative games is one maximizing the product of the player's utilities. This amounts to an analytic explication of the common sense notion of balance of power in negotiated bargains. It indicates each person's willingness to risk conflict and how much cost each can inflict on the other in the conflict, i.e. how dependent each participant is on the other for his self-interested benefits. The riskier one is and the more able to use threats, the more favorable bargain one negotiates (Harsanyi, 1961).

Of course, the determinate relief provided by this formal approach is purchased at a price. Just as Popper emphasized only the situational factors in 'the logic of the situation', so Harsanyi and game theorists capture at most its abstract logic. The theory takes as given all the important empirical parameters of choice situations: the player's utility functions, the strategic structurally defined alternatives open to them, and the information available to them. The hope, nonetheless, is that this non-parametric explication of the logic of rational choice can provide an analytical framework for positive purposes.

In accord with usual economic methods of theory construction, this framework is depicted as the uninterpreted calculus of a hypothetico-deductive system. For the classical economist, reliance upon psychological theories of motivation were minimized to consumer utility maximization and to entreprenurial profit maximization. Similary, Downs' non-economic extension to a theory of political parties and voting participation postulates maximizing only voter utility and party vote getting in quest of election as primary goals. The added task of empirical social scientists then becomes that of supplying a further interpretation of this rational calculus rich enough to predict much of social behavior. This requires, for instance, a set of bridge principles capable of transforming the normative rationality postulates into behavioural laws by linking them to empirical observations.

Thus Harsanyi earlier claimed that "game-theoretical analysis can be used only if we have a sufficiently specific theory of human motivations specifying the objectives people are actually seeking by engaging in various types of social behaviour, i.e. a theory specifying the main variables entering into people's utility functions" (1968, p. 321).

Likely he would now agree that the social scientist must also present empirically supported theories about the alternative options available to social agents, the institutional constraints governing interaction between agents in various contexts, the information available, the belief sets of diverse groups of agents, and the psychological computational capacities of the agents.

By indicating how such empirical theories are to be utilized in general theory construction, Harsanyi provides the specific guidance to social scientists lacking on Popper's view. More importantly, he avoids severing the link between normative and positive uses of rationality by following the classical economic methodology of hypothetico-deductive systematization. Evidently, the reason why he finds no puzzle about the connection between them must be of the following sort. Given an abstract uninterpreted calculus with its postulates of rationality, one can interpret them in either way merely by plugging in evaluative parameters to yield a normative, perhaps even ethical, theory of choice on the one hand, or by substituting empirical parameters to produce a positive theory of social action on the other.

If this construal accurately represents his position, then we have a case of what I referred to as a proto-economic theory, since the basic mechanism underlying both the normative and positive social theory is an economic logic of choice characterizing 'economic man' or 'rational man', in what Weber called static ideal-typical terms. The social scientific general theory results from the rationality postulates merely as an interpreted application of the relatively autonomous basic economic logic. Harsanyi concedes, of course, that these basic models of rational behaviour might not yield adequate explanations or realistic predictions in many social situations. They might therefore require substantial modification as suggested by Simon's theory of 'bounded or limited rationality'. But he reads such modifications as merely approximations to 'pure' rationality and dismisses the objection as an issue to be settled by empirical research.

However, the current research results (of Simon, Edwards, Rapopport, Fouraker and Siegal, Caplow, Shubik and Deutsch) seem to argue rather convincingly against this proto-economic position. This suggests still another reason for social scientific neglect of rationality theory, despite the important work done by Downs, Olson and the others mentioned above. I will spare you the details of this empirical research along with the prevalent philosophical criticisms of positivist methodology. Instead, let me indicate why I think this position offers misguidance to the social scientist. Again the point is twofold.

First, by supplementing the proto-economic view with the reductionism of methodological individualism, Harsanyi rejects the use of social norms and other social structural or interactional concepts as "basic explanatory variables in analyzing social behaviour" (1968, p. 321). I take this to mean they cannot be introduced into the vocabulary of the theory as primitive terms. Rather they themselves must be "explained in terms of people's individual objectives and interests" (1968, p. 321). But if my earlier assumption was correct, that an adequate interpretation of the rationality calculus for non-economic purposes requires not only a theory of individual human motivation but also social structural theories about alternative available options, institutional constraints on choices, and socially established belief and value sets; then social scientists constructing such theories are left in a cyclical dither. They cannot construct such theories legitimately without assurance that terms like roles, rules, rituals and norms are themselves explained by an adequate individual rationality theory; yet, they cannot construct the latter without the former.

This is simply to amplify the common criticism that "In general, it is not possible to explain norms without reference to other norms and institutions. If a given norm is regarded as the outcome of a game, other norms will generally figure among the rules; and even then the outcome may not be unique" (Foldes, 1968, p. 336). Put in terms of a recent social exchange theorist (Emerson, 1969, p. 381), such reductive heuristic advice misleads sociologists into conceiving of society as the sum of individual behaviours and thus failing to offer any theory of society. By placing the overwhelming burden of social explanations on 'given' parameters, it also misleads them into taking social structure, the very dimension sociology attempts to comprehend, as given.

Clearly, Harsanyi, and Homans e.g. are correct in insisting that social structural notions must be explained, that functionalism is inadequate to the task, and that group minds or other holistic metaphysics must be eschewed. However, every individualistic account Homans provides eventually assumes some other social structural concepts and premises. A better strategy of inquiry would, therefore, allow some such terms to be introduced into the theory as theoretical primitives, undefined by reference to individual properties. One such possibility might be Emerson's definition of 'social sanction' and 'group norm' in terms of individual attributes plus the notion of 'acting in concert', i.e. "the behaviour of a social coalition or a person representing a coalition" (1969, p. 383). As he suggests, and we are all painfully aware, a parent's sanctions are often ignored quite rationally when the child learns to form more rewarding coalitions with others not acting in concert with the parents.

But having swallowed the camel by admitting social structural terms as non-observable theoretical primitives into the theory in non-reductive fashion, there seems little reason to strain at the gnat by persisting in a proto-economic mode of individualistic theory construction. Particularly vulnerable at this juncture is the relative autonomy, the non-empirical character, of the rationality postulates or the basic economic logic of choice. Instead of restricting the postulates to the idealized characteristics of a basically non-social 'rational man' – such as a perfectly consistent utility function, perfect information, and perfect computational capacities – a wiser option dictates admitting a more realistic social-psychologically informed set of postulates. These would characterize rationality as a bounded 'strategy of conflict' along the social interactional and interdependent lines proposed by Simon and Schelling. But now these empirically revisable and informed postulates are no longer mere approximations to the pure case. As themselves the fundamental postulates of rationality, they serve both the normative and positive functions.

The second type of misguidance stems from this latter point. While agreeing with the economist's general sense of the connection between the normative and positive functions of rationality, I reject the non-empirical, as well as the non-social, character of the underlying logic of choice. For this seems to give priority to the normative over the positive function concerning the appropriate criteria for appraising the epistemic acceptability of the postulates. It leaves the decision theorist to his own intuitive devices for needed revisions in the basic postulates instead of relying on the direction of social empirical theories. This also tends to reinforce the aversion of social scientists to normative dimensions of their theorizing by suggesting that the latter are the relatively autonomous domains of logicians, mathematical economists and philosophers, and that this domain is therefore immune from the criticism of their own social empirical theorizing.

Mainly because of this normative priority I also find the dual functioning of rationality more puzzling than does Harsanyi. The issue is surely more complex than acknowledging the goal-directedness of human behaviour, even along proto-economic lines. However, since the issue bears mainly on the connection between the normative and positive functions and thus only indirectly on the surprisingly fruitful unifying features of rationality for the current social research programme, let me delay further comments about it until the concluding section of my paper. Meanwhile, let me sketch some of the important features of the empirically oriented theory of rationality in completing this part.

7. SOCIAL PSYCHOLOGICAL RATIONALITY

Although Schelling has made important contributions from this perspective to bargaining theory and coordinating games, I will restrict my comments to Simon's more elaborate and more empirically grounded theory of 'bounded or limited rationality'. Starting from the phenomenological viewpoint of the individual agent (with all of his limited knowledge and capacities) as the basic action and choice frame of reference, rationality is defined relative to this contextual 'definition of the situation', to the subjective perspective of the agent. Simon insists, in our terms, on linking the logic of choice to empirically determinable situational factors and to common sense notions, rather than to mathematically tractable idealizations. Hence, he finds the objective game-theoretic approach "fundamentally wrongheaded ⋯ in seeking to erect a theory of human choice on the unrealistic assumptions of virtual omniscience and unlimited computational power" (1957, p. 202).

Working mostly with the positive function of rationality in organization theory, as opposed to the normative function of classical economic theories of the firm, he advances simplifying models of rationality designed to capture only the main empirically limited features of the agent's 'life space', or situational problem solving. The central move in thus applying his simplifying principle of bounded rationality is "the replacement of the goal of *maximizing* with the goal of *satisficing*, of finding a course of action that is 'good enough'" (1957, p. 202), for the particular agent given his subjective limitations in coping with the vast complexities of the world. Actual rationally adaptive behaviour is then modelled as a simplified approximation to the game-theoretic global sense of rational choice. In most cases the model requires postulating neither the agent's utility function nor his ability to calculate marginal rates of substitution among different wants.

Firms, e.g., often fail to maximize anything in particular, including profits. Instead they establish minimal standards of satisfaction, levels of aspiration, in hopes of achieving enough profits to assure continued growth. Concentration on such limited objectives has led many organization theorists to construct a 'behavioural theory of the firm' more in accord with 'satisficing' procedures. Generally, then, the model contains the following simplifications:

(1) Optimizing is replaced by satisficing – the requirement that satisfactory levels of the criterion variables be attained.

(2) Alternatives of action and consequences of action are discovered sequentially through search processes.

(3) Repertoires of action programs are developed by organizations and individuals, and these serve as the alternatives of choice in recurrent situations.

(4) Each specific action program deals with a restricted range of situations, and a restricted range of consequences.

(5) Each action program is capable of being executed in semi-independence of the others – they are only loosely coupled together (March and Simon, 1958, p. 169).

Finally, all of these features reflect the fact that there are social structural and cognitive 'boundaries' to adaptive rationality. That is, there are situational elements in an agent's psychological field at any particular time that must be treated as constraints on strategic rational calculations. They are not derivable from individual decision making itself but are essential parts of it. As such they provide empirical guidance for possible extension and revision of the basic postulates. Hence, while Simon does not talk directly to the issue, I take it that he would oppose the principle of methodological individualism.

Further clarification of this empirically oriented approach to rationality theory can best be achieved by considering the two major objections to it. The standard and most tenable criticism rejects Simon's sharp distinction between maximizing and satisficing by interpreting the latter as merely a more complex version of the former. Rationality remains a matter of maximization on this critique, but what precisely is being maximized may well need drastic revision. Firms, e.g., may fail to maximize profits, or sales, or even growth. Likewise, if political candidates pursued only sufficient votes for election, then the Goldwater and Wallace and Stephen Lewis strategies remain inexplicable. For they, like other individual agents, may, and usually do, fail to maximize any one thing in particular. Yet this does not preclude maximizing expected utility in a wider sense which combines these factors in their utility function. At most this points to the need for less simplistic postulation of goals, for more empirically reliable goal attributions made on the basis of social as well as psychological considerations.

Individual expected-utility maximization also clearly fails to establish determinate decision rules for bargaining situations between interacting agents or groups, whether of a cooperative or non-cooperative sort. Game theorists, accordingly, make additional intuitive assumptions to increase the scope of the logical calculus. But this only requires extending the notion of maximization, as in the Nash-Zeuthen-Harsanyi model, to the product of individual utilities of the participant bargainers.

Here again, however, the need is for more social-psychological information about interacting situations, which in turn suggests more refined rationality postulates and maximization decision rules; not for a different kind of rule altogether. My claim, therefore, is that the satisficing rule serves better as a heuristic principle guiding further developments in rationality theory along maximizing lines than as a "replacement" for the latter.

Moreover, Simon's important stress on both the limited computational capacities of the agent and the need for search procedures to discover, sequentially, appropriate alternatives and consequences of action – i.e. his insistence on something less than omniscience – can also be accommodated by an optimality model. One required modification is to give up the assumption of perfect and costless information. Harsanyi and most other game theorists are, I believe, quite prepared to make this concession. Witness his present proposal about the "*agent normal form*" as intermediate between the standard extensive and normal forms of games in order to capture the essential information about the game. Witness, as well, his insistence on clarifying "what institutional arrangements (negotiation rules) each particular solution concept is meant to assume about the bargaining processes among the players", and his invoking the "*tracing procedure*" to represent the psychological processes (Harsanyi, 1977, pp. 334, 335, 340). Other important work has also been done on calculating the opportunity costs both of collecting additional information and of searching for alternative options.

Needless to say, such calculations call for highly complex and difficult conceptualization and empirical support to preclude adhocness. Nonetheless, Simon's point fails to sustain the conclusion that maximization must be replaced by the vague standard of satisficing, of achieving subjectively satisfactory or 'good enough' solutions.

However, I've always been dubious that this is in fact the conclusion Simon intends by advancing his alternative model of rationality. He clearly accepts a maximization rule in cases where the conditions of application are met. He seems to be insisting only, and properly, that such ideal conditions are seldom met; and consequently that the proto-economic model, which invokes omniscience, cannot account positively for most social behaviour. In addition, he often refers to his own satisficing model as an approximation to the global game-theoretic model. Clearly, e.g. he would not contend that it is rational, perfectly or imperfectly, to choose a worse option to a recognizably better one, even if the former were good enough to satisfy one's present level of aspiration. Precisely this choice of the better option in such situations is what causes the agent to raise his future level of aspiration.

In other words, Simon seems not to be opposing the maximization rule so much as the proto-economic model of rationality and the accompanying heuristic principle of methodological individualism, along the lines I suggested above. Satisficing is simply what maximization looks like when set in the complex social-psychologically informed context of interacting individual's 'life-space'.

Finally, without some such maximizing rule, Simon's theory of satisficing would fall prey to the same indeterminacy objection as Popper's rationality principle. By placing undue importance on the positive and behavioural situational dimensions of the 'logic of the situation', it would lack a determinate normative logic of choice (Michalos, 1974). By emphasizing only psychological or structural considerations to the exclusion of a set of normative rationality postulates, it would fail to provide a satisfactory adaptive mechanism capable of generating social behaviour. All the more reason, I think, to read his theory as an alternative to a proto-economic model. This latter, rather than the goal of maximizing, is indeed the "fundamentally wrong-headed" part of game-theoretic rationality.

Unfortunately, critics of the satisficing model attribute this stronger, but more untenable, 'replacement' conclusion to Simon and then object to it on one of two grounds. One group claims satisficing is identical to maximizing and thus contributes nothing additional to a proto-economic model, while the other group maintains that it is not so assimilable and therefore lacks an appropriate determinate logic of choice. In either case the main thrust of Simon's position is overlooked. The importance of his contribution stems not from an inadequate alternative decision rule but from an immensely fruitful heuristic principle. Considerably more specific and less misguided advice accrues from it than from either metholological individualism or a proto-economic model. Most importantly, it guides the construction of richer more realistic rationality postulates and thereby the development of a more adequate theory of rationality in both its positive and normative functions.

Consequently, despite the current disarray within rationality theory, and the resulting need to construe it as part of the ongoing research programme, it nonetheless constitutes the essential unifying core theory of a social interactional programme by supplying the adaptive mechanism generating social behaviour. In this context, the rational causal mechanism seems best explicated by the third methodological formulation considered above, by Simon's more realistic social-psychologically informed analysis of the basic rationality postulates. When properly interpreted as a heuristic principle, it links together the logical and the situational aspects of the 'logic of the situation', while sharing none of the objectionable features of either the Popper or Harsanyi proposals.

Being itself part of an interactional theory of rationality, it also satisfies the directional demands of the current research programme more adequately than either of the alternatives. Instead of sharply separating the normative rational calculus from the positive social psychological theory of cognitions and motivation, e.g., it advances a more integrated theory by linking them together in an interactional context which offers important heuristic guidance of a feedback sort to both dimensions. The resulting theory of motivation and cognition is, in effect, a rationality theory of strategic interaction.

Only when Popper's fundamental heuristic rationality principle is supplemented by game-theoretic logical postulates of rationality, and these in turn are governed by empirical theories of cognitive, symbolic, and social interaction – only in this systematic type of unified research programme – will we have a general theory of social action capable of accounting for the vast complexities of social behaviour.

8. Conclusion: connecting the dual functions of rationality

On another occasion I have offered detailed analyses of a number of social scientific theories, some in an operant behavioural framework and others from a social structural perspective, in order to further defend my unification thesis. These theories range from social psychology and sociology to anthropology and sociopathology. Since time does not permit such additional elaboration here, let me consider instead some of the philosophical implications of this research programme concerning the nature of justification and the relation between the positive and normative functions of rationality theory.

Earlier discussion of the methodological aspects of this two-fold functioning revealed three main positions: Popper severs the connection between them; Harsanyi connects them via hypothetico-deductive systematization but upholds the autonomous priority of the normative; and Simon links them in a unified empirical interactional framework allowing mutual dependence. In the latter case, normatively rational mechanisms constitute the core theory guiding construction of positive theories, which then feed back guidance for revision of the basic logic of choice. For instance, the rationality postulates and associated decision rule prescribe many types of social behaviour as rational which were earlier thought to be at best non-rational. The studies on voting behaviour and individual actions in large groups by Downs and Olson are clear cases. The attempt to formulate positive social theories which adequately account for such behaviour, in turn, suggests needed revision in the basis postulates and their social psychological presuppositions.

Detailed examples are also needed to elaborate this position. But for the moment I want only to elicit a simple argument implicit in this view, one with major philosophical implications. If rational choice postulates and decision rules do indeed function dually as both normative and positive mechanisms, to prescribe and explain social action, then it follows that failure to function adequately in *either* role may be good presumptive epistemic grounds for revising one's theory of rationality.

This argument parallels another I invoked in a different context about rational belief (1968). Namely, if beliefs function both as structural mappings of the world and as guides to action, then failure to serve reliably in either role may be good presumptive epistemic grounds for rejecting the belief as irrational. Both arguments are, needless to say, highly controversial since they link normative and empirical considerations together epistemically. Let me therefore pursue the former case further.

Ordinarily, of course, philosophers and decision theorists find no issue with the normative side of the argument. Since rationality postulates and decision rules were initially intended to provide a normative theory defining rational action and prescribing for choice situations, however conditionally, one would obviously expect them to be appraisable on just such grounds: in terms of whether or not they adequately capture and codify our intuitions about subjectively prudential behaviour. In other words, as with any normative theory in ethics, epistemology or logic, failure to be reliably responsive to our intuitions constitutes good grounds for revision, whatever theory of justification one may hold about prescriptive judgments.

Another dimension is added to the usual issues surrounding the nature of justification, however, when one considers the positive side of the argument. On the one hand, it is also uncontroversial that any positive empirical social theory is justifiable only when reliably responsive to empirical data. As such, any rationality-based causal theory or explanation must be appraisable minimally on empirical grounds. But, on the other hand, the nomological connections between wants, beliefs, situational factors and actions, i.e. the basic causal laws of our positive social rationality theory, are merely empirical interpretations of the rationality postulates and the corresponding decision rule. The same decision rule which normatively guides action is here invoked as a basic positive causal law explaining behaviour. Consequently, if the rule fails to serve this explanatory function, fails as a core part of our social theory to account for behaviour, then this empirical failure also constitutes good presumptive epistemic grounds for revising our basic rationality theory, and hence our normative theory.

The upshot of the argument, then, is that justification of a rationality theory requires a mutual adjustment between normative and empirical considerations (Goodman, 1965, Chap. III). Just how the adjustment procedure works in any given case is, of course, another matter. Nonetheless, we seem driven, by accepting the dual functioning of rationality theory, to a naturalistic and empirical theory of normative justification. This seems to me to follow from the social psychological theory of rationality presented above. It is also the spirit in which I earlier paraphrased Bentham to the effect that nature has indeed placed mankind under the governance of rationality in its dual normative and positive functioning.

One final comment. This general line of reasoning about justification as mutual adjustment seems also to bear on recent discussions which emphasize the role of the history of science in epistemically appraising scientific theories. One of the central issues is whether the historian of science merely positively describes and explains the course of past theorizing or whether in so doing he also prescribes certain methodological norms of scientific inquiry. And if he manages the latter, how are his dual positive and normative functions related?

The above argument suggests one answer to this problem. For the historian, as any social scientist, must invoke a rational account of scientific developments, both intellectually and socially, both internally and externally. But to do this he needs a general theory of action with its core rationality theory. One might, in fact, want to distinguish between internal and external history, on this basis, as the difference between invoking purely epistemic utilities or including pragmatic utilities as well. It would be instructive, e.g., to consider how a decision theoretic epistemology, such as Levi's or Churchman's, might function as a positive explanatory account of the history of science (Levi, 1967; Churchman, 1961).

In any event, for the historian to invoke a normative decision theoretic view of rationality to explain positively the historical developments of science, would lead, by way of the above argument, to a view of his methodologically normative function as a case of mutually adjusted justification. Some such theory of rationality seems not unlike what Kuhn, Lakatos and Popper have been suggesting. It would surely provide some explication of the latter's 'principle of transference'. For now the 'logic' referred to in "what is true in logic is true in psychology" via situational analysis becomes a decision theoretic logic. Moreover, if sociological explanations are properly construable as rational ones, in accord with the current research programme, then the above comments begin to sound like a revival of the 'sociology of knowledge'. But this also is a topic for another paper.

The University of Western Ontario

NOTE

[1] The present discussion of this debate is limited to the. version as it appears in a social scientific context. The wider, and deeper, issues not pursued here concern the challenge of Romantics to this very context, to the very notion of scientific rationality as 'single vision'. These issues require separate treatment.

BIBLIOGRAPHY

Bandura, A.: 1971, *Social Learning Theory*, General Learning Press.
Churchman, C. W.: 1961, *Prediction and Optimal Decision*, Prentice-Hall.
Emerson, R.: 1969, 'Operant Psychology & Exchange Theory', in Burgess and Buskell (eds.), *Behavioral Sociology*, Columbia U. Press, New York.
Foldes, L.: 1968, 'A Note on Individualistic Explanations', in Lakatos and Musgrave (eds.), *Problems in the Philosophy of Science*, North-Holland Publishing Co., Amsterdam.
Goodman, N.: 1965, *Fact, Fiction and Forecast*, Bobbs-Merrill Co., Second edition.
Harsanyi, J.: 1977, this volume, pp. 315–343.
Harsanyi, J.: 1961, 'On the Rationality Postulates Underlying the Theory of Cooperative Games', *Journal of Conflict Resolution*.
Harsanyi, J.: 1968, 'Individualistic & Functionalist Explanations in the Light of Game Theory', in Lakatos and Musgrave (eds.), *Problems in the Philosophy of Science*.
Harré and Secord.: 1972, *Explanation of Social Behaviour*, Blackwell, Oxford.
Leach, J. J.: 1968, 'Explanation and Value Naturality', *B.J.P.S.* **19**, 93–108.
Levi, I.: 1967, *Gambling with Truth*, A. Knopf, New York.
March and Simon,: 1958, *Organizations*, Wiley & Sons.
Michalos, A.: 1974, 'Rationality Between the Maximizers and Satisfiers', in Schaffner and Cohen (eds.), *PSA 1972*, D. Reidel Publishing Co.
Popper, K.: 1972, *Objective Knowledge*, Oxford U. Press.
Popper, K.: 1967, 'La rationalité et le statut du principe de rationalité', in Classen (ed.), *Les Fondements Philosophiques des Systèmes Economiques*, Paris.
Simon, H.: 1957, *Models of Man*, Wiley & Sons.
Watkins, J.: 1970, 'Imperfect Rationality', in Borger and Cioffi (eds.), *Explanation in the Behavioural Sciences*, Cambridge U. Press.
Wilson, B. (ed.): 1970, *Rationality*, Harper & Row, New York.

INDEX OF NAMES*

---

*Names listed in the bibliographies are not included in this index.

THE UNIVERSITY OF WESTERN ONTARIO
SERIES IN PHILOSOPHY OF SCIENCE

A Series of Books on Philosophy of Science, Methodology, and Epistemology published in connection with the University of Western Ontario Philosophy of Science Programme

Managing Editor:

J. J. LEACH

Editorial Board:

J. BUB, R. E. BUTTS, W. HARPER, J. HINTIKKA, D. J. HOCKNEY,
C. A. HOOKER, J. NICHOLAS, G. PEARCE

1. J. Leach, R. Butts, and G. Pearce (eds.), *Science, Decision and Value.* Proceedings of the Fifth University of Western Ontario Philosophy Colloquium, 1969. 1973, vii + 213 pp.
2. C. A. Hooker (ed.), *Contemporary Research in the Foundations and Philosophy of Quantum Theory.* Proceedings of a Conference held at the University of Western Ontario, London, Canada, 1973. xx + 385 pp.
3. J. Bub, *The Interpretation of Quantum Mechanics.* 1974, ix + 155 pp.
4. D. Hockney, W. Harper, and B. Freed (eds.), *Contemporary Research in Philosophical Logic and Linguistic Semantics.* Proceedings of a Conference held at the University of Western Ontario, London, Canada. 1975, vii + 332 pp.
5. C. A. Hooker (ed.), *The Logico-Algebraic Approach to Quantum Mechanics.* 1975, xv + 607 pp.
6. W. L. Harper and C. A. Hooker (eds.), *Foundations of Probability Theory, Statistical Inference, and Statistical Theories of Science,* 3 Volumes. Vol. I: *Foundations and Philosophy of Epistemic Applications of Probability Theory.* 1976, xi + 308 pp. Vol. II: *Foundations and Philosophy of Statistical Inference.* 1976, xi + 455 pp. Vol. III: *Foundations and Philosophy of Statistical Theories in the Physical Sciences.* 1976, xii + 241 pp.

8. J. M. Nicholas (ed.), *Images, Perception, and Knowledge.* Papers deriving from and related to the Philosophy of Science Workshop at Ontario, Canada, May 1974. 1977, ix + 309 pp.

9. R. E. Butts and J. Hintikka (eds.), *Logic, Foundations of Mathematics, and Computability Theory.* Part One of the Proceedings of the Fifth International Congress of Logic, Methodology and Philosophy of Science, London, Ontario, Canada, 1975. 1977, x + 406 pp.

10. R. E. Butts and J. Hintikka (eds.), *Foundational Problems in the Special Sciences.* Part Two of the Proceedings of the Fifth International Congress of Logic, Methodology and Philosophy of Science, London, Ontario, Canada, 1975. 1977, x + 427 pp.

11. R. E. Butts and J. Hintikka (eds.), *Basic Problems in Methodology and Linguistics.* Part Three of the Proceedings of the Fifth International Congress of Logic, Methodology and Philosophy of Science, London, Ontario, Canada, 1975. 1977, x + 321 pp.

12. R. E. Butts and J. Hintikka (eds.), *Historical and Philosophical Dimensions of Logic, Methodology and Philosophy of Science.* Part Four of the Proceedings of the Fifth International Congress of Logic, Methodology and Philosophy of Science, London, Ontario, Canada, 1975. 1977, x + 336 pp.